SAFE

THE RACE TO
PROTECT OURSELVES
IN A NEWLY DANGEROUS WORLD

SAFE

Martha Baer
Katrina Heron
Oliver Morton
Evan Ratliff

HarperCollins*Publishers*

HarperCollins books may be purchased for educational, business, or sales promotional use. For information, please write: Special Markets Department, HarperCollins Publishers Inc., 10 East 53rd Street, New York, NY 10022.

FIRST EDITION

Designed by Elliott Beard

Printed on acid-free paper

Library of Congress Cataloging-in-Publication Data
    Safe : the race to protect ourselves in a newly dangerous world / by Martha Baer ... [et al.].
      p. cm.
    ISBN 0-06-057715-0
    1. Terrorism—United States—Prevention. 2. Security systems—United States. 3. Internal security—United States—History. I. Baer, Martha.
    HV6432.S23   2005
    363.32—dc22   2004053935

05 06 07 08 09 ❖/RRD 10 9 8 7 6 5 4 3 2 1

ACKNOWLEDGMENTS

THIS BOOK WAS conceived and edited by David Kuhn and Katrina Heron as a collaborative project with co-authors Martha Baer, Oliver Morton, and Evan Ratliff. We wish to thank all the sources quoted within, without whose time and patience the book would not have been possible, and also to express our gratitude to the following people for their support: John Arquilla; Paula Balzer; Peter Belhumeur; John Bennet; Shoshana Berger; Jay Boris; Art Botterell; Warren Braunig; Roger Brent; Mark Bryant; John Caffin; Sarah Chalfant; Morgan Cole; William Crowell; Brian Dalton; Richard Danzig; Megara Doorly; Eric Eisenstadt; Rich Eisner; Claire Fraser; John Gage; Chris Giovinazzo; Bill Goggins; Tory Haljun; Cathy Hemming; Bruce Hicks; Danny Hillis; Martin Hugh-Jones; Ian Schrager Hotels and Anda Andrei; Russ Johnson and the people at ESRI; Shaun Jones; Bill Joy; Ali Kazemi; Douglas Keeve; David Kehrlein; Chip Kidd; Billy Kingsland; Robert Koenig; Stuart Krichevsky; Scott Layne; Sarah Lazin; Howard Lipson and the people at CERT; Emily McDonald; Tom McNichol; Peter Merkle; Sara and Katie Miles; Rachel Miller; Thomas Monath; Ryan Moore; John S. Morgan; Randy Murch; Nathan Myhrvold; Jeff O'Brien; Tom O'Rourke; Tara O'Toole; Maureen Orth; Rebekah Parker; George Poste; Ken Powell; Mark Rasch; Jacques Ravel; Don, Glynda, Warren, Elta, Carrie, Hayden, and Owen Ratliff; Seth Reubenstein; Paula Scalingi; Linda Schacht; Sam and Dana Schaffer; Orville Schell; Brad Smith; Alex Stock; Kip Thomas; Kevin Thompson; Lt. Michael Thompson; Lindsay Traylor; Keegan Walden; Susan Weinberg; and Andrew Woods.

CONTENTS

SAFE

Introduction

ON THE MORNING of September 11, 2001, virtually all of us, from ordinary citizens to supposedly well-informed leaders, were shocked to discover that a very small, secretive, and incredibly destructive enemy we knew little about could truly hurt us; that this enemy understood more about the routine functions of our world than many of us did; and that this enemy could infiltrate our physical space with utter ease. We were shortly to learn that this enemy had been diligently studying a host of technological developments we tended to take for granted in our lives, looking for ways to create catastrophe. Suddenly our naïveté came face-to-face with the enemy's deeply informed intentions. It made for a rude awakening.

The question *How could this have happened?* has still not been fully answered, and what answers there have been are genuinely troubling. We have learned that our intelligence agencies were woefully ill equipped to safeguard us against terrorist attack; that our homeland-defense strategy was a muddled collection of separate, and often conflicting, directives; that our emergency-response capabilities were tragically lacking; and that our entire national infrastructure—from our water supply and electrical grid to transportation and industrial and telecommunications networks—remains vulnerable to future harm.

The fact that passenger planes were turned into missiles was a horrific reminder, on the cusp of the twenty-first century, that technology's potential for good is coupled with the possibility for immense destruction. At the same time we were whipsawed by the revelation that terrorists had passed unimpeded through all the security checkpoints and used weapons no more sophisticated than box cutters. But a larger realization was hovering uneasily at the edges of our consciousness: We were used to seeing our technologies as discrete objects (jetliner, skyscraper), when the more accurate way to see them is as parts of linked and complex systems—systems capable of many possible interactions and outcomes, systems full of possible glitches, systems in need of constant vigilance. The enemy, while not better or smarter than us, had studied our technological systems far more creatively than we had ever anticipated.

Many of our systems failed that day in a cascade of unpreparedness—air travel security, emergency radio and dispatch networks, the stock exchanges. In the first hours, the fundamental breakdown in information flow shackled those on the ground and those in the corridors of power. Firefighters and emergency personnel had charged headlong into the two doomed towers, while, as Richard Clarke recounts in the spellbinding opening chapter of his book *Against All Enemies,* a skeletal team at the White House struggled to somehow coordinate a response. Paul Wolfowitz, Deputy Secretary of the Department of Defense, called in from a remote location: " 'We have to think of a message to the public. Tell them not to clog up the roads. Let them know we are in control of the airways. Tell them what is happening. Have somebody go out from the White House. Paul,' " Clarke writes he responded, " 'there is nobody in the White House but us and no press on the grounds.' "

Other systems would fail in the days to follow, and many more would be found to have failed long before September 11. The more we probed our national vulnerabilities, the weaker our defenses appeared, and the more elusive the answer to our most pressing question: *How do we prevent this from happening again?*

It's a commonplace to observe that technology makes things better while at the same time making things worse. We now have ready proof that the openness and interconnectedness our technologies provide—and on which we increasingly depend, not only in a practical sense but also as an expression of modern life—have also created the potential for great danger. Put another way, innovations as magnificent and progressive as the Internet and its myriad applications trail evil twins. For all their popular applications, for all their potential to enrich our lives and our society, these technologies also make possible such dangers as untraceable terrorist communications, guerrilla cyberwarfare, and the seamless transfer of assembly instructions for known biological, chemical, and nuclear weapons as well as new genetically modified pathogens. As *The 9/11 Commission Report* noted, "Terrorists [could] simply buy off the shelf and harvest the products of a $3 trillion-a-year telecommunications industry. They could acquire without great expense communication devices that were varied, global, instantaneous, complex, and encrypted."

Back before the unthinkable happened, we weren't accustomed to seeing our technologies in light of their dangers. Whether as corporate CEOs or individual consumers, most of us were used to thinking of that cumbersome word *technology* just as a blizzard of ever-newer tools at our disposal, if only we could figure out which ones we really needed and wanted. If we felt like victims it was as the victims of marketing campaigns, exhortations to upgrade and add on, problems with equipment incompatibility, the nuisance of bugs and computer crashes, and the overall annoyance of finding that offerings supposedly designed to make our lives more efficient were often doing the opposite.

The world beyond our immediate grasp—beyond the keys and screens of computers and cell phones, behind the cash from ATMs or even the water from the kitchen faucet—was veiled, invisible, or completely alien to most of us. This was in a way the apogee of a historical trend that took root at the beginning of the twentieth century, when developments in science and technology started moving with a speed and sophistication never before witnessed. An ever-widening gap grew up between the ordinary things most people knew and the extraordinary things a few people knew how to make. As advances in these areas accelerated and became increasingly specialized, with both positive and negative implications, it became harder to understand, discuss, or debate their profound effects.

We tolerated this growing knowledge gap in part because we couldn't see any handy alternative; it seemed an inevitable byproduct of the fact that an astonishing new frontier was accessible only to gifted inventors and visionaries. This cadre of experts seemed to speak their own, virtually untranslatable language. The stuff was just too damn complicated for most of us to follow. And the businesses that turned the new knowledge into products for us to consume were quite happy doing so in ways that allowed us to leave our ignorance intact. As long as it worked, we didn't need to know why it worked; when it didn't work, increasingly we just needed to know whom to call. (And if it couldn't be fixed, well, something newer was prepared to take its place.) There was dissent and argument, to be sure, but the learning curve in these areas was so steep that only a few could be said to know what they were talking about, and they didn't always have the skills, or the interest, needed to fill us in. Over time, a cultural rationale for being in the dark about technology took root: It was boring.

The authors of this book have spent a great deal of time thinking about the knowledge gap—the rift between the excluded and the exclusive, so to speak. We have confronted it regularly as writers and editors at *Wired* magazine, which was the first mass-market publication devoted to the cultural relevance of science and technology. We have explored this hidden, invisible, alien domain as enthusiastic storytellers—fascinated, curious, aware of being alive in a time of extraordinary change and intellectual ferment, mindful that the technological future holds both promise and peril for the human race, and eager to find ways to bridge the knowledge gap.

With September 11 and its aftermath, we confronted it anew. The knowledge gap was no longer just a cultural divide. It was now also the source of widespread fear. People were frightened by what they didn't know.

We were spurred to write this book by two conflicting truths: first, that the mass of Americans naturally reacted to the attacks by questioning whether our faith in the power of technology was simply misplaced; second, that very few Americans (indeed, very few people anywhere) have sufficient knowledge of this arena to be able to make this, or any other, critical judgment about it. What we were really

looking at, in other words, was an epidemic-scale anxiety of ignorance, an insidious disease that made us think the cure—a commitment to understanding our technologies—was tied up with the very
causes of our sickness. Our political leadership was as sorely afflicted
as any of us, offering a mix of policy proposals and cynical half-
measures for enhanced security whose consequence has been to exploit public fears without actually improving protections.

SAFE aims to be a treatment for this ailment. Before we can have an
intelligent debate about how technology might secure us, we must
first understand what technology is capable of: We have to know how
things work, and not just in the near-term. As riveting as the government's postmortems have been, they have stopped short of addressing
the full spectrum of technological realities and potential. In terms of
the practical failures of intelligence collection, for example, recommendations have relied on old-fashioned organizational fixes. As
Richard Posner, the Federal Appeals Court judge, wrote in a review of
the *9/11 Commission Report* for the *New York Times,* "The commission
thinks the reason the bits of information that might have been assembled into a mosaic spelling 9/11 never came together in one place is
that no one person was in charge of intelligence. That is not the
reason. The reason or, rather, the reasons are, first, that the volume of
information is so vast that even with the continued rapid advances in
data processing it cannot be collected, stored, retrieved and analyzed
in a single database or even network of linked databases. Second, legitimate security concerns limit the degree to which confidential information can safely be shared, especially given the ever-present
threat of moles like the infamous Aldrich Ames. And third, the different intelligence services and the subunits of each service tend, because information is power, to hoard it."

In other words, it's not that simple. Quite the contrary—and here
is where the collective eye of the nation typically glazes over.

Lost in the recursive non-debate is the fact that technology is anything but boring. It's easy to see why many would find it so, however.
So much writing on the subject focuses on the gadgets and, increasingly, the balance sheets while leaving out the subject's vital, com-
pelling core: people. Technology does have a face—indeed, it has
many faces, ranging from those of the experts in these pages to our

own in the mirror. We are all, to a greater or lesser extent, on an unprecedented learning curve, whether we are intimately engaged in research, rushing to the scene of an attack as first responders, or going about our daily lives. We are all part of the story of our technological systems, their vulnerabilities, and the means by which we can remake them.

This book tells the story of people who are looking at these complex, interconnected systems—people exploring questions such as how they work; what their implications and possibilities are; what makes them fail and what will prevent those failures; how important the human element is in a technology-driven world; how human intelligence and machine intelligence can partner well and how and why problems can arise; and what specific kinds of tools and approaches are likely to ensure our greater security. It is also the story of people working in government agencies and in emergency and relief capacities, and of ordinary citizens, who today have unprecedented access to information and resources.

To a large—and perhaps, to some, surprising—degree, the search to understand our most promising new technologies leads us backward, into the history of optimistic starts and dead ends, accretions of fact, epiphanies, and unearthed connections that mark all scientific inquiry but are often left unrecorded by the doers. The aim of this book is to connect the dots—to show the continuum of research that spans decades and even centuries, as well as the vital links that can exist between diverse and at times seemingly opposed lines of inquiry.

Our aim is also to identify and explore, through the lens of the new threats we face, interlocking themes that illuminate the inner logic, versatility, and constraints of machine—and human—nature. Perhaps the most important of these is the power of unintended consequences, the way in which each new discovery or invention is tantamount to a move in a board game in which only some of the rules are known. Each move is influenced by what came before and each will elicit a response, but not necessarily the one we expect.

Another central idea can be summed up this way: Although technology has made lots of things easier, it has rarely made anything simpler. The problem lies in assuming that it should. Technological exploration is the process of initiating ever more contact with the unknown. Put another way, the more we learn, the more we don't yet know. As a practical matter, meanwhile, the most cunning solutions inevitably give rise to new problems. Create tools to, say, amass and store huge amounts of information and you are handed back, among other problems, the challenges of learning how to make that database actually useful to you (without overestimating its usefulness) and finding a reliable way to safeguard your new asset. Create automated systems to manage enormous real-world infrastructure networks such as transportation lifelines or public utilities and you confront the task of preserving the functionality of each constituent part while getting the whole to progress smoothly.

Another overarching issue is the dual-use quality of technology, which is to say the manner in which it can be put to good or bad purposes—which we are only too aware of at present, but which has a gripping backstory of its own. This issue, unavoidably, hits close to home for the authors of *SAFE*. To write about technologies, about their capabilities and possibilities and vulnerabilities, as we have done, is to provide material that might be of use to malefactors. When scientific papers about how to grow various microbes were found in a terrorist camp in Afghanistan, popular books on biowarfare were found with them. As authors we have labored to be sensitive to this and have taken care to not reveal specific vulnerabilities of operational importance. But we have to acknowledge the possibility that this book, too, will be rifled for clues as to promising modi operandi.

The reason for risking this possibility is that the knowledge gap is asymmetric. Small, relatively poorly resourced networks of sociopathic fanatics have a very powerful motivation for understanding the weaknesses in our systems. The citizenry at large and the forces that protect us are far more numerous and far better resourced, but not as focused on these issues. As authors we can't keep the bad guys ignorant, but we can help to bring the good guys—with their far greater resources—up to speed. We can add to people's perception of the threats and, crucially, augment their ability to assess the countermeasures on offer. We can contribute to an open debate that would be impossible if all such assessment were confined to classified discussions in windowless rooms. And we believe that, in facilitating such debate, we will help strengthen our defenses far more than we will aid any potential attacker.

The role played by the human spirit is yet another part of the technological saga. Technology can make us feel newly disempowered, and we need to overcome both our intimidation and our frustration. We need to be reminded that we're still essential to the picture. To know how essential, just think of the passengers on United Airlines Flight 93 who, empowered by a basic utilitarian technology, devised the only effective defense against the hijackers that September morning. As we write, "Their behavior is a powerful illustration of the advantage of human adaptability. The longstanding rules regarding hijackings stipulated that the passengers and crew should follow the instructions of the hijackers. The best way to ensure their safety, the conventional wisdom said, was not to antagonize the terrorists. Those on Flight 93, however, discovered through cell phone conversations with people on the ground that the terrorists had no intention of landing the plane, so in a heroic act of self-sacrifice and human resourcefulness, they developed a new rule: Fight back."

At the same time, where we fail to understand our own limitations, we cause our technologies to fail. Human error is perhaps the most common source of machine distress. Technologies can also fail without our help, of course, but it turns out that we frequently have a hand in undoing our own handiwork. The idea that creation implies control is one of our most cherished beliefs, cherished all the more because, deep down, we know it isn't true. (Think of the Garden of Eden. Or a teenager.) With the advent of systems so complex they exceed the bounds of human comprehension, we have to accept the reality that at times the best we can do is oversee the technologies we have brought into being; creation gives way to participation. While we might crave a retreat to simpler and more manageable scenarios, we're going to have to learn to live with the fact that mind-boggling complexity is now a permanent fixture of our world. We have to reframe our own perceptions of it so that we can recognize its inherent challenges and capitalize on the core strengths it embodies. We have to get out of our own way.

How can we balance the clear benefits our technologies bring with the risks and vulnerabilities to which they expose us? How can we stay in the picture but get out of our own way? And, while we're at it, is the water supply safe? Is the power grid secure? Is it okay to get on an airplane? Is there a way to spot a terrorist before the terror is unleashed? The good news is that, for every time we ask one of those questions, the experts in these pages have already asked it many times over and are positing solutions that take the swath of aforementioned complications into account. The bad news is that to implement solutions requires reasoned, open debate and political will—and few, if any, of these experts have forums in which their debates can be heard, let alone privileged access to political leadership.

George Dyson, the renowned historian of technology, has said of the age-old quest for artificial intelligence that "anything simple enough to be understandable will not be complicated enough to behave intelligently, while anything complicated enough to behave intelligently will not be simple enough to understand." In the context of present-day technology research and present-day politics, it's safe to say that anything simple enough for decision-makers to encapsulate into a sound bite won't be complicated enough to solve the problems we face, while anything complicated enough to solve the problems won't be simple enough for decision-makers to squeeze into a sound bite.

It's here that the yawning knowledge gap becomes a dangerously stark liability: Our political leaders are put off by all that complexity. They want a fix, something clear-cut, something they can easily understand. Better yet, they want tech support to just show up and take care of it.

And, to be fair, don't we all. But it's not that simple.

Do we have reason to feel empowered in today's world? Is it naive to have hope for a safer future?

Read on.

ONE

Lifelines

H OW WOULD YOU take out the Holland Tunnel?" asks Tom O'Rourke. This courteous 54-year-old engineer stopped updating his look somewhere around 1980—outsized square wire-rims, turtlenecks, a herringbone sports jacket that's more architectural than sartorial at the shoulders. His gray-white hair is parted at the distant left. In a crowded San Francisco restaurant, he looks across the table at three companions, attendees at a lecture he gave several hours ago, who have tagged along to dinner. But they're too shy to venture responses to his question. Visions of giant U-Hauls parked mid-tube and filled with explosives take shape in their minds and hang there.

O'Rourke answers the question himself. "Disable the ventilation," he says. "No human being could get from one end of the tunnel to the other without asphyxiating. The energy of an explosion wouldn't blow out the walls of a tunnel," he adds. "It would be funneled out the ends." This would be deadly in the moment, but not structurally harmful. "Destroying the ventilation towers," he explains, "would make the tunnel unusable for months."

Despite its dark import, O'Rourke finds a certain satisfaction in this little quiz. It embodies a truth he cares about passionately: that countless ordinary, invisible, disregarded systems are what make modern life possible. Disrupt them and a routine day can come to a stop, workers flooding suddenly out of offices into the streets, television news anchors interrupting regular programming, neighbors gathering and speculating.

The capabilities of these systems are both marvelous and daunting. From the geometry of columns that make buildings stand up to the elimination of bacteria in food-processing plants, from 9-1-1 dispatches to ventilation ducts—it's impossible to fully comprehend all the mechanisms and processes we rely on every day. Yet all these systems are vulnerable—to accidents, to internal failures, and to the relentless specialization that a technologically advanced society demands. And they are vulnerable to terrorism, both directly through targeted sabotage and indirectly as the result of attacks aimed nearby.

O'Rourke is a master of this invisible realm. He has spent his life in places most people don't think about, much less visit: excavation pits that prefigure buildings, cisterns that hold cities' rainwater, dirt passageways soon to become public transit lines, pipe networks that deliver natural gas. He lives, in a sense, in the land of omission, where buried conduits make the news only rarely and only if they fail, and he has spent his career making sure they don't. You might say he's like those systems themselves: hard at work accomplishing specific tasks, influential across distances, ingenious, unglamorous, connected.

In fact, O'Rourke is one member of a sprawling network, made up of thousands of engineers much like him, who spend their lives protecting the hidden machinery of daily life. They oversee those elements that the rest of us have mythologized and forgotten: air, water, earth, fire. They build roads and design ports, track diseases and program computers, fix aqueducts, monitor farms, and anticipate disasters. And they tend to see the world in terms of its systems: collections of elements that comprise a functioning whole, which can be assembled, dismantled, reconfigured, and improved. Engineers think this way about all sorts of phenomena, from tiny biological systems to hulking, man-made transportation systems. They see systems inside what might look to others like single, fixed entities—buildings, for example, or genomes—and they imagine ways to isolate, alter, or exploit individual components.

Long before terrorism made it to the top of the national agenda, this network of experts, so vast that many members don't know of many others' existence, was pondering the weaknesses and strengths of the systems that support us. Their careers have evolved and their ideas have flourished in tune with the nation's worries and hopes—and science funding. Many worked on nuclear weapons and defense during the cold war (O'Rourke himself "took a pass" on an opportunity to work on missile launching pads tunneled so deep in the ground that they could withstand attack, their weapons used to retaliate); they moved on to various interests such as natural hazards or environmental ills. And today, as the United States turns its attention away from past fears, pledging to heighten its defenses against terrorism, this community of specialists is already on the job. You just have to know where to look for them.

Born in Pittsburgh, Pennsylvania, Tom O'Rourke played sports, and read Dickens and James Fenimore Cooper, while his father worked as a salesman for a chemical company. It wasn't until attending Cornell University, in Ithaca, New York, that he discovered civil engineering. Since then he has cleaved loyally to upstate New York, returning from earning a Ph.D. at the University of Illinois, Champaign-Urbana, to Cornell. Today, after 20-odd years, he continues to teach in Cornell's Department of Civil and Environmental Engineering (recent courses include "Retaining Structures and Slopes" and "Rock Engineering") while logging at least 100,000 miles of travel every year. A lean, 6-foot, 4-inch figure whose permanent slouch is a gesture of inclusion to anyone looking at him from below, O'Rourke will show up at a congressional hearing in Washington, D.C., or the offices of the multinational construction company Bechtel in San Francisco, or at the site of the Big Dig—a massive urban reconstruction project—in Boston, or at the scene of an earthquake in Turkey. And these days, he's making a concerted effort on behalf of dozens of engineers like him to get the attention of the Department of Homeland Security, where he's certain his expertise is relevant. O'Rourke's peripatetic life, however, is a function not just of his energetic disposition (his voice-mail greeting encourages callers to "have a productive day") but also of the complexity and scope of his field.

In 1998, the Clinton administration issued a manifesto of sorts, calling on business and government to take heed of the systems that sustain them. Titled Presidential Decision Directive 63 (PDD 63), its focus was physical infrastructure—ordinary, dingy, ubiquitous—and specifically six types of it: telecommunications, energy, banking and finance, transportation, water systems, and emergency services. In the parlance of civil engineering, these are called "lifelines," and their tremendous importance has made them an object of interest for civil engineers as well as a target of warring armies and terrorists throughout history. In the 1940s, the Zionist Irgun attacked railroad sites in Palestine. The Shining Path in Peru has consistently gone after the power grid, leaving Lima without electricity or under rations, sometimes for more than 40 days. In the 1980s, the contras in Nicaragua attacked the food supply, planting weeds in cultivated fields, and not only destroyed a hydroelectric facility but also executed its designer. In his *Minimanual for the Urban Guerrilla,* Brazilian Carlos Marighella, a Marxist organizer killed by police in 1969, explicitly directs comrades to target infrastructure: "The derailment of a cargo train carrying fuel is of major damage to the enemy. So is dynamiting railway bridges," he writes. And: "As for highways they can be obstructed by trees, stationary vehicles, ditches, dislocation of barriers by dynamite, and bridges blown up by explosion. . . . Telephonic and telegraphic lines can be systematically damaged, their towers blown up, and their lines made useless."

Yet to most of us, and to the numerous government officials who read PDD 63 in the late 1990s, an awareness of these ambient structures was quite new. It was the emergence of the Internet, and its reminder that *connection* can yield amazing conveniences but also insidious surprises, that made the old systems linking us together— some of them for centuries—become more conspicuous. A fresh fear for the stability of this substratum of civilization emerged. PDD 63 warned rather desperately, "Any interruptions or manipulations of these critical functions must be brief, infrequent, manageable, geographically isolated and minimally detrimental to the welfare of the United States."

In part, these new worries were a reaction to decades of neglect. Much of the infrastructure in the United States was built early in the twentieth century and has undergone little maintenance. In New

York, the two main conduits that convey water to Manhattan, serving 90 percent of the borough, were commissioned in 1917 and 1938. Both have been in continuous use ever since; neither has been dewatered for inspection. (This failure is not because authorities are careless, but because they can't let half the borough go dry.) In many communities, water officials don't even know where their pipes are, any relevant maps having been lost long ago. In the power sector, the vicious competition unleashed by deregulation in the 1980s and 1990s has squeezed out not just ancillary functions such as research and development, but even some fundamental upkeep. As one Department of Energy engineer puts it, "The best way to bring down a big infrastructure is just like in the Roman Empire: Let it rot from the inside. That's about where we are."

Decay alone might not be so disquieting, but add to it the sheer scale of our lifelines, and the potential problems can seem overwhelming. The transportation category in President Clinton's directive, for example, contains a massive number of components: roads, runways, rails, harbors, docks, bridges, overpasses, signs, signals, railings, bus stops, and benches, to name just a dozen. The New York State Department of Transportation manages 456 aviation sites and 12 ports. Detroit maintains 472 public buses and approximately 9,000 bus stops. Ohio has 42,000 highway bridges. And so on.

In the energy sector, the word "complexity" is an understatement. The United States counts about 10,400 generating stations within its borders. Texas power provider CenterPoint alone owns 84,000 miles of distribution cable. Con Edison workers in New York access cable through 250,000 manholes and service boxes. Following the disastrous ice storm of 1998, Canadian power companies replaced 35,000 poles.

In recent years, a new breed of scientists has begun to plumb this type of complexity by focusing on the nature of networks—dynamic, interwoven systems made up of nodes (such as train stations) and links (the tracks that connect them). Again, inspired partly by the Internet—and more broadly by the power of computers to simulate some of their ideas—these theorists have made a series of discoveries about the behavior of networks such as the power grid. Perhaps the most interesting revelation to emerge so far from this vigorous new field is that networks from wildly disparate arenas can have closely related properties. The Internet, the economy, the proteins in a cell, social circles, infrastructure, and even fads behave in strikingly similar ways.

Yet while promising, these nascent studies have yet to produce any explicit new measures for strengthening our lifelines, and since Clinton's call to arms, concerns about what's now persistently referred to as "critical" infrastructure have steadily intensified. In the Bush administration's sleek blue 2003 directive, *The National Strategy for the Physical Protection of Critical Infrastructures and Key Assets,* the writers increase the number of lifelines in focus from Clinton's six to eleven, adding food and agriculture, public health, defense, chemicals and hazardous materials, and mail and parcel delivery. The United States, the report reads, is home to 66,000 chemical plants and 137 million postal delivery sites, and the nation's food supply network includes 1.9 million farms—all part of intricate flows that involve supplying, processing, production, packaging, storage, distribution, and, it is hoped, tonight's untainted hamburgers.

Despite these systems' ubiquity and geographical sprawl, they have never been our primary worry in the face of disaster. For obvious reasons, human beings are the objects of our most visceral concern: the real horrors of coordinated blasts in two Istanbul synagogues at the close of 2003 were the deaths and injuries and blood-spattered streets, not the loss of surrounding electricity. For the most part, damage to critical infrastructure is felt later, secondarily, when the body count is already a given. This injury is indirect, imposed on the safety or health mechanisms we depend on, but not on our physical selves. Power outages don't often kill people, though they can, for instance, kill cows, which produce key elements in our diet: without their electrically powered milking machines, dairies can't keep up, and the animals contract mastitis, a condition that has killed hundreds and incapacitated thousands at a time. The destruction of roadways won't so much send people to hospitals as prevent the already sick and injured from getting there.

Critical infrastructure is like the Holland Tunnel's ventilation system—not the thing itself but what makes the thing possible—and the power of PDD 63 was its will to bring these hidden essentials into the sphere of acknowledged concern. In a complex society, the importance of interconnections is paramount: data networks that confirm bank transactions; phone lines that dispatch ambulances; roads, airports, TV signals—all these vigorous linkages comprise what we think of as order. Orchestrate prolonged disconnections and you can create monumental chaos.

You also create ripple effects. Case in point: The loss of use of the Holland Tunnel might primarily cause deaths by explosion and simultaneous traffic disruptions—but it would also slow the delivery of goods to Manhattan and keep millions of people from going to work. And ripple effects can include policy shifts and mass psychological changes: As a consequence of 9/11, for instance, the number of foreign academics coming to teach or study in the United States in 2003, for the first time in recent history, plateaued; no doubt a combination of immigration restrictions and the altered social climate kept them away. Oddly enough, as the 2001 story of "shoe bomber" Richard Reid makes clear, even foiled attacks can cause major ripple effects. It doesn't take a successful bombing to get a whole nation of travelers to walk sock-footed through the airport.

Meanwhile consequences spread laterally as well. An attack on a single port would mean not just the crippling of that single shipping node; it would also mean port closures on every coast, creating a self-imposed embargo across the country—whether merited or not. The expectation of copycat assaults swells the shutdown of services well beyond the circumference of one actual event.

Tom O'Rourke internalized these truisms long ago. He and other engineers and researchers have been scrutinizing these secondary entities and their trails of effects all their professional lives. These men and women are the sentries of *systems,* and for decades, beginning long before PDD 63, they have been developing an understanding of how "extreme events" affect the stability of their wards.

Not surprisingly, natural-disaster specialists have detailed, proven, and transferable expertise. Today, they are our deepest resource in understanding how to fortify and restore lifelines in the face of terror. When the World Trade Center towers collapsed and officials had no way to determine whether nearby structures were safe, they turned to earthquake engineers, whose protocols for evaluating the viability of damaged buildings had been tested and refined for decades; New Yorkers ventured into dust-covered downtown edifices carrying the inspection forms developed by seismic experts. Hurricane specialists know how to evacuate 100,000 citizens from an urban center without jamming intersections; this would be useful wisdom in the event of a chemical attack. Planners at the National Forest Service can rapidly establish command structures for teams of thousands of individuals responding to wildfires, a helpful skill if a dirty bomb were to draw dozens of rescue agencies to an incident.

Like floods and earthquakes, terrorist attacks are low-probability, high-impact events; any hope of defense requires methods of prediction, concentrated investment, informed planning, and elaborate, speedy response. Earthquakes especially resemble terrorism, because they erupt without even the slightest warning, while hurricanes and tornadoes tip off meteorologists beforehand. And, like natural disasters, man-made ones are powerful connection breakers. They can rupture physical links such as water mains and power lines, and they can upset less obvious flows, causing, for instance, refineries to stop operations or emergency responders to flock to one site, depriving other areas of services. No one can anticipate these effects with the acuity of disaster engineers. As at home in the scenarios of forethought as they are in the land of afterthought, they are masters at quantifying the risks of disruption and recovering from its upheavals.

Tom O'Rourke is, as it happens, an optimist, a spirited man who uses adjectives such as "can-do" and dispenses praise for his favorite colleagues in near swoons of appreciation. (His alma mater, Cornell, also gets lavish treatment: "It's beautiful, progressive, sensitive, and caring.") O'Rourke harbors a combination of a scientist's pragmatism and constitutional good cheer, which recently prompted him to conduct a somewhat bizarre experiment. Although he doesn't believe that when something goes wrong, it was somehow meant to be, he began, not long ago, to wonder whether dashed expectations were necessarily a bad thing. If he was so disappointed when the waiter announced that there was no more petrale sole, then why did he thoroughly enjoy the panfried cod he ordered instead? Hey, there are so many positive outcomes! Why should one thwarted objective give way only to inferior ones? So, with typical scientific attention (and an endearing obliviousness to human psychology), he began an empirical study, using his own life experience. He kept count each time he experienced a letdown. The data, it turned out, looked great. "So far," he reports, "I'm finding about 50 percent good alternative outcomes!" In this way, O'Rourke is logically invalidating the expectation of regret.

He is also exposing a small marvel that dwells along the undersides of the scientific method: the power of the unexpected. Great experimenters understand that being wrong can lead to triumph, as long as they maintain the will to know. Scientific rigor, picayune attention to detail—these traits may look like rigidity to the outsider but, on the contrary, they can be the very roots of flexibility. The truth frequently propels change. In an age when advances in computing and chemistry and, perhaps most of all, biology are accelerating, the best engineers can learn, evolve, and adjust their thinking in decidedly unrigid ways.

O'Rourke's little study illustrates another principle as well—that to prove a scientific proposition requires observation and repetition. Which is why he must sample many apparent disappointments (and perhaps many fish dishes). It's also why earthquake specialists have a certain morbid appreciation for disaster. "Disaster researchers don't want disasters to happen," he says apologetically, "but you've got to take advantage of them if you're going to learn."

O'Rourke's own repeated observations of infrastructure began with the soil. In the 1970s, when most scientists were studying the direct effects of earthquakes on underground structures, he saw that even more detrimental to pipelines and tunnels was the lasting deformation of the ground. O'Rourke bypassed the collapsed homes caused by quakes, and he even bypassed the pipelines that deliver water and gas. He chose as the core aim of his career to understand the earth in which the pipes were buried. But how could you quantify or—even harder—predict the form of something so massive and elemental? It may seem inert, but the ground can rotate, lurch, rupture, crack, liquefy, oscillate, extend, compress, shear, consolidate, and dilate in the most mind-boggling ways. And here's where O'Rourke's rigor and flexibility—both at once—kicked in. He thought, Well, let's give it a shot. With some research money from Cornell, he began trying to parse the earthly confusion—and proceeded to revolutionize the way quakes were studied. His great contribution to the field has been to help figure out, as he describes it, "how the ground behaves."

The funding kept coming. With the possible exception of flooding, seismic events have been the most underwritten subject of research among the natural hazards, with the federal government doling out about $100 million each year to earthquakes. (In these disciplines, individuals refer to groups of professionals by their specialty, as in "I don't think wind would agree" or, "Floods are very well organized.") One of the fruits of this support is "a dossier of real-world conditions," as O'Rourke puts it, amassed observations that reveal not just how strong the jolts are or how deep the fissures in a tremor, but also the ways buildings fall and tunnels cave in, which pipes bend and which break, and at what frequencies houses tend to shake. O'Rourke has been one of the chief contributors to this documentation.

Sometimes he makes his observations right there in Ithaca. In 2000, for instance, he and partners from Tokyo Gas Company and a Tokyo university built the largest full-scale replication of ground effects on pipes ever conducted inside a lab. They constructed two boat-sized boxes, the first anchored to the floor and the second rigged so that it could slide alongside the first. Then they poured 60 tons of sand from a three-story bin into the two containers, burying a steel pipe 3 feet beneath the surface. With the innocent strand of pipe running through both containers beneath the sand, the team shoved the movable box at 4 miles per hour a distance of 3 feet. The buried pipe writhed and bent. Twice more they hurled the movable box across the flooring, each time with wetter sand. Each time they meticulously read their results from 150 strain gauges stationed throughout the containers. From this effort, O'Rourke learned that certain steel pipes under certain conditions bend in the ways he expected, and that his mathematical model was sound.

But many of O'Rourke's observations are made far from home. His first trip to the scene of an emergency was to the site of the magnitude 8.1 earthquake that devastated Mexico City in 1985, where he actually had no official business—"I was doing post-earthquake reconnaissance *tourism*," he says. But soon after, his work produced critical data when he perused the damage and measured the flooding caused by the Ecuador earthquake of 1987. (In that event, landslides destroyed a single 26-inch-diameter oil pipeline, which resulted in the loss of 60 percent of the country's export revenue. In response, one index of U.S. oil prices jumped 6.25 percent and Ecuador ceased exports for five months, sending the economy into a devastating spiral.)

O'Rourke also observed earthquake consequences in Armenia in December of 1988. Ten days after the country suffered a magnitude 6.9 shock, he boarded an Ilyushin 76—the Russian-built cargo plane with a colossal gut and double chin hanging from under its wings— that was headed for a Moscow energized by glasnost. "It was the first time since World War II that the USSR had accepted aid from outside the Soviet bloc," he says. Having welcomed outside search-and-rescue teams a week before, the Soviets now asked a group of American engineers to survey the damaged infrastructure, anticipate aftershocks, and make recommendations for the longer-term future. O'Rourke slept through the long flight and refueling stops, curled atop a 20-foot-tall pile of hardware in the Ilyushin's hold: digital seismometers and dialysis machines (the Armenians were suffering from "crush syndrome," in which damaged muscles release toxins that collect in the kidneys; victims were leaving the hospitals, thinking they were okay and dying several days later).

"The whole trip was a space odyssey," O'Rourke recalls. By the time the team had been briefed in Moscow and flown into Armenia, they were in a deep haze. Through banquets and Armenian brandy (the welcoming protocol), O'Rourke would watch for his team leader to nod out and then immediately let his own head drop to the table. "But nobody had jet lag when we got to Leninakan," he remembers, reliving the wakefulness that desolation induced. "It was leveled, destroyed. Eight-story houses looked like they'd been detonated. The weather was freezing cold, which was good, because the minute we had a thaw, the smell of bodies was overwhelming. I remember all the clocks were stopped, all showing the same time.

"The first night we camped out in a damaged building," O'Rourke goes on, "where it was well below zero, and then we headed to Spitak, which means 'white' in Russian, and I could understand why: The mountains were covered with snow, antiseptically white. When we got up high to survey the town, it looked like it had been hit by an atomic bomb. I couldn't even figure out where the city was. I remember scanning the horizon and realizing *that* was the town. It was virtually 100 percent razed." He attempts to recapture the horror. "You're transfixed, you're not part of this world."

Such events put the relatively minor effects of quakes in the United States in perspective. In Bam, Iran, in late 2003, the death toll hit 30,000. Only once has a U.S. earthquake and its aftermath taken more than 500 lives, and that was the infamous 1906 temblor in San Francisco, whose death toll has been contested over time, ranging from 700 to 3,000. The second most deadly quake in the United States took place in 1946, when the sea floor ruptured near the Aleutian Islands off Alaska and triggered a tsunami throughout the Pacific, causing a death toll of 165.

O'Rourke's team did a technical assessment of Spitak building designs (these had been dictated out of Moscow, where the quake that architects had anticipated, he says, "was more like a strong wind") and the ways the local masonry tended to vibrate; they analyzed pipelines and the two nuclear plants (unharmed), and they wrote reports that future Spitak engineers would refer to. The team brought ideas for retrofitting—how to do it and to which buildings, based on risk and cost—as well as credence the Soviet engineers could use to make their case for funds.

O'Rourke's travels led him to a number of observations, which he carried back to the United States and used to help protect comparatively deluxe American homes and infrastructure more proactively. The work that he's most proud of bore fruit in California 10 months after his journey to Armenia. Before the Loma Prieta earthquake, which hit the San Francisco area in 1989, O'Rourke was able to combine and manipulate the observations he'd accumulated throughout his career and put them to use in order to make predictions. His main technique for transforming that dossier of real-world conditions into actionable wisdom was to layer old and new maps, pictures, and measurements, in order to better understand San Francisco's geography. At the time, O'Rourke and his colleagues used their hands and their brains; today, this technique is accomplished on computers using geographical information systems, or GIS.

The nature of this technology is hard to get a grip on. The acronym, in its own right, is a weird term of art. You can refer to "a GIS," as in "a water cooler"—a self-contained database and set of functions that lives on a computer and is shared by a bunch of coworkers. Or you can refer to GIS in the abstract—not a brand name, not a technique, but a blurry grand term, less analogous to a water cooler than to all of H2O. Used in this way, the term is as broad as cartography itself, as in "GIS has revolutionized the world."

Like many intellectual revolutions, this one can be said to have changed the way we see things. In the case of GIS, though, the claim is literally true. The GIS revolution is based on a remarkably simple idea—that software can be made to arrange the data in a database according to the places in the world it applies to, and to display that data on a screen. At one level, a GIS is just a computerized map: it tells you what is located where. But whereas maps tend to deal with only one kind of data at a time—a street map tells you routes and names, a census map tells you who's who, a geological map tells you where to find iron ore—a GIS can present you with whatever data set happens to be of interest, all in one place, as long as it's available in the right format.

This might seem interesting only to geographers and city planners—people traditionally concerned with places. But though most of our everyday data doesn't come from maps, that doesn't mean it's not mappable. In some way or other, most data can be tied to a place. Information about people and companies can be linked to home addresses and business premises. Crime statistics can be linked to precincts and, if collected the right way, to specific sites. According to Russ Johnson, a manager at the leading software company in the field, if you look at it the right way, "Ninety percent of the entries in a typical database are geographical."

But GIS software is not only a convenient way of looking at different data sets. It's also a way of seeing connections between them. It provides you with a way of seeing data arranged by space, making connections visible in ways they never would be in mere columns of data. Web services such as Mapquest use GIS techniques to link the latitude and longitude of specific addresses to maps of an area. GIS applications might allow a bank to find out where to install ATMs in order to be close to as many office workers as possible, or to help a fast-food chain put new franchises where its competition hasn't got any. What route should a convoy carrying hazardous waste take across the country so as to expose the least number of people to risk? Ask a GIS.

The geographical connections Tom O'Rourke made before San Francisco's 1989 quake were the kind of connections GIS is especially good at. At the request of then-mayor Dianne Feinstein, O'Rourke and his colleagues had already begun a study of San Francisco's auxiliary water supply system, which was built after the 1906 quake. The fire that was set off by that temblor had blazed for three days and destroyed 490 city blocks. It had made the need for a water system dedicated purely to fire fighting painfully clear, and between 1908 and 1912, San Francisco responded by building the only high-pressure water supply of its kind. Eighty years later, however, officials weren't sure how the infrastructure would hold up in the wake of another Big One, and O'Rourke had a plan for finding out. Using calculators, pencils, and rulers, he merged maps of San Francisco in 1985 with repair data from records of 1906, along with information about the city's massive areas of fill (at three different times over the century, the city had leveled its terrain and expanded its boundaries by dumping fills along the shores of San Francisco Bay) and the lengths, types, and diameters of the city's underground piping. Then he searched the city for photographs, accumulating about 60 old images (he found one of them on the wall of a bar and made a copy of it with his own camera) that depicted the damaged sidewalks and structures in 1906. "Every picture," O'Rourke explains, "is a measurement. If you know that the cobblestones were 2 inches by 4 inches by 6 inches and the telegraph poles about 100 feet apart, you can use that as a basis for estimation." With all this material, he generated a rendering of the city that applied observations from across history and from across disciplines to each meter of turf in the present. Next, he ran a computer program that could take input describing a pipe break here or a leaky hydrant there and, based on all that data, calculate the overall effect on the auxiliary system. With a breach that lost water at 20,000 gallons every minute, say, you could compute in your head that you'd have an hour before your 1.2-million-gallon tank ran dry. But what would happen if there were ten breaches and five tanks, plus a few valves you could shut off within 10 minutes to spare flows? O'Rourke's maps and computer program could tell you.

What they told O'Rourke and his colleagues about San Francisco was ominous: the auxiliary water system would not hold up in the event of a serious jolt. In November of 1986, the city responded by voting to put $46.2 million into upgrading its fire- and emergency-response capabilities; in addition, Deputy Fire Chief Frank Blackburn ("Aaaah, he's a hero," says O'Rourke) insisted on purchasing several trucks, packed with hosing that could lock into giant pumps on a fireboat and suck extra water from the bay. The vehicles could then race to the blazes with endless supplies.

On October 17, 1989, the lurches of Loma Prieta caused 160 water main breaks in San Francisco, and, indeed, within about 40 minutes, in the neighborhood most vulnerable to fire, the firefighters' system was empty. Without O'Rourke and his colleagues' recommendations and Blackburn's portable hoses, San Francisco would have burned once again.

For several years after Loma Prieta, O'Rourke studied its effects, adding to that ever-fatter dossier of real-world conditions. He updated his maps with new data, bored 50-meter holes around town to better understand the city's soils, and checked the quake's reality against what his models had predicted. "Then, by 1992, I thought, 'OK, this is old. What's next?'" He answers his own question as if drawing the most logical conclusion in the world. "I thought, 'It's a good time to be in LA.'" In 1994, when the Northridge quake hit, O'Rourke had already begun studying the configuration of pipes beneath Los Angeles, and soon afterward he began building a true, computerized GIS that could protect the inhabitants above.

GIS begins here: a leafy campus tucked into a hot, industrial valley in Redlands, California. The buildings are painted earth tones in the contemporary architecture style, circa 1969. In fact, that's the year that the campus was founded. Originally called Environmental Systems Research Institute and now known formally by its acronym, ESRI is the uncontested front-runner in developing software for geographical information systems. With revenues of about $500 million a year and customers in every industry from natural resource management to banking and finance, real estate, and defense, this quiet, privately held company has an unyielding grip on the business of GIS. In fact, according to 2002 data, about 35 percent of all GIS software in the world—with closest competitors straggling at 10 and 9 percent—is sold by this quirky company, still owned by a guy who drives a Ford Taurus station wagon and his vice president wife. Today, ESRI has its own three-story research and development center and its own press, which pumps out booklets and volumes of maps—gorgeous, colorful maps that resemble chemical spills and CAT scans, candy wrappers and the paintings of Richard Diebenkorn.

The company's founder, visionary, and resident workaholic is Jack Dangermond, a sizable man in his sixties, with round eyeglasses and a droopy face, who was trained as a landscape architect. (He's responsible for the shady, peaceful feel of the company's 20 buildings.) In the 1960s, at Harvard, he'd been doing geographical market research using punch-card machines, and after nursing a successful consultancy for a decade, he began selling his own software. For the last 15 years, his company has grown 20 percent annually. Today it has a solid customer base—300,000 businesses and institutions use its software—and 1,500 developers currently create tools that snap into its platform. Prices range from hundreds of dollars for a simple desktop program to $10,000 for a high-end system.

ESRI's products—GIS software programs of varying sophistication and capabilities (some programs, for instance, allow you to put your maps and geographical data on the Internet)—is useful to scads of professionals ranging from campaign consultants to archaeologists, but they're particularly valuable in the world of infrastructure. Presumably, no contractor would try building an electric power system without several good maps, and good maps are just what ESRI software creates. Over the past decade, large and small utilities—including power, water, gas, transit, and telecommunication companies—have been scrambling to build computerized depictions of their networks using ESRI's software and its consulting and technical services. One Texas electric company has a staff of 200 "GIS editors," whose job is the care and feeding, correcting, updating, and enhancing of that company's system.

Some of the advantages of GIS software are simple and obvious.

The problem of maps being drawn to different scales, for example, goes away; a GIS can be consulted at any number of scales. If topographic data about the landscape, say, and practical data about rights-of-way are on maps of different sizes, combining them becomes easy once the two are in a GIS format. The different data sets form "layers," all of which can be seen separately or on top of each other. The software is like the transparencies in old anatomy books: turn a page, see the skeleton; turn another, see veins; and one more, see muscles wrap the bones. Then do surgery.

Not only can maps at different scales be combined; so can maps made with different "projections." Here, the disparity isn't simply size; it can be in the very curves, angles, shapes, and square mileage of the two pictures' individual elements. To make a map, cartographers translate the three-dimensional curvature of the earth into a two-dimensional picture, and they do this using a projection. The trouble is that there are nearly as many possible projections as ways to peel an orange. Score the fruit with straight lines, pull off wedge shapes, and lay those flat across a plate, and you won't end up with anything like a rectangle. Unwrap the fruit in one long curlicue or break off the husk into a few sturdy cups—however you approach the task, you can't end up with a perfectly flat piece of peel. And so all map projections distort the landscape they represent, one way or another. Continents get pulled one way or another. Oceans and coastlines tilt. Places with the same area don't end up looking the same size, features with similar shapes end up looking wildly different. To GIS software, though, projections are just mathematics. Changing projections in order to combine two data sets is a simple matter.

The harder work is in making the GIS in the first place. It's not just a matter of scanning existing paper maps into a computer—though that's where a lot of GIS building starts. The result of a scan is a mere picture, an array of dots—pixels—with no other information attached. For a GIS to work, it needs information; the dots and the patterns need geographical meanings. Typically, transforming an old paper map into a meaningful image on a screen is done by an operator with a mouse who defines specific sites on the map and tells the machine what they are. It's grunt work, meted out by government agencies, utilities, and such to college students and entry-level graduates.

ESRI's software makes it easier; it will automatically draw in lines from highlighted point to highlighted point and expedite the process of characterizing them (street, path, border, stream). But for the most part, computers still need guidance to see a picture as an actual place.

Then there's the business of collecting the data that will fill up the GIS, tables and tables of it, list after list, and integrating all that material. Data itself is a product today, sold by data companies. Geotrac, for instance, based in Norwalk, Ohio, will sell you data about flood zones. Geographic Data Technology, in Lebanon, New Hampshire, will sell you average daily traffic counts for 400,000 points around the United States. From the WhiteStar Corporation, you can get data on the location and productivity of millions of oil and gas wells in the United States, and from the Janus Group you can purchase data about cable television markets. These companies and a slew of others sell data that's already formatted and ready to load into ESRI's software; for thousands of other sources, once you've collected the facts and figures, you still have to reorganize them so that your ESRI GIS recognizes what they are.

And last there's the most potentially difficult task of all: making sure that the world in your maps corresponds to the world outside your window. The data in a GIS needs to be "georeferenced"—to be defined with respect to a fixed set of coordinates in the real world. If your GIS is not properly georeferenced, it's not much good, and the different forms of data in it may not be lined up correctly—it would be as though our anatomy textbook's binding were bad, and the spleen appeared to be tucked up into the ribcage.

One way of georeferencing is to use images taken from aircraft or satellites as base maps. (Problems of perspective—the geometrical distortion of the more distant things in the picture—can be removed by the computer in the same sort of way that a map in one projection can be changed to match a map in another.) Such image-based maps are very good for showing you a real-world picture of what's where. For other purposes, though, you need a more sophisticated satellite-based approach: the global positioning system (GPS), operated by the Department of Defense. GPS is arguably the most transformative military technology since the invention of the ICBM. The system uses a constellation of 24 satellites in six different orbits, each of which constantly transmits a very accurate time signal; from anywhere on the earth at any time, it should be possible to pick up the time signals from at least four of these satellites. The time signals will all be slightly different, because the distances between you and the four satellites will be different, and those slight discrepancies allow the circuitry in your GPS receiver on the ground to work out exactly where you are.

For the military, GPS offers the immense tactical advantage of knowing exactly where forces are deployed, regardless of whether they are tanks, infantry soldiers, aircraft, patrol boats, or smart bombs. And in the case of the smart bombs, it is also a way of telling them where to go: bombs are given the GPS coordinates of their targets and use that knowledge to trim their trajectories as they fall. For civilian users, GPS is a way of not getting lost when hiking, sailing, or driving around a new city. As the cost and size of the receivers fall steadily, they are appearing in more and more places—such as in cell phones. In the long run, this technology will do for people's sense of place what cheap wristwatches have done for our sense of time, giving us easy, reliable, and precise updates whenever we want them. (Timex, as it happens, already sells a GPS-enabled wristwatch.)

A GPS receiver makes it possible to get reliable georeferencing data anywhere you go without all the traditional mapmaking paraphernalia of theodolites and other surveying instruments. This makes it much easier for GIS developers and users to tie their data to the real world. In fact, the two technologies are as mutually supportive as their initials are similar. GIS displays are by far the best way of showing GPS data—a receiver that shows you an electronic map, visually indicating your actual location, is much more helpful than one that simply flashes numeric coordinates. And given your GPS coordinates, a GIS map not only shows you where you are; it also shows you where you've been and what you're near.

Dangermond, the passionate father of GIS technology, thinks that what you're near matters a lot. He likes to point out that spatial information is what connects us. It's a language that most, if not all, modern cultures speak. To recognize that the school is "across the street," and later that the supermarket is "on Howell," and eventually that the streets of Paris "form a star" and Pakistan is "just north of India"— these functions are virtually universal. What's more, they are capable of nearly infinite extensions: Pakistan is also east of Iran, counts a population of more than 160 million, and has a 46 percent literacy rate. Dangermond believes that its ability to reveal data and connections in this way makes geography itself a "unifying metaphor" for society—and the GIS revolution is making that metaphor ever more powerful. And he's an idealist, to boot: he believes that geographical information, if cleverly arranged and interpreted, can help us better secure and govern our world. It's key to protecting our environment, conserving natural resources, and even negotiating peace when borders are in dispute. It brings together disparate interests, constituencies, and even cultures, he says, rendering a telephone pole not just a wooden rod planted at such-and-such latitude, but simultaneously a conveyor of kilovolts, a reason for an easement, a lessor to cable TV outfits, an asset, a risk, and a maintenance node.

Russ Johnson is flesh to Dangermond's spirit—and he has direct experience of working where the twin questions of where you are and what you're near have life-and-death importance. Johnson, the top contact at ESRI for customers using GIS for homeland security and public safety, worked as a smoke jumper in the 1970s; he and three or four other guys would parachute into fires raging in the middle of nowhere. To parachute into a fire, he needed a good idea of where to land—in this case an area of young vegetation with no steep drop-offs, upwind of the blaze but close enough to get his work done. Until ESRI's software came along, Johnson and his team would fly overhead and eyeball the terrain, jotting down what they saw. Then they'd drop crepe paper streamers with weights attached, to see how much drift to expect for the real plunge. At best, in those days, a cartographer back at dispatch might have scared up topographic and vegetation maps, prayed that the two were drawn at the same scale using the same projections, and, if so, traced one onto acetate to lay atop the other. At worst, Johnson took his best guess and jumped.

Today, the terrain of a fire exists on a hard drive, available in myriad forms, accessible by ESRI technology. Lay wind directions over topography over vegetation type over latitude-longitude, and today's smoke jumpers know exactly where to drop before they're up in the air.

A short man with perfect posture and a muscular neck, Johnson comes off as someone who has little fear, not so much out of bravery but as a matter of habit. For 30 years, he worked for the federal government in emergency response, "chasing fun things" from Mexico to Alaska. At storms in Florida, floods in southern California, the great fire in Yellowstone National Park in 1988 and the Inniki hurricane in Kauai in 1992, Johnson served on teams called incident commands (ICs), which fly into locales in crisis "to add a management layer for command and control," he says. "The IC brings in the weight and checkbook of the feds." Its leader establishes a headquarters and names a committee, usually including the fire and police chiefs, that reports to the governor or mayor. This incident commander handles the press and the politicians. He also brings in top-of-the-line mapping resources. "I wouldn't take my team to an incident," says Johnson, "without a GIS specialist."

Johnson had a bumpy transition to ESRI when he arrived five years ago to help public safety customers such as fire departments exploit the technology. "I went from being big chief of ops to being one of 3,000 employees. The first year was all about eating crow. I liken it to throwing a newborn into a swimming pool." But by 2001, a couple of years in, he was finally assimilated into this insular, idiosyncratic enterprise—just in time for his area of expertise, emergency aid, to spin into a frightening new form.

"Homeland security is all about geographical data," Johnson says, as he glides through a PowerPoint recap of September 11, 2001. "Everything that occurs during and after an incident has a spatial component. The first image you need is a search-and-rescue grid. Then you want to bring in construction equipment. Well, where do you put it? A massive crane needs a staging area."

Johnson was one of some 40 ESRI employees who helped run a mapping station during the weeks after September 11. When the city's state-of-the-art Emergency Operations Center (EOC) vanished into the rubble of World Trade Center Building 7, the city's GIS was destroyed with it. For three days, Alan Leidner, New York's director of GIS, dragged his meager replacement equipment from place to place looking for a home. By September 14, the new EOC was established at Pier 92, housing 20 GIS workstations and 5 large-scale map printers called "plotters." That took care of the hardware; still, all the relevant data and the hard drives that held it were either deep under the rubble or scattered among government offices, many of them shut down by the disaster.

So the mappers reached out to dozens of agencies and enterprises, some of whom even delivered data by hand: Space Imaging Inc. for satellite pictures, the Department of City Planning for lot numbers, Finance for property records, and the Department of Buildings for floor plans. Utility companies and the health department contributed. Command posts in the field submitted daily updates. There were thermal images and light detection and ranging (LIDAR) pictures and three-dimensional renderings that allowed viewers to "fly through" the wreckage.

"For the first few days," Johnson recounts, "it was slow, until people realized the value of what we were doing. Then we'd have people standing around at all hours yelling for maps. They had all sorts of questions: Where was the underground water wall that was keeping out the Hudson? Where were vacant buildings that could hold supplies and provide shelter? Where should we run a fiber-optic cable to bring the stock exchange on-line without tying up vital streets? Which bridges are strong enough to take transported debris?" The crucial information was culled from between the layers, where data met data—where the coordinates of fuel tanks met floor plans. When phones went down, mappers checked the affected areas against the addresses of homebound residents—matching utility data with health department data—in order to send visitors to help.

The vision of many in homeland security—and in fact the vision the Bush administration articulated in its sleek blue book—is of "an integrated critical infrastructure and key asset geospatial database for access and specific use by federal, state, and local government officials, and the private sector." Which is to say: *GIS for everyone*. The federal government is in the process of developing a "national map" that will serve as a single foundation for public and private organizations to build on and refer to; as part of that common base of information, the map will include exquisitely detailed satellite pictures of 133 major American cities.

•　　•　　•

While GIS can create an image of an infrastructure system, using gorgeous hues and endless detail, scientists and engineers charged with protecting that system from attacks must also study how it behaves. They have to grasp the system's overall dynamics: Which configurations, for instance, operate most efficiently? Which are most robust? How many elements can the system spare and still achieve a particular goal? And most important, what makes it susceptible to disruption? "To understand the vulnerability of infrastructure," says O'Rourke, "you have to understand the behavior of its components *and* the configuration of the network."

This big-picture thinking is particularly crucial when it comes to terrorism. One of the key discoveries so far in the emerging field of network science is that connected systems of all kinds perform differently when responding to random error than when confronted with deliberate attack. While an onslaught of accidental failures might have almost no effect on many networks, a concerted, selective attack could bring an entire system down. Albert-László Barabási, a physics professor at Notre Dame and a prominent figure in the discipline, has pinpointed the reason for this. In most networks, he and his colleagues have found, individual nodes have widely varying numbers of links; specifically, most networks exhibit an arrangement in which a few nodes have many links, while the majority of nodes have only very few links. In social circles, for instance, typical members may have 9 or 10 acquaintances while a handful of especially popular members might have 50. Because of their many connections, these ringleaders are key to the network, binding all the less popular members together.

What this means for terrorists is disturbingly significant: by identifying only a few well-linked nodes—like Chicago's O'Hare versus the Greater Peoria Regional Airport—they have a chance at damaging an entire system. As one Department of Energy engineer puts it, chillingly, "Give me the nineteen 9/11 terrorists and put them all to work on the power grid, and they could bring down the whole country for months." With vital hubs compromised, failure would travel rapidly and unpredictably throughout the network. In such a scenario, elements and interrelationships across the entire picture become pertinent. Now the great tangle of connectivity is in play, and suddenly the world is irredeemably, irreducibly complex.

This is a scary thought. But, as Barabási points out, there is a good side to the complex configuration of these networks, too. Complexity gives us options. Complex networks, in the unintended poetry of mathematics, can have more "degrees of freedom." The Internet is vulnerable to targeted attacks, but its fundamental architecture—that the vast number of its nodes are link-poor, meaning few other computers depend on them, and that the critical nodes are hidden among billions of dispensable ones—makes it also very resilient: hard to damage, near impossible to crash. The complexity works *for* us, not against. As in nature, diversity makes us stronger, not weaker; homogeneous systems—gene pools, monocultures—are more vulnerable than heterogeneous ones. Or take the way that New York City—as complex a network of interests and lives and infrastructure as the world can boast—responded to September 11. Hundreds of thousands of people were housed, fed, kept safe, helped home, and put in touch with those who feared for them. The resilience was remarkable—and it was achieved because of the complex makeup of the city, not in spite of it.

The road to safety is not to simplify things. It is to try to arrange the world in such a way that its complexity keeps options open, offers alternatives, provides opportunities to adapt. Sometimes this is a matter of finding the possibilities for resilience that a complex system has hidden within it. Sometimes it's a matter of redesigning a brittle system so that it can offer us more room to maneuver.

New theories seeking to define the ways in which styles of networks behave differently will begin to help here. But one approach to network health has been obvious for centuries: redundancy. Backup systems—from polygamy to the spare tire—are age-old forms of security. In infrastructure, they abound. In fact, the geometry that describes so many collections of power lines, underground conduits, and roads—those uniformly spaced horizontal and perpendicular lines we call a "lattice" or "grid"—is a form of redundancy itself. This is one of the beauties of cities based on blocks. With four links at every intersection, you've got multiple paths toward every node. "The square lattice is a very redundant system," says Albert-László Barabási. "The grid is very error-tolerant."

When necessary, utilities add more redundancy on top of the mesh. Alex Tang, a former engineer at the Canadian phone giant

Nortel, who now consults for telecommunications and energy companies, helps local Bells serve financial institutions. Banks make millions of transactions each day, and lost minutes, even lost seconds of connection time, cost money. So they buy premium-priced redundancy from the phone companies. "Banks are supported by *two* telecom networks, though they use only one," Tang explains. "The other is a standby, which we call the 'slave.' The two systems are running in sync, but information flowing into the second system isn't actually going anywhere—unless there's a problem. Then the system makes a millisecond decision to switch on the slave."

Networks rarely maintain a grid structure for long, however. That kind of regularity only occurs where a centralized design authority is making choices about how a network should look, and for infrastructure, which covers large expanses of territory, that sort of authority rarely exists or lasts over time. "Most supply networks out there are consumer-driven and designed locally," says Barabási. "Wherever people decide to build their house, that's where the water pipes go." Instead of ending up with a grid where each node radiates the same number of links, in most networks we end up with the much less consistent arrangements we see in the Internet or the social circle. Barabási and his colleagues refer to these link-and-node configurations as "topologies" and believe that the nature of various topologies—whether some nodes have many links and others very few, for instance—can tell us a great deal about how the networks behave overall. Understanding which topologies create brittleness and which create resilience is proving a tremendously valuable area of research.

When it comes to large infrastructure, the lattice disappears in response to other conditions as well: namely, a network's extending geographically beyond populated centers. Outside cities and suburbs and towns, long transmission lines are strung across acres of open country, with no crisscrossing paths. The trans-Alaska pipeline runs 800 miles, passing through three mountain ranges, without backups. The Colorado River aqueduct runs 242 miles, portions of it through barren desert. A parallel aqueduct, built to take over if the original were damaged, would be unthinkably expensive and wouldn't do the job anyway: to protect against malicious intentions, you need *dispersed* redundancy; otherwise, attackers can destroy the first and second lines simultaneously. Instead, the fix here is a blunt instrument—like cement. Utilities "harden" these parts of their systems, wrap or reinforce them with impervious materials, sometimes literally pouring concrete over pipes. Steve Grise, an ESRI designer and product manager, considers this a pretty feeble solution to a nearly impossible problem. "A person, a lawn chair, and a shotgun would do a lot more," he remarks. And it's true: outstretched miles of our critical connectors, hardened or not, remain exposed. In fact, the purpose of most security implemented throughout the electrical system has been to keep people out of facilities for their *own* safety, not to protect facilities themselves. Juan Torres, an infrastructure engineer for the Department of Energy, makes the point this way: "If you think of telecommunications as the nervous system and energy systems as blood, all our organs are hanging *outside* our body."

The trouble with redundancy, meanwhile, is that free markets don't like it. Sometimes called "surge" or "reserve capacity," surplus is the enemy of efficiency. Profit makers want to boil out the excess, pull up the slack. Supply chains, marketplaces, networks—all are considered most successful when processes are taut, transfers immediate, and inventories minimized. American infrastructure, of which 85 percent is owned and operated by the private sector, is subject to the same forces. Spend little, generate much, and don't let anything sit idle for even a second. The average oil refinery in the United States today operates at 93 percent of its maximum capacity. The larger the portion of capacity in use on a normal day, the less there will be available, and the more profound the losses will feel, in an emergency. As another DOE engineer puts it, "Everything you do to an infrastructure to make it more efficient makes it more vulnerable."

When it comes to infrastructure, one of the more troubling realities for the homeland security establishment is the lack of "excess" equipment critical to the power supply. For instance, the U.S. power grid is dependent on transformers—mechanisms that shift voltage levels up so that electricity can travel long distances and down so that it can arrive safely at your stereo. Among the hundreds of thousands of transformers operating throughout the system, about 1,100 are referred to as "extra high voltage" (EHV). These are rare, massive, and brutally expensive mechanisms. Utilities do store extra EHV transformers and sometimes share them with other companies, but it is doubtful that spares would be accessible in the event of a terror attack for a number of reasons. Transmission companies take a risk when they lend out equipment they may need themselves or that will benefit competitors; also, because the production of these transformers is highly unstandardized, very few of the giants are the same, so very few swaps are possible. More thorny than these issues, though, is the problem of transporting the equipment. Today, the only conveyance for EHV transformers—which look something like the twisted guts of a Navy gunship—are special railcars or barges. Scheduling the right freight car to travel to what is usually a remote location is already a logistical ordeal, on top of which nearly all manufacturers are based overseas. Meanwhile, building one of the beasts can take more than a year. One initiative, driven by the energy industry's joint R&D institution, is looking at stockpiling some number of these transformers, designed specifically so as to be movable by cargo plane. The cost of such a stockpile is likely to be a few hundred million dollars. That's nine figures' worth of surplus—an affront to the doctrine of efficiency—but it might be a good investment in resilience.

Even harder to counter than shortages in huge systems is the tendency for one problem to trigger another, and another, each time sending danger into ever more pathways and spreading further injury—called "cascading." Tom O'Rourke recounts a devastating example of cascade effects in a paper he coauthored with the support of the National Science Foundation and the Natural Hazards Research Center, titled "Lessons Learned from the World Trade Center Disaster About Critical Utility Systems." This particular incident took place in New York's garment district in 1983, and it started with a water main break, which tangled traffic throughout the area and flooded an underground electric facility. The deluged power station caught fire, burning some transformers, which released PCBs. Simultaneously, the destruction of the substation knocked out electricity and phone service for thousands of residents and businesses. "As the damage cascaded from one level to another," the authors noted, "losses were sustained in increasingly larger and complex systems."

This example traces failure from one type of infrastructure to another, but cascading can bring down a network independently. The term itself is used commonly in the power industry, where instances of it are particularly dramatic. And because the energy apparatus is, in a sense, the infrastructure for infrastructure, the prospects of attacks on it that produce cascading are especially disturbing. As O'Rourke puts it, "Once failure gets into the electrical system, it spreads very rapidly."

One striking example of a cascading power failure was the blackout that began in Ohio in August 2003 and spread north to Ontario and east to New York City and parts of New England. Ultimately, 50 million people lost electricity, more than during any other power failure in U.S. history. A year later, analysts were still not certain about the step-by-step causes of this complex cascading, but they had determined with some confidence that the chain of events was set off by trees. That some sagging wires and minor arboreal interference could lead to a crisis of this magnitude goes a long way in explaining why the Pentagon is so concerned with "asymmetrical warfare": In an extremely networked society, seemingly minor bad acts can have alarmingly wide impact.

The electrical grid's sensitivity to cascading is due in part to electricity's audacious preference for moving at nearly the speed of light. (Compared with that rapidity, the 1989 depletion of San Francisco's auxiliary water supply—after 40 minutes—is a bit less stirring.) The perfect energy waves that travel the planet from turbine to toaster will react to the tiniest, most imperceptible disruptions. For U.S. energy companies, a "long-term interruption" in service is one that lasts more than one minute. Problems are frequently corrected within one-hundredth of a second, and a really drawn-out fix takes between 100 and 1,000 seconds. That's when you have a lumbering human being involved. To make split-second adjustments requires circumventing the bottleneck of human reflexes and using systems calibrated electronically, instructed to do just the right thing at the right time. The trouble is that today's automatic fix-it devices are fairly simple-minded, capable of handling just a few rules. Maximum flows, drops, noise, temperatures, and other aberrations are set conservatively, so mechanized repairs kick in fast. And those repairs—usually in the form of a circuit breaker that shuts off a flow altogether—create their own minidisturbances. Down the line, another eager circuit breaker might respond. There are tens of thousands of these mechanisms on alert at any moment.

John Hauer, a dedicated, if beleaguered, engineer at the Department of Energy, is one of those infrastructure guys who spends his time pondering the big—really big—picture. His bailiwick is electrical utilities, and his goal is to better tune their controls so as to prevent cascading, to give all those simple-minded protective devices—as well as the sluggish humans who help out—better instructions for handling anomalies. While Tom O'Rourke is shoving sand against a pipe in the lab, Hauer is keeping watch on the entire expanse of one of the most complex networks in the world, the U.S. energy-delivery system.

In fact, in recent years, as the industry has gone private and more energy is flowing greater distances among more numerous parties, the complexity he wrestles with has increased. The North American grid has grown both more interconnected and more compartmentalized. In today's deregulated environment, utilities are free to trade with distant generators and myriad players across regions. This makes the whole more interwoven, whereby the failure of a generator or a disturbance along a transmission line might have effects hundreds, even thousands, of miles away. Simultaneously, however, with no more public monopolies to feed off, utilities have become more viciously competitive and protective of whatever advantage they can get. There's little interest in sharing information, which could tip off competitors about deals in the making or weaknesses in their positions, so there's no shared data bank for investigating the nature of the whole. "Everybody operates their piece of the system," says Hauer. Why share data that might end up benefiting their rivals? John Hauer is one of a dwindling number of engineers who is actually still struggling to see beyond regions and corporate boundaries. "All the data for the entire Western U.S. system is integrated in the computer on my desk, and there's no one formally assigned to analyzing it," he says.

It's not just policy, though, that makes the system so difficult to view holistically. There's also some physics involved. To read the condition of the grid in Miami and compare that with the condition of the system in Des Moines requires having a standard approach at each location. Just as an analysis of geography depends on a common map projection, a study of how energy is moving through wires in different locations needs a shared context. In this case, shared timing. When data to be read is measured in milliseconds, clocks had better be well synchronized. The trouble is that there's latency—or lag time—in everything. Data takes time to get from collection points to control centers, and even the voice of a person on a phone line saying "Set your watch to 12:02—now" has a lag.

So how do you know if the voltage in Florida at .082 seconds past the hour is the same as the voltage in Iowa? John Hauer's answer is to use the 24 GPS satellites orbiting 11,000 miles above the planet. The GPS satellites broadcast time signals that GPS receivers compare in order to derive their position. But if you know your position, you can use those signals as a straightforward way of telling the time with extreme accuracy. Like an infinitely large clock hung so high that everyone can refer to it, the satellites provide the same time reading for all of Hauer's measuring devices wherever they are on the continent. So Hauer can compare voltages reliably, not just within the local electric network serving Des Moines but potentially across the nation, seeing, for instance, a tiny discrepancy before it upsets the balance in Cheyenne. As Hauer's colleague Jeff Dagle puts it, "Before you had x-rays. Now you have a CAT scan. The time stamp is the key." Measurements of flows once taken every 2 seconds across a local network can now be recorded 30 times a second from utilities hundreds of miles away.

The peculiar irony of cascading effects is that they feed off the very connectivity we rely on. The solution to cascading is to cut off, go cold, withdraw. In fact, that's just what those electrical relays are trying to do when they zealously break a circuit: contain the problem by isolating it. Although in the tenuous world of electrical flow, this prevention measure can actually worsen matters, in most cases, treating the cascade with isolation is the right solution. Consider a biological analog: quarantine. It was difficult to argue with the edict to keep people home if they'd been exposed to SARS in Toronto. Despite healthy fears of being stranded (dark house is empty; phone goes dead), despite legitimate *Lord of the Flies*–style renderings of social separation, too much connection can make us sickly; often, solitude is the cure.

*Dis*connection—called "islanding" in the energy trade—provides a certain safety, even if it scrambles daily life. Following the attacks on the World Trade Center, every utility used islanding to protect areas of Manhattan. Tom O'Rourke cites half a dozen examples in his paper "Lessons Learned." Among them, water: "Isolating the broken mains restored pressure in the intact system outside the perimeter of the closed gate valves." Gas: "To ensure stoppage of gas flow into areas of greatest damage . . . isolation was performed by cutting and capping mains." Steam: "[Authorities] shut down the steam distribution network . . . to preserve service in areas unaffected by damage." While the city recovered, these islands of decommissioned infrastructure gradually shrank in size, as the officials reopened valves closer and closer to the site and gradually restored services. This approach is one way of using the complexity of the system to its own advantage: a complex mesh allows you to create islands while leaving the network largely functioning, sacrificing a small portion of service for the good of the whole. In fact, a U.S.-Canadian task force report on the August 2003 blackout asserted that islanding some Midwestern customers— that is, shutting off their power—could have prevented the outage from spreading.

In the PowerPoint opus he has assembled on the Loma Prieta earthquake, O'Rourke likes to pause over the strange contradictions of connectivity. He doesn't offer a solution, and he doesn't exactly smile, but there's some sort of amusement, or perhaps some awe, glinting behind his wire-rims. The slide is a map of San Francisco, with leaks and tanks and pump stations labeled with arrows. O'Rourke is gesturing before the screen like a weatherman, the stiff shoulders of his sports coat climbing up toward his chin. He speaks as if this subject alone matters—as if all the world's libraries contain only this one book, the one about a marvelous and captivating, beautiful and mysterious . . . auxiliary water supply. And he has reached the part of his story in which prescient engineers back in 1912 included a shutoff valve at Sixth Street to protect the water supply up on the higher terrain of the city from pouring out through leaks in the lowlands where pipes were more likely to succumb to a quake. Now O'Rourke's tale gets rolling, and he's using his long white hands to suggest impending crisis. The Loma Prieta earthquake has occurred—it's early evening and sirens are roaring—and on Seventh Street, a block from that clever valve, a 12-inch pipe has burst. Meanwhile, Pacific Gas & Electric, the city's utility, shuts down the power. "Gas *and* electric," O'Rourke repeats. "PG&E kept the E off for fear of G leaks igniting; if its sparks and its gas caused fires, the liability would be all theirs." And now the broken main on Seventh, along with a number of leaky hydrants, are gushing more than 5,000 gallons a minute, and sure enough, the Jones Street tank, perched on the hill, is also losing water fast. In 15 short minutes, the level in the tank drops by half. The valve! If the valve were shut between the tank and the leaks, water would stay in the system! The operator throws the switch in order to island the bedlam occurring beyond Sixth Street. Cut off the flow before the big break and prevent the Jones Street tank from hemorrhaging. But nothing happens. "The operator threw the switch," says O'Rourke, "but there was no electricity. The switch didn't close." Cut off from its power source, the valve couldn't cut off the water supply. The tank bled dry within an hour.

O'Rourke pauses. It's not frustration or fatalism or doom he wants to convey. It's simply complexity. How it's awesome and challenging. How the effects we anticipate are never the whole of it. How the systems we need are not the ones we're worried about but the ones we forget about, because we forget about the complexity.

Engineers and civil servants around the room hold their Styrofoam cups of lunch ramen in their laps and stop eating. They're waiting for direction. Connect? Disconnect? How do you know? The answer, in principle, is both. Connections are what make systems so powerful *and* so vulnerable.

In practice, the right answer in any given instance remains elusive. But there are a few members of that other system, that human network, committed to the quest. Grid mastermind John Hauer, network scientist Albert-László Barabási, subterranean explorer Tom O'Rourke—they are nodes on a network of engineers and thinkers and technologists that, when cleverly connected, may just figure it out.

TWO

Behavior and Betrayal

A T 6 P.M. ON A DAMP, cold mid-December day in 1999, Diana Dean, a U.S. Customs inspector in Port Angeles, Washington, three hours northwest of Seattle, was finishing up her shift at the port's border checkpoint. She watched as the MV *Coho,* a car ferry from Victoria, British Columbia, chugged across the Strait of Juan de Fuca on its last run of the day. Dean walked over to the port's holding lot, where she and two other inspectors stopped each car as it pulled off onto U.S. soil. She asked every driver a few simple questions—where they were from, the purpose of their trip—and then sent them on their way.

Dean approached the last driver, a trim man in his thirties with close-cropped black hair in a green Chrysler 300M. Speaking in stilted English, the man told her that he was from Montreal, headed to Seattle, and was planning to stay in a hotel. As he answered her routine questions, however, Dean noticed something suspicious about him. "He was agitated," she recalls, "jittery, nervous." He kept rummaging tensely through the car's center console and made little eye contact with her as they spoke. It was odd behavior.

Dean, who has large round glasses and medium-length sandy blond hair, is not an intimidating presence. She had 19 years of experience at Customs, the last 10 at Port Angeles. She had seen thousands of cars come off the ferry, asked countless passengers the same questions. She had learned to trust her sense that something about the man just wasn't right. "I didn't know that we were going to find something," she says, "but I did know that he was somebody we would want to look closer at." Suspecting that he might be carrying drugs, she decided to conduct a secondary examination, asking him to fill out a customs form. He did so quickly, listing his name as Benni Antoine Noris, the same name as on his Canadian passport and driver's license. Dean asked Mark Johnson, one of two other Customs officers who had finished their duties, to take over the questioning, knowing that he spoke French. Noris remained evasive, however, and eventually Dean asked him to exit the car, which he did only after prolonged coaxing from her and Johnson. When he did step out, Dean was struck by the sight of his oversized camel hair coat. "It was probably six sizes too big for him," she recalls, laughing at the memory.

Two other agents, meanwhile, opened the Chrysler's trunk. Inside they at first found only a suitcase full of clothes. When one of the inspectors unscrewed the spare-tire compartment, though, Dean heard him call out, "Uhhhh, Diana?" In the wheel well he had found something considerably more sinister: bags of powder, pill bottles, and two olive jars of a strange brown liquid.

Realizing that they had uncovered his secret, Noris slipped out of the coat, leaving it in Johnson's grasp as Noris made a run for it. He led officers on a five-block chase, but when he tried to commandeer a car, the alert driver sped through a red light and knocked him off balance, enabling the agents to catch him and place him under arrest.

Noris turned out to be an Algerian whose real name is Ahmed Ressam. His Chrysler, the officers would later find out, was carrying 118 pounds of urea crystals, 14 pounds of sulfate powder, and 48 ounces of nitroglycerin. He would eventually confess to authorities that he had been trained in Al Qaeda camps in Afghanistan, where he studied explosives and chemical warfare, and that he had planned to detonate his explosive mixture at the Los Angeles airport on the millennium. He is now serving a 130-year term in a federal prison as part of a deal in which he informed on other Al Qaeda operatives.

The capture of Ressam, one of the most dramatic terrorist arrests in

U.S. history, did not benefit from any technological sleuthing. The Customs officers at Port Angeles did not employ finely tuned explosive sensors or penetrating x-rays to select his car for a search. They did not find his name through an automated search of government computer databases, nor did advanced facial-recognition cameras snap his mug and match it to a file. Ressam, in fact, wasn't under surveillance at all, and his identity didn't appear on any U.S. government terrorist watch list. Instead, his capture was based on Diana Dean's hunch, the sum of her experience and observational powers (and perhaps a bit of luck: Ressam was likely groggy from malaria he had caught overseas, although Dean says that he wasn't "sweating profusely," as some accounts later claimed).

Dean now supervises a border crossing in North Dakota, where she moved in 2003 to be closer to her family. After the arrest, she and her fellow officers were lauded as heroes, and she was interviewed countless times by the media. Her answers were always the same: She just saw something strange in Ressam and decided to act on it. "I saw so many people come through that port," she says. "And pretty much they all start looking alike. I don't care what country they are from, I don't care what language they speak. When they drive up to you, you pretty much know this guy is on business, or he is out for a pleasure trip. So when somebody comes up who is out of the norm, those are the people you are going to look at."

What does a terrorist look like? Ask most Americans this question, and you will undoubtedly receive answers still clouded by the attacks of September 11 and the ensuing "war on terrorism." They will very likely say that a terrorist is a male, and perhaps that he is Arab, Muslim, young, and single. But examine the broader history of terrorist threats—from the Irish Republican Army to the Japanese Aum Shinrikyo cult to Oklahoma City bomber Timothy McVeigh to Ted Kaczynski to female Palestinian and Chechen suicide bombers—and that profile quickly falls apart. The common, self-evident thread among terrorists is not their ethnicity, age, or gender; it is their behavior. They stand out because of the brutal and sometimes suicidal acts of violence they commit—and the actions that precede that violence.

Much of our defense against such attacks typically focuses on the weapons terrorists might employ, from firearms, to explosives, to viruses, to nukes. Billions of dollars have been spent developing high-tech ways to seek out these physical representations of danger, with x-ray machines, explosive sniffers, and radiation detectors. But all such attacks, whether a hijacking or suicide bomb, consist of two elements: the weapon and the person who delivers it. The nature of the weapon changes, but the basic nature of the terrorist's intent does not. Is it possible, then, to uncloak terrorists based on their intentions, the subtleties of behavior that separate them from ordinary citizens?

Raphael "Rafi" Ron is prepared to bet your life on it. Ron, the former director of security at Ben Gurion Airport in Tel Aviv, Israel, has been traversing the United States since September 11, 2001, preaching the virtues of the Israeli model for airline security. With short gray hair and deep-set eyes, he has a perpetually composed demeanor even while discussing the grisliest terrorist assaults. He's sitting in the lobby of the University of California faculty club, where he's come to give a presentation at an invitation-only conference on transportation security. When he speaks, his hands move in graceful synchrony with his words. Ron was a paratrooper in the Israeli special forces in the late 1960s, and then became one of the first air marshals for El Al, the Israeli national airline. After a stint in diplomatic security, protecting Israeli embassies and its UN mission, he spent 20 years in the prime minister's Office of Intelligence and Special Operations. He took over as director of security at Ben Gurion in 1996, where, he says, he gave the same speech to each batch of security trainees. He told them that every morning on their way to work, "you should repeat a mantra: 'It's going to happen today. It's going to happen on my shift. It's going to happen to me.' If you find it difficult to believe that, don't show up." Anticipating attacks on Ben Gurion, he says "is not a question of if, it is a question of how. That's the way I perceived my job."

Ron was a month away from his retirement on 9/11. Since coming to the United States, he has helped Boston's Logan Airport revamp its defenses on the basis of a program called "behavioral pattern recognition." The central tenet of Ron's philosophy—and to some extent the Israeli security philosophy overall—is that whether they are in the subway, entering a crowded café, or passing through an airport checkpoint, "terrorists are there for a purpose which is completely irregular. It must have some type of expression in their behavior." For Ron, "behavior" means something more general than just the way people act. In some cases, it means something as simple as people being somewhere they shouldn't be—such as loitering around the perimeter of a nuclear facility—or doing something they shouldn't—such as shooting video. In others, the clue might be more subtle, such as a back story that unravels under questioning or contains seemingly minor elements that simply don't fit.

Then there are cases such as Ahmed Ressam, where a person's demeanor itself provides the evidence of his or her intent. Attackers preparing to execute their plans, says Ron, "are going to be involved in an extreme act of violence, very soon. This is something that affects your mentality. Nobody is completely indifferent to the idea that very soon he is going to initiate an extremely dramatic event with extreme violence—even if you are trained for this. And they are training for this." The stress of carrying out an attack, he says, inevitably manifests itself physically: "It is basic human nature that when you are going to do something that is extreme, something happens to you. You are not the same person that you were a week before when you were going to do your shopping in the supermarket."

Terrorist behavior in our society is an anomaly, and anomaly detection is something that—as Diana Dean's example shows—trained human beings can do well. If we know what is normal in a situation, we can often pick out what is not. But terrorists are, needless to say, also actively trying to conceal exactly the stress that Ron describes, creating a kind of arms race between the detector and the deceiver. Ron's solution is to train humans to understand the subtle clues, whether a jittery appearance or a dubious story, connected with threatening behavior. Researchers, meanwhile, are racing to apply technology to the task of detecting suspicious intentions, building automated tools that could fill the gaps in human perception of behavioral evidence. The hope is to let machines carry some of the load in detecting terrorism, but the challenge is to use technology without creating a system that will flag innocent bystanders, or that terrorists—who are always seeking vulnerabilities—will easily exploit. How we are able to meld the twin approaches of human ability and technological muscle lies at the heart of our attempts to identify terrorists at or before the moment of their attack.

Perhaps as much as anyone in the history of science, Dr. Paul Ekman understands how a person's physical cues provide a window into his or her intentions. Ekman, the world's foremost expert on facial expression recognition, is a renowned psychology professor and director of the Human Interaction Laboratory at the University of California at San Francisco (UCSF). He has spent a career dissecting the subtle physical manifestations of human emotion and has authored or edited 13 books on the topic, including the groundbreaking *Telling Lies* in 1985, and, in 2003, *Emotions Revealed,* a guide to detecting and understanding the surface cues to emotion.

In five decades of research, Ekman advocated and then proved the notion that humans share a universal catalog of facial expressions. The idea dates to Darwin, who wrote a book about it, *The Expression of the Emotions in Man and Animals.* But before Ekman's research most anthropologists discredited the notion, believing that expressions were instead culturally determined. To prove them wrong, Ekman traveled the world and examined reams of film to compare the expanse of human expressions among a cross section of cultures. He then set about recording and classifying them. Ekman and a colleague, Wallace Friesen, developed a program called the Facial Action Coding System, or FACS, to catalog the facial movements of humans. They created a kind of GIS of the face, exhaustively mapping the motions made possible by the 43 facial muscles, and discovered 3,000 combinations of movements—such as a scarcely lowered eyebrow or a faintly upturned mouth—that signified some underlying emotion. All emotions create these subtle but perceptible physical reactions, which provide revelations about human feeling. Emotional reactions can also generate what Ekman calls "micro expressions," involuntary movements that last only a fraction of a second. Micro expressions can reveal the emotions of people even when they are trying to conceal them—a revelation that he has termed "leakage."

In the course of his research, Ekman found that certain highly adept people could naturally read those cues and discern the underlying emotion. Others, by studying FACS, could develop the skill.

Ekman himself, after years of scrutinizing thousands upon thousands of faces, is a master of reading expressions. "My eyes are very educated," he says, when asked what it felt like to be trained in the science of emotion detection. Sporting sweatpants and running shoes, he sits in a leather easy chair in his living room, high in the hills of Oakland with a vista of the fading sunlight over San Francisco Bay. "It's really a very strange talent. When I was a kid in the 1940s and 1950s there was a comic book character—I think it was called Plastic Man—who had x-ray vision." The Plastic Man effect, he says, isn't unique to himself. "Everybody who learns FACS gets the same effect."

Having the ability to perceive subtle expressions, however, amounts only to recognizing emotions that others might not. The antecedents of feelings can often be complex and elusive. The skill, he says, is "only to know how people feel. It's not to know what they are thinking. When I meet people at some social event, and they either recall what I do or they find out what I do, they often get concerned. And the first thing I tell them to reassure them is: I don't know what you are thinking. The most I can tell is what you are feeling. But then, emotions don't tell you their source. So if I see the fact that you are looking annoyed, I don't know if you are annoyed at yourself, or annoyed at me, or annoyed recalling what happened this morning on the freeway."

Nonetheless, the ability to detect emotion is valuable in other tasks, in particular as a tool for detecting lying or unease. Ekman has consulted with thousands of law enforcement officers, training them on a combination of the basic components of FACS and an understanding of voice and body language. He has studied every available piece of public footage showing assassins captured just before they struck— some 40 to 50 clips that were assembled by the Secret Service— deducing common themes and training Secret Service agents to identify them. Lately the FBI and the CIA have increasingly enlisted him for counterterrorism training.

Even in tutoring law enforcement officers, however, Ekman is careful not to confuse the awareness of emotion with an understanding of its origin. "The main thing I teach when I teach law enforcement or counterintelligence—it's really important to pick up on the emotion that the other person is feeling. But you've got to find out what the source is. Don't presume that you know the source." Ekman calls that presumption "Othello's error," after Shakespeare's famous Moor. In the play, Othello accuses his lover Desdemona of infidelity and informs her that he has ordered her lover Cassio murdered. Othello then interprets her distress, as Ekman wrote in *Telling Lies,* "as a reaction to the news of her alleged lover's death, confirming his belief in her infidelity. Othello fails to realize that if Desdemona is innocent she might still show these very same emotions: distress and despair that Othello disbelieves her." Othello kills Desdemona, offering Ekman a handy example of the consequences of the error. As he later summarized the lesson in *Emotions Revealed,* "the fear of a guilty person about being caught looks just like the fear of an innocent person about being disbelieved."

Ekman has not only trained law enforcement personnel; he and a colleague at UCSF, Maureen O'Sullivan, have also studied their natural ability to detect lying. What they found, to their surprise, was that the vast majority faired poorly. We tend to think that we are good at detecting lies, and there is a lot of evolutionary biology that says we should be. But among all the groups they tested—using an exam that shows subjects lying or telling the truth about having watched either a violent film or a nature film—only Secret Service agents performed better than random chance. Ekman and O'Sullivan speculated that they did so partly because in their jobs they scan crowds searching for assassins, which might give them a sharpened awareness of human expressions. Interestingly, though, the two also noted that Secret Service agents, who typically interrogate people who have only threatened harm to public officials, encounter *fewer* liars on a daily basis than those in law enforcement. "Secret Service officials told us that most of these people are telling the truth when they claim that their threat was braggadocio, not serious," Ekman and O'Sullivan wrote. Officers in the criminal justice system, on the other hand, tended to believe that "everyone lies to them. Thus, the Secret Service deals with a much lower base rate of lying and may be more focused on signs of deceit." The lesson, while still speculative, is an intriguing one: Constant contact with lying may in fact desensitize officers to its cues, rendering them continually susceptible to Othello's error.

Ekman and O'Sullivan did, however, discover a tiny group of

people—about 20 out of a pool of 12,000 law enforcement personnel, judges, arbitrators, and psychotherapists tested—that they call the wizards of deception detection. "The wizards rarely if ever make mistakes," Ekman says. "They are 80 to 100 percent accurate when you show them a piece of videotape, able to discern in one viewing, in real time, whether the person is lying or telling the truth. It's a very rare skill."

Ekman and O'Sullivan studied the wizards more closely and found that they relied not only on facial expressions but also on body language and voice to identify deception. "Everyone does it a little bit differently," says O'Sullivan, the lead investigator on the study. "And one of the things that is very important is their life experience." The wizards, she says, seem to fall into three categories: those that seem to be naturally talented; those who have unusual backgrounds (for example, their parents were alcoholic or mentally ill, making them "sensitized to emotion as a coping mechanism"); and those who have made explicit midlife decisions to improve their ability to detect lying. To detect deception, they employ multiple strategies, sometimes switching from reading the face to vocal intonation depending on the lie being told. Diana Dean's hunch, O'Sullivan observes, was likely a combination of reading facial expression, body language, and voice—specific techniques that Dean herself may not even be aware of. "She has just seen lots and lots of people in that context," O'Sullivan says. "So she has probably developed particular types of templates for that situation."

Deception is a skill as well, and there is a parallel population of its best practitioners, whom you might call the anti-wizards. In the process of creating tests to rate the deception detectors, Ekman has come across some spectacularly good liars. Such natural deceivers avoid the mistakes that usually trip up deceivers: the inability to plan a coherent lie, the apprehension over being detected, and "duping delight"—the excitement, often betrayed by a facial expression, that people sometimes feel when successfully deceiving someone. Being a master of deception has its advantages, particularly in the worlds of sales, con men, poker, and politics. In *Telling Lies,* Ekman offers the example of Hitler and his deception of Neville Chamberlain in 1938 concerning Germany's intention to invade Poland. Hitler likely betrayed little fear of being caught, Ekman writes, in part because he "knew exactly when he would need to lie and what he would need to say, so he could prepare and rehearse his line. There was no reason for Hitler to feel guilty or ashamed about his deceit—he considered deceiving the British an honorable act. . . . The one emotion that Hitler might have felt that could have leaked is duping delight. Reportedly, Hitler took pleasure in his ability to mislead the English. . . . But Hitler was a very skilled liar, and apparently he prevented any leakage of these feelings." Nor did it hurt that Chamberlain wanted to believe the lie Hitler was offering.

Terrorists may present a similar problem for intention detection. They are often meticulous in their plans and may also feel they are committing "an honorable act." Organizers, meanwhile, have a strong incentive to recruit skillful liars. "If I were running a terrorist organization," Ekman says, "that's who I would be after." Terrorists also train to conceal their emotions: Al Qaeda training manuals confiscated in Britain make multiple references to how terrorist operatives should handle their demeanor in public places. In some cases, terrorist organizers may even conceal the final goals of the attack from the operatives, to prevent them from inadvertently or purposely revealing the plot. According to a Bin Laden videotape issued after the attacks and confirmed by reports quoting intelligence officials, some members of the 9/11 hijacking teams—the ones who were not trained as pilots— may not have known the ultimate intention of their mission until they boarded the planes, so as not to compromise the plan.

Yet on the day of the mission, all the 9/11 hijackers had at least some knowledge that they were about to commit an act of terror— even if they did not know how dire an act. Washington Dulles surveillance footage of the hijackers shows them looking serious and calm, and passengers who encountered them in the airport claimed later that they looked like harried businessmen. But to trained human observers asking probing questions, their demeanor might have betrayed some sign of their intentions. According to the report issued by the independent 9/11 Commission, a ticket agent at Dulles selected two of the hijackers for additional security because they appeared suspicious.

Even assuming terrorists will go to great lengths to conceal their feelings, minimizing one's natural expressions of emotion is an extremely difficult task. Difficult, but not necessarily impossible. Ekman describes how he recently reviewed tapes of a Buddhist monk who was attempting to squelch all reaction to an unexpected 110-decibel noise—135 decibels will damage the eardrums—through two forms of meditation. "He's very good," says Ekman. "I have to look at it really carefully 8 to 10 times in a row to see what's happening, because he is able to prevent almost everything." The point, he says, is that "informed countermeasures" against emotion detection are not out of the question, one reason he won't talk about the specifics of his counterterrorism training.

Aware of the potentially nefarious uses for his research—by foreign intelligence agencies conducting interrogations, or even terrorists themselves—he says he refuses to do training in countries that are not constitutional democracies, despite being offered many times his usual consulting fee. Still, FACS is available to anyone who wants to purchase it, as are Ekman's less in-depth CD-based training tools, and *Emotions Revealed* offers many of his tips on detecting emotion from facial expressions. He maintains that the social benefits of sharing his findings to this degree outweigh the antisocial costs; that, regardless of the circumstance, whether in law enforcement or everyday relationships, people benefit from having more information about emotion. "I operate under the basic assumption—which is a convenient one for me—that in general the better we understand people, the better off we and they are."

In mid-April 1986, a 32-year-old Irish woman named Anne Marie Murphy walked into London's Heathrow Airport with a single carry-on suitcase and a ticket to Tel Aviv on El Al. Murphy, who was pregnant, thought she was on her way to visit the parents of her London-based boyfriend, a Jordanian named Nezar Hindawi. Hindawi had packed Murphy's bag for her and dropped her off at the airport, promising to catch a later flight and meet up with her in Israel. He told her that she should avoid telling anyone about him or their plans, lest their unconventional relationship raise eyebrows in Israel.

After Murphy passed through Heathrow security, and her bag was cleared through the airport x-ray system, she arrived at an El Al security checkpoint. The security officer began by asking her basic questions: Who was she? Where was she going? What was the purpose of her trip? Not wanting to mention Hindawi, she replied that she was traveling to Tel Aviv for pleasure and would be staying at the Hilton for two weeks. When asked her profession, she said she was a chambermaid. Murphy told the officer she was carrying 50 pounds sterling in cash to pay for the hotel. That, according to former El Al security director Isaac Yeffet, was the clue that first roused the agent's suspicion. April was high season for hotel bookings in Tel Aviv, and in 1986, 50 pounds would certainly not cover two weeks at the Hilton. She then proffered a credit card, but given her profession, the security officer remained dubious. She also said that she planned to visit holy sites in Israel, among them Bethlehem. Where, the officer asked, did she plan to stay in Bethlehem? Again, she named the Hilton. There was only one problem: There was no Hilton in Bethlehem.

The officer flagged Murphy for a full screening, and security personnel took her to a separate area to search her bag by hand. Inside, they found a high-end Texas Instruments scientific calculator that Murphy said she planned to use for currency conversions. When they emptied the bag, they located a false bottom that concealed three pounds of plastic explosives, to be triggered by the calculator. Nezar Hindawi turned out to be working for the Syrian intelligence service, and the bomb was set to go off in flight. He was later convicted and sentenced to 45 years in a British prison for the plot. Murphy, authorities later established, had never known about the bomb.

The type of inquiry that tagged Murphy, which Rafi Ron calls a "targeted conversation," has its roots in the genesis of the Israeli air security system. Until 1968, flights on El Al were largely free of any security measures. In July of that year, members of the Popular Front for the Liberation of Palestine hijacked an El Al flight from Rome to Tel Aviv and forced it to land in Algeria, where they kept the passengers and crew hostage for five weeks. That attack spurred Shin Bet, the Israeli security service, to place armed air marshals on all El Al flights.

Israeli flights, however, had become high-profile targets. When it became clear that air marshals would stop a simple hijacking, terrorists upped the ante, attempting to smuggle bombs and weapons on the planes. Shin Bet started examining more comprehensive security measures. "Our tendency was to go and look for the weapons," says Ron, who at the time had recently finished his military service and become an El Al air marshal. "Because it is something concrete, and perhaps it's easier to detect a weapon than to detect an intention." In trial runs, however, they discovered that in order to detect every type of weapon, they would be required to spend an hour or more hand-searching each passenger, an impossibility even for the small El Al.

So instead they began to focus not only on the weapons, but on the people themselves, creating a passenger profiling system to narrow the focus of the weapon search onto a small percentage of potential high-risk passengers. "We developed the system by saying, Let's try to see the pattern of those people," says Ofer Einav, a former El Al security director and now CEO of Ganden Security Services Solutions (GS-3), a Virginia- and Israel-based security consulting firm. "If you are taking the one bad guy that is trying to hide between the 100 innocent people, what will be the type of passenger that we are looking for?" Under the system created in the early 1970s, El Al airport security personnel receive a several-week training course in which they study previous terrorists attacks, their methodologies, and the perpetrators' motives. Trainees, many of whom are college graduates, are taught how to conduct targeted questioning, to begin with simple questions and then trace back the passenger's story, asking for verification at various stages. They are instructed to observe the eyes, face, and body language for signs of nervousness or doubt. Passengers deemed suspicious are selected out for more rigorous questioning and a full hand search of themselves and their belongings.

The system, by any practical measure, has been extraordinarily successful. The Israelis haven't suffered a hijacking since 1968, and, according to the former El Al security officials, numerous attempted bombings—including one in 1979 in which a Palestinian terrorist organization paid a German to smuggle a bomb on board an El Al flight from Switzerland, telling him it was a drug shipment—have been thwarted in the act. Like Anne Marie Murphy, the German was caught by the suspicions of an El Al questioner.

Such cases reveal both the potential flaw and the inherent advantage in the Israeli method. Anne Marie Murphy offered a particular type of challenge to a system bent on detecting the emotional signs of unease: She did not know about the bomb in her luggage. She betrayed herself only because her boyfriend had told her not to say she was visiting his parents, and she proved to be a poor liar. If some of the 9/11 hijackers did not know their true mission, they would have offered a similar dilemma for behavior detection, had it been used. Yet both examples also illustrate the advantage of including face-to-face human contact in security. Without targeted questioning, Murphy's bomb would have remained undetected, just as the 9/11 hijackers and their box cutters passed through security. Because terrorists alter their attacks to coincide with vulnerabilities, the unintended consequence of any security system is to shift them to new tactics. Part of the Israelis' success has come from using better technology while still emphasizing the role of humans, who offer a flexibility to counter the deadly creativity of terrorists. Contrast that with the history of x-ray machines and explosive detection systems (EDS), which, while they serve as an excellent deterrent for potential bombers, have never alone uncovered a terrorist bomb in the history of aviation.

For any attempt to replicate the Israeli success in the United States, however, the problem is making the system scale. El Al (which is highly subsidized by the Israeli government, spending around $100 million annually on security) is Israel's sole airline, serving 4 million passengers a year. The U.S. airline system comprises dozens of carriers serving 800 million passengers a year. Questioning each passenger to the depth of El Al—which requires that fliers arrive three hours before their departure time—simply isn't feasible.

There is also a cultural divide between the United States and Israel when it comes to security, one forged by Israel's longer exposure to a constant threat of terror. "Israel, not by our own free choice, has served as something of a kind of a laboratory for combating terrorism, because we have been targeted for so many years," says Dan Meridor, a former Israeli legislator, minister of finance, and minister without portfolio overseeing the country's security agencies. "Living with that means that you are ready to allow the state to do more things than people who don't live under that condition would allow. People are more patient. They are more permissive."

Paul Ekman says that he has been unable to interest the Transportation Security Administration (TSA), the U.S. Department of Homeland

Security agency that oversees aviation security, in his training—a fact that he attributes to a culture that promotes technical fixes over training humans when it comes to security. A drive for efficiency through technology, as September 11 showed, can often be the enemy of security. "We are a country that is hooked on technology," Ekman says. "People tend to think that if a machine can't do it, it isn't worth doing. Contrast that with the Israeli approach, where they are really using the best people they can get, and then training them like mad."

Ekman's observation points to a particular challenge in combating terrorism, that of partnering humans and machines. The United States is indeed a nation that tends to look to our own inventiveness to tackle our thorniest problems. In the case of terrorism, technological tools do offer the potential for extending human capabilities—allowing us to sense things we cannot, track potential dangers, verify identities, and process vast amounts of data. But technology also introduces its own flaws into any defense. Our tools tend to be inflexible and unvarying, relying on preset rules rather than the elasticity of human judgment. They are often brittle—when they fail, they fail completely. Humans, by contrast, excel at identifying patterns, applying knowledge to new problems, and adapting to changing circumstances. At the same time, humans can become tired and distracted, and fall victim to tricks of perception or failures of memory. The ideal defense, then, is one that combines the best attributes of both people and technology, using one as a redundancy for the other—the same kind of redundancy required for the security of the power grid or water supply.

Finding the proper balance between humans and our machines is a concern that dates to the beginning of the industrial revolution. But the academic study of how we and our tools interact, an area known generally as "human factors," began after World War II. It was born in Air Force and Navy labs, which began to study how military personnel worked together with increasingly advanced technologies, particularly radar. Radar was extending human senses, creating new possibilities for detection while also suddenly generating massive amounts of information to process. In the 1950s and 1960s, thousands of studies were conducted on how humans and machines could best cooperate to improve military performance.

As computers began to advance in the middle part of the twentieth century, the question of what and how many human tasks might be shouldered by technology grew more pressing. Machine intelligence seemed poised to surpass its human masters. In 1950, the brilliant World War II code breaker and computer pioneer Alan Turing proposed what later became known as the Turing test: a machine would have to be considered truly intelligent if it could convince a human interrogator that he or she was communicating with another human. Many computer scientists believed that computers would soon pass the Turing test. "It seems entirely possible that, in due course, electronic or chemical 'machines' will outdo the human brain in most of the functions we now consider within its province," wrote psychologist J. C. R. Licklider, echoing the concerns of the time in his seminal 1960 essay "Man-Computer Symbiosis." Before machines took over those functions, Licklider proposed, there would be a period of human-computer partnership in solving problems. "It seems worthwhile to avoid argument with enthusiasts for artificial intelligence by conceding dominance in the distant future of cerebration to machines alone," he wrote. "There will nevertheless be a fairly long interim during which the main intellectual advances will be made by men and computers working together in intimate association."

In the 1960s Licklider headed the Information Processing Techniques Office at the Advanced Research Projects Agency (ARPA), the Pentagon outfit chartered to investigate technologies too far out for other people to try. ARPA—which later added "Defense" to its title, becoming DARPA (and then ARPA again, and then once more DARPA, which finally stuck)—was formed in 1958 in response to the Soviet launch of the Sputnik satellite. The agency funds research intended for both military and commercial uses. Over four decades it has incubated the technologies behind the computer mouse, the stealth bomber, and, most famously, the Internet—which grew out of a communications network called Arpanet that Licklider spearheaded. But the vision that Licklider cited in his essay—that of a certain triumph of machines over humans—turned out to be further off than many of his contemporaries envisioned. The field of artificial intelligence (AI) faltered when it failed to find a way to categorize and replicate the complexity of human intelligence. AI has advanced significantly over the last few decades—by 1997, the IBM supercomputer Deep Blue was able to defeat the world chess champion Garry Kasparov. The world today, though, is still one of Licklider's human-computer symbiosis, where computers handle certain prescribed tasks under the direction of humans but remain unable to tackle some problems intelligently. We happily offload to machines tasks of solving mathematically difficult problems, piecing together gene sequences, and flying surveillance missions. Other responsibilities, such as flying commercial airlines and running factory assembly lines, are cooperative efforts in which technology plays at least an equal partner. But there are many judgment-intensive activities, such as driving cars and questioning witnesses, that we can't or don't want to let computers perform on their own—at least not yet.

As machines have increasingly become equal partners in complex endeavors, researchers in human factors have examined high-risk systems such as airlines, nuclear power plants, and chemical refineries—places where a failure in the interaction between humans and machines can result in disaster. Such was the case during the 1979 meltdown at the Three Mile Island nuclear plant near Harrisburg, Pennsylvania. The worst nuclear accident in U.S. history was a combination of human and equipment failure—most importantly among a string of breakdowns, technicians misinterpreted a warning signal alerting them that a valve was stuck in the open position, allowing the reactor's coolant to leak out. The Chernobyl meltdown, the Bhopal chemical disaster, and some of the most devastating airline crashes in history have all resulted from the dangerous intersection of human and machine fallibility.

When it comes to counterterrorism, then, the critical question is how much of our perception to automate. How much responsibility should we entrust to our technologies? Outsourcing the thinking to machines is often necessary, but it is also rife with dangers. Because our tools tend to be rigid, they look only for the things we have told them to look for, and carry out only the tasks for which they are programmed. They often cannot adapt to new circumstances and rapidly changing environments—technology's version of Othello's error. As one human-factors expert wrote in 1963, "Men are flexible but cannot be depended upon to perform in a consistent manner, whereas machines can be depended upon to perform consistently but have no flexibility whatsoever."

Human inconsistency is potentially even greater when it comes to terrorism, given that very few people have had any exposure to it, even among law enforcement personnel. As a result, they have very little sense of what they might be looking for. Despite the best efforts of Paul Ekman and Rafi Ron, many will also lack the kind of training that would let them maximize their natural human talents. "The people doing the terrorism work," says Ekman, "are mostly people who six months or a year ago were doing vice or drugs. That's a lot easier, and they know way more about how to do it. They didn't want to have to do terrorism. It's a lot harder, the success rate is much lower, no one is teaching them much how to do it. They are trying."

Airport security, in particular, is an arena that calls for a careful balance of human skills and technological tools. Since September 11, 2001, the TSA has beefed up its technological capability, requiring that all checked bags are screened using EDS. Security personnel use x-ray machines to search diligently for any sharp object, from a nail file to a pocket knife. There are even more sophisticated detection machines under development, including ones which will employ tera-hertz radiation—penetrating light rays that can better detect explosives, conventional weapons, or even the signatures of biological agents. Relying on such detection alone, however, is an invitation for terrorists to simply develop more sophisticated weapons that will pass through undetected. "Once you have the access to that technology," says Ron, "you can study its weaknesses—and every technology, especially those being used in airports, has weaknesses. Once you have studied its weaknesses, you can trust that they will always be in the same place at every given moment." The x-ray machines at airport security checkpoints are operated by people, but they are people doing what humans do poorly: repetitive tasks that require finding a rare signal—in this case, a weapon—among a sea of noise that is the vast majority of benign bags. Screening machines alone are a classically brittle defense. And when they fail, as they did on September 11, they fail completely.

Which is not to say that they should not be used. But recall that the only effective defense that September morning turned out to be the actions of the passengers on United Airlines flight 93, which crashed in Pennsylvania. Their behavior is a powerful illustration of the advantage of human adaptability. The longstanding rules regarding hijackings stipulated that the passengers and crew should follow the instructions of the hijackers. The best way to ensure their safety, the conventional wisdom said, was not to antagonize the terrorists. Those on flight 93, however, discovered through phone conversations with people on the ground that the terrorists had no intention of landing the plane, so in a heroic act of self-sacrifice and human resourcefulness, they developed a new rule: Fight back.

The attacks on the the Pentagon in Washington and the World Trade Center in New York created an instant demand for—and a lucrative business in—counterterrorism expertise, especially in aviation. Israel was a natural source for such experience, and Rafi Ron was among the Israelis who came to offer advice. Unfortunately however, the Israeli system is not easily portable. "It was obvious to us that what we were doing in Israel was not compatible with the system here," he says, "for many reasons—cultural, political, and operational."

Ron started a firm called New Age Security Solutions—so named, he says, because September 11 changed the face of aviation. (The company was later acquired by the American security training company Advanced Interactive Systems.) When Boston's Logan Airport hired Ron and his firm to help revamp its security system, the first task was to study its weaknesses. Two of the hijacked flights had left from Logan, and although its selection by the hijackers was probably based more on their operational plan than on the airport's security relative to other, equally lax, airports, officials were understandably anxious to make changes that would instill public confidence. More than 400,000 flights a year take off and land there, carrying some 20 million to 30 million passengers. So Ron and a team of 10 other experts helped conduct a vulnerability assessment—the final report ran to 500 pages—covering everything from perimeter security, to passenger screening, to stray vehicles on the tarmac. Among their recommendations was a strategy to preserve the essence of the Israeli method without trying to simply scale it to size. At the strategy's core is what the Israelis call profiling but what a Logan official named "behavior pattern recognition."

Using Ron's philosophy, the Massachusetts Port Authority (MassPort)—the independent public agency that manages Logan Airport, the Port of Boston, and the city's Tobin Memorial Bridge—trained several hundred state police officers who patrol the airport, both in uniform and undercover, in techniques for identifying suspicious anomalies. MassPort officers were taught how to conduct targeted questioning without provoking the ire of innocent passengers and to look for the same visual and logical clues used by Israeli security officers. They learned how to turn a simple "Hi, how are you?" conversation into more severe questioning if the situation warrants. "Seasoned cops use it without even knowing what to call it," says Dennis Treece, MassPort's director of corporate security, of behavior pattern recognition. "The bottom line is it gives the average law enforcement officer the confidence to approach someone on just the hint of suspicion."

Among the factors that Ron pointedly excluded from his training, however, are race and ethnicity. For many in the United States, the word "profiling" automatically raises the specter of race. "I'm against racial profiling, not only because it's against my values," says Ron. He goes on to explain that, in the case of behavioral detection, race is not an anomalous—that is, relevant—factor. There are an estimated 3 million Arab Americans living in the United States, and targeting them only distracts from recognizing truly suspicious behaviors. "Racial profiling is not an asset," he says. "Quite the contrary."

"You have to be careful that you don't take the easy way and start looking for people who meet a profile," agrees Treece, who had to have the state's attorney general approve Logan's system, and who brought in the American Civil Liberties Union to examine it. (He ultimately gained their grudging approval—although they later lodged a complaint about potential racial profiling under the program.) "What we are trying to teach people is, It's not who they are. It's what they are doing."

In Boston, Treece says, the system has already shown results. He cites the example of a TSA inspector who came to Logan to test the airport's checkpoint screeners, arriving unannounced and carrying a briefcase full of phony bomb components. Even playing a terrorist, it turns out, can cause one to give off detectable behavioral cues. "He knew that if he was stopped by a state trooper, and they looked in there, and it looked like he had bombs in his case, that would not be good," Treece says. "Especially at Logan, where they carry submachine guns. So he was acting a little nervous." A veteran MassPort officer spotted his demeanor and approached to start a conversation, which Treece says "quickly developed into, 'Hey, I think we should take this downtown.'" It was then that the inspector showed his badge, thereby avoiding a nasty scene. But in this case, Othello's error—misreading the fear of an innocent person—had served the officer well, and foiled the inspector's ruse. TSA got the message: In October of 2004, *Time* magazine reported that the agency planned to test a behavior-based training called Screening of Passengers by Observational Techniques, or SPOT, based on the Logan system.

Such behavioral screening techniques alone won't solve the dilemma of airline security, but they can be layered together with technology. One possibility might be to reduce the number of potentially suspicious passengers by using trusted-traveler programs, under which some passengers voluntarily submit to a full background check. These low-risk fliers would then be verified by biometric technologies, which use biological traits such as fingerprints or irises as a form of unique identification. Such a system has been in effect since October 2001 at Amsterdam's Schiphol Airport, utilizing an iris-scanning technology built by the New Jersey–based company Iridian Technologies. Passengers who join the program carry an identification card containing a digital representation of their iris structure. "It's sort of an iris fingerprint," says Jim Cambier, Iridian's chief technical officer. "Those structures are unique in every eye. Your right and left eye are completely different. Identical twins have different iris structures." When they arrive at the security checkpoint, passengers insert their card and step up to a camera, which takes a picture of their iris and matches it against the one on the card. If the irises correspond, the passenger is cleared through minimal security. Such systems are under consideration by the TSA.

Another technological layer involves the TSA's Computer Assisted Passenger Prescreening System (CAPPS), a software program designed to identify high-risk fliers. The original CAPPS system, which relied on weighted rules to flag passengers—such as those who purchased a one-way ticket, traveled abroad frequently, or paid in cash—was deemed too easy for potential attackers to figure out. CAPPS II, the controversial successor program developed under the TSA and then killed in mid-2004 due to privacy concerns, was meant to overcome those flaws. The plan was to use reservation data, combined with information from private databases, to assess a "score" for each passenger based on his or her background and details of the reservation; this score would in turn determine whether the passenger should be targeted for additional screening. A new system called Secure Flight—which the TSA says may or may not access public records—is under development.

The quandary of trusted-traveler and software profiling systems—in addition to the privacy issues that eventually sank CAPPS II—is that they also create new vulnerabilities for terrorists to exploit. By necessity such systems require that someone define in advance the parameters of who is considered dangerous and who is not. Trusted-traveler programs and their associated biometrics, because they allow for less scrutiny of the passengers in them, suddenly become inviting targets. Find a way to become a trusted traveler, the terrorist reasoning would go, and you are guaranteed less scrutiny. Such systems are also vulnerable to trial runs, like those the 9/11 terrorists conducted on flights well in advance of their attack, that expose their weaknesses.

Even when they succeed in their predefined goals, our best technologies may still fail to protect us. In one of many devastating post-9/11 could-have-beens, it was revealed that 9 of the 19 hijackers *were* in fact flagged by the CAPPS system, identified as high-risk fliers. The tragic failure occurred in what happened next: their selection meant only that their checked bags were screened and matched to their boarding passes. Under CAPPS, the terrorists never had to answer a question or encounter a security agent trained to ask questions and follow a hunch. There was no chance for an intuitive officer such as Diana Dean to save the day. Even when four of the five hijackers at Washington Dulles were taken aside at the security checkpoint—after setting off metal detectors or being selected by ticket agents—the screeners didn't question them. Instead, they merely examined their tickets and searched their bags.

Despite its high-tech advantages, whatever software program the TSA eventually deploys may be similarly vulnerable to terrorists clever enough to game the system. Two students from MIT, Samidh Chakrabarti and Aaron Strauss, demonstrated as much in 2002, in a journal article entitled "Carnival Booth: An Algorithm for Defeating the Computer-Assisted Passenger Screening System." The problem with any CAPPS-like system, they observed, is its obviousness. Anyone flagged by the software would immediately know their "CAPPS status," since they'd be subject to an additional search. Chakrabarti and Strauss showed the danger of this with what they called a "carnival booth algorithm"—an algorithm is a procedure or formula for solving a problem, and it's often used in computer science to describe a piece of software code that accomplishes a goal. Terrorist cells could feed various members through the system repeatedly, find the ones who weren't flagged, and then use them for hijacking missions. Like a carnival booth barker, Chakrabarti and Strauss wrote, the transparency of a CAPPS II-like system "entices terrorists to 'Step Right Up! See if you're a winner!' "

Rafi Ron maintains that such vulnerabilities will persist if our security concentrates on flagging passengers and searching for weapons, rather than detecting behavior and arranging person-to-person interaction with high-risk fliers. "The problem with everything we are doing in American airports today," he says, "is at the end of the day if somebody becomes a selectee in the process, the only thing that happens is that he is asked to take off his shoes. And that's ridiculous."

Israeli experts insist that biometric systems and profiling software could easily be layered with behavioral detection. Since passengers already interact with a variety of airport personnel—from ticket agents to checkpoint screeners—those people could be trained to recognize subtle behavioral clues. Ron—whose company has trained, among other law enforcement groups, New York City's counterterrorism bureau—also sees applications for behavioral expertise in other areas of security. Behavior detection in Israel, he says, is not confined to the airport. The security guards that are required at restaurant and café doors are trained to watch for similar signs in suicide terrorists. According to Ron, in roughly 40 percent of the suicide bombings in Israel, the bomber is stopped before he or she can detonate the weapon. The same techniques, with trained police, could be used here. A May 2004 FBI bulletin warning of suicide bombers in the United States, reported by *Time* magazine, instructed law enforcement officers to watch for people wearing heavy jackets on warm days, smelling of chemicals, or having tightly clenched fists that could indicate they are holding a detonator. "The main thing you have is your ability to watch people," Ron says, "and take decisions on who do you want to pay more attention to, and be in a position to react to situations immediately."

Such observation has its limits, of course, but even 9/11 provided evidence that alert agents can play a role in thwarting terrorist plans. A month before September 11, an immigration officer in Orlando turned away a man later confirmed to be a potential hijacker. The man, Mohamed al-Kahtani, was unable to explain to the officer what he was doing in the country. As the 9/11 Commission reported, "The inspector relied on intuitive experience to ask questions more than he relied on any objective factor that could be detected by 'scores' or a machine. Good people who have worked in such jobs for a long time understand this phenomenon well. Other evidence we obtained confirmed the importance of letting experienced gate agents or security screeners ask questions and use their judgment. This is not an invitation to arbitrary exclusions. But any effective system has to grant some scope, perhaps in a little extra inspection or one more check, to the instincts and discretion of well-trained human beings."

A few years ago, Ioannis Pavlidis, then a pattern recognition expert in the Minneapolis research office of the high-tech conglomerate Honeywell, received a "request for proposal" from DARPA. The agency was looking for research projects to fund in the growing area of biometric technologies. The term "biometrics" refers generally to the application of statistical and mathematical models to biology, but in the last decade it has also come to describe technologies that attempt to uniquely identify people based on biological traits. All areas of the latter form of biometrics—from fingerprint to face to iris to hand geometry recognition—rely on software-based methods of identifying unique patterns, the same type of algorithms that Pavlidis was developing at Honeywell.

Born in Greece, Pavlidis emigrated to the United States in the early 1990s to earn a Ph.D. in computer science at the University of Minnesota. He wrote his thesis on methods for optical character recognition (OCR), which involves using software algorithms to translate text from a paper document to a digital, editable form. The essence of OCR entails teaching computers to "read" and distinguish text—in the same way that biometrics requires that they read and distinguish human features. After completing his doctorate in OCR, Pavlidis had gone straight to Honeywell, where his focus was on creating software that could automatically identify handwriting and signatures.

Upon receiving the DARPA request, Pavlidis surveyed the biometrics landscape and quickly realized that all the burgeoning technologies fell into the same category: whether fingerprint or face recognition, they were designed to find a distinguishing physical feature and match it to a person's identity in a database. It occurred to him that all these biometric technologies required already knowing who you were looking for—for example, making sure that the right person is accessing a secure door. "The question in my mind was, What happens in the case of a bad guy who is not in the bad-guy database?" he recalls. "He appears normal but he still means bad things. How do you figure out what is happening in these cases? That is how the whole thing started." The ultimate biometric, Pavlidis reasoned, would be not one that verified your identity but one that—like the airport security personnel and counterterrorism officers trained by Paul Ekman and Rafi Ron—read your behavior without having to know it. "The idea was not, 'Tell me who you are,'" he says, "but, 'Tell me what you are about to do.'"

Pavlidis started looking around for a way to do this, and eventually concluded that he might be able to measure stress by applying pattern recognition to a thermal imaging system that would detect heat signatures produced by the body. To do so, he needed a partner who understood physiology. So he hopped on the Web and found a renowned endocrinology researcher at the nearby Mayo Clinic named Dr. James Levine. Levine, considered a world authority on measuring the body's use of energy, was investigating ways that different behaviors increased human metabolism. He was enthusiastic about Pavlidis's biometric idea, and they developed a series of experiments to try to determine whether stress might create a unique thermal pattern in the face—one that Pavlidis could use thermal imaging cameras and software to detect. Though temperature seems like an obvious indicator of stress, no effort had ever been made to detect and measure it at a distance. The researchers obtained funding from the Department of Defense Polygraph Institute to investigate.

During one of the early experiments at the Mayo Clinic, Pavlidis and Levine were in the lab, discussing what method to use to induce stress in their research subject, a loud noise or a difficult math test. While they were debating this, a piece of ceramic tile they had brought along fell and crashed to the floor, startling the subject waiting in the other room. "All the equipment was on," recalls Pavlidis, and suddenly the monitors around the room lit up, showing a quick but prominent flash of heat around the subject's eyes. They had found it: a thermal signal of the human "fight or flight" response. Levine began working on the physiological causes behind the heat flash, while Pavlidis applied his pattern recognition skills to interpreting three-dimensional temperature information. By using the temperature to infer things such as blood flow in the face, they concluded that they could detect psychological conditions such as nervousness and fright, and possibly even deception.

Pavlidis and Levine decided to put the theory to the test. First they randomly assigned 20 volunteers to commit a mock crime—stabbing a dummy and robbing it of $20—and then proclaim their innocence under questioning. A control group was kept isolated, with no knowledge of the crime. Pavlidis and Levine then asked each of the subjects a series of questions about the crime, measuring the thermal signature of their face and having Pavlidis's software interpret it. The results of their first clinical study, published in the journal *Nature* in January 2002, showed that the thermal imaging system was just as effective as a polygraph in determining whether subjects were being truthful. Eight-three percent of the "criminals" were flagged by the system, while 90 percent of the "innocents" were not. It was the biometric technology that Pavlidis had envisioned, one that could potentially read a person's intentions.

Pavlidis shifted his work from Honeywell to the University of Houston, where a large part of his research is devoted to civilian applications for the technology, employing it to monitor patients' temperature signatures—and thus medical indicators such as their pulse—in their hospital beds, at home, or even at work via the Internet. For DARPA, the researchers are trying to perfect a system that would give questioners an additional tool to read their subjects' truthfulness. Unlike a traditional polygraph, which measures a combination of respiratory activity, fingertip sweat, and blood pressure, the thermal system doesn't require strapping in the subject. It can be conducted noninvasively and even without the subject's knowledge (and with heat signatures, the physical response is even harder to control than facial expressions). Also, it doesn't require a specialist to interpret the results; the software simply spits out a score indicating the subject's truthfulness.

Pavlidis and Levine's work is just one of many similar attempts to automate behavior detection. Researchers at the University of Pittsburgh and the University of San Diego are working to automate Ekman's FACS using video. Such systems attempt to replicate and even improve upon the basic human ability to read the cues of nonverbal behavior. Automated behavior detection could provide aids to human questioners in visa application interviews, border control, and other areas where investigators are already questioning people about their intentions.

But every test or detector can go wrong in two ways: It can fail to spot something that it is supposed to spot (a false negative), or it can think it has found something when it actually hasn't (a false positive). If you are infected with HIV and a blood test says you aren't, that's a false negative. If you are HIV-free and the test says you have it, that's a false positive. Which type of error you care about more depends on the context. In the realm of airport security, letting hijackers slip through is a false negative with tremendous consequences. But false positives, in which innocent people are deemed suspicious, are also more than just inconvenient. In great numbers, they can doom a system to irrelevancy—and undermine democracy in the process.

Consider Pavlidis and Levine's system, if it were employed on its own at an airport such as Boston's Logan (a possibility raised by some media accounts of the research, but one about which Pavlidis is skeptical). Assuming for the sake of argument that the system's 90 percent accuracy rate carried over to the airport, the false negative problem could be endured (only 1 chance in 10 of missing a terrorist is not perfect, but a lot better than nothing for a single layer of security). But 20-plus million passengers pass through Logan each year, and the system would flag millions of airline passengers who were simply stressed out over flying—another version of Othello's error. With so many false alarms, the system would quickly be rendered virtually useless. "That was never our method, pointing it at people and saying this is a good guy and this is a bad guy," Pavlidis says. "The major secret is in what questions you ask and the way you ask them. You measure the response to a stimulus. If you don't have a stimulus, you don't have anything." But even with questioning, thermal imaging might still be vulnerable to Othello's error—fooled by innocent subjects who are simply stressed out about the intense questioning process.

Which is why Paul Ekman favors a multitiered security setup, using trained behavior detection personnel. "Tier 1 are people who are doing screening at airports or visa interviews, where they've got four minutes," he says. "In a country of our size, there are thousands, tens of thousands of such people. What's really important for tier 1 interviewers is to train them as best you can, because you don't want any false negatives. You really don't care about false positives, because all that happens is that they get shifted into tier 2, and you get a more serious interviewer who is better trained." The tier 2 interviewers, potentially, could be the wizards that Ekman and O'Sullivan identified in their research.

In such a system, technology such as thermal imaging could potentially be used as an appendage to human observation, to fill in the gaps where expertise isn't available. "Ideally you would have the very best psychologists attending the interviews, next to the consular officer in an embassy that grants visa applications, or next to the immigration officer at a border checkpoint," Pavlidis says. "The effectiveness of a top-of-the-line psychologist cannot be replicated with any machine. But of course this is not feasible." (In fact, another type of thermal imaging—to measure the surface temperature of skin—has already been used in airports around the world, to try to detect SARS patients during the 2003 outbreak. Such detectors faced their share of false-positives.)

"The idea," Pavlidis says, is "to provide a machine tool that will make these people a little bit more effective in what they are doing anyway." By way of example, he mentions Diana Dean and Ahmed Ressam at the U.S. border in 1999. "That was a classic case of a very good interviewing agent," he says. "The guy felt a little bit uncomfortable, so she ordered a search of his car and that's how they found the explosives. If we are not lucky enough to have a good interviewing person, like in that case, the guy might have slipped through and what would have happened?"

Early one October evening in 2002, as the last daylight faded over Boston Harbor, a man walked purposefully along the restricted waterfront area that abuts Logan Airport. Logan sits along the northern shore of the harbor, flanked on three sides by the open water. Following 9/11, MassPort had established a no-traffic zone along the airport's waterfront edge, preventing any boats from approaching within 250 feet of the shore. The new regulations were meant to ensure that the security system that Rafi Ron helped establish wasn't rendered useless by gaping holes in the airport's perimeter. MassPort set up security patrols to periodically monitor the water and shorefront, but soon discovered a problem: with 5 miles of shoreline, there simply weren't enough officers or resources to watch the area 24 hours a day. The lack of manpower was now evident as the man strode toward the airport's perimeter. As he approached the outer fence, he could see no sign of the security force.

On a nearby tarmac, however, sitting at a bank of laptops, Alan Lipton and his engineers were already aware of his presence. Lipton's video cameras had tracked the man as he approached a virtual "trip wire." Within seconds an alert went out to authorities to investigate the intrusion. Fortunately, the man was himself a member of the state police, testing a new automated video system from a company called ObjectVideo, where the 33-year-old Lipton is the chief technology officer. Over several days the system was able to detect—without human intervention—almost 90 percent of the staged perimeter incursions by boats and people into the safety zone.

The premise behind automated surveillance technology is that suspicious behavior is not confined to darting eyes or trembling hands.

Because carrying out a terrorist attack usually involves executing a series of intricate steps, terrorists' behavior "is also dictated to some extent by their operational plan," observes Ron. "If they want to attack many people, they have to be in a place where many people are present. If they have to have a place to prepare themselves for the attack without too many people, you need to look for them at that place." When terrorists are probing their targets for information, or turn up somewhere they are not supposed to be, they become anomalies.

Video cameras are a natural—and often controversial—tool for trying to detect these macrobehaviors. The use of video surveillance is ubiquitous in many public places and critical facilities, particularly in the United Kingdom. A researcher at the University of Sheffield reported in 2003 that there were more than 4 million closed-circuit television (CCTV) cameras in use in Britain, roughly one-fifth of the number used worldwide. In the United States, camera surveillance is often the first step taken to beef up defenses of critical infrastructure. The nation's 66,000 chemical plants, 103 commercial nuclear power plants, and 10,000 or so power plants are now viewed by authorities as inviting targets; to take one example, more than a hundred chemical plants are in areas of sufficient population density that an attack on them would threaten a million or more people.

The growth of video surveillance of public spaces, particularly in urban areas, raises important privacy issues for policy makers and the public to debate. Civil liberties advocates in the United Kingdom have increasingly questioned the implications—and the uses—of the exploding number of CCTV cameras. But when it comes to protecting facilities such as chemical plants, the issue is not privacy but capability.

Because of the difficulty in having a human being monitor large numbers of cameras, CCTV is rarely used to watch areas in real time. Instead, the presence of cameras is usually intended first as a deterrent and second as a forensic tool for law enforcement to use after an event. The problem, wrote Lipton and three colleagues in a recent white paper, is that "no matter how highly trained or how dedicated a human observer, it is impossible to provide full attention to more than one or two things at a time. A vast majority of surveillance video is permanently lost without any useful intelligence being gained from it. The situation is analogous to an animal with hundreds of eyes, but no brain to process the information."

The attempt to create machines that will automatically analyze video has its roots in the field of computer vision, a branch of artificial intelligence research. The goal in computer vision has been, essentially, to create machines that understand what they see in an image. The field traces its history to the 1960s, when the Pentagon began developing signal processing software to interpret the two-dimensional photos returned by satellites. The military funded successful research in automated target recognition, in which a computer looks at shapes on a map and picks out potential targets such as tanks or ammunition depots. On the civilian side, the field branched out into medical image processing, such as automated analysis of CT scans and x-rays for cancer detection.

Attempts to use similar analysis on video—particularly to try to develop robots that could process visual information—were met with dramatic failures. Marvin Minsky, the artificial intelligence pioneer, once speculated that he could have a graduate student solve the "computer vision problem" in a few months. "Thirty years later," says Lipton, "we are still almost nowhere nearer to solving the problem." The greatest obstacle was the sheer density of visual information, which includes both a visual component and a time component. The calculations required to establish even basic elements of a scene, to distinguish inanimate objects, say, are incredibly complex. "It seems really easy, because we do it as human beings," says Lipton. "If we look at a video stream, or we just look at a scene, we instantly understand what is happening. It's really hard to teach a computer to do that. Computers are, for all their intelligence, very dumb. And vision is one of the toughest problems there is."

It was only in the mid-1990s that scientists began to crack it. In 1997 Lipton was a newly minted 26-year-old junior professor from Australia, working at the Carnegie Mellon Robotics Institute in Pittsburgh on computer vision problems in robots. Just as he arrived, the lab received a grant from DARPA to develop what they called video surveillance and monitoring technology—the ability for machines to interpret visual data and identify suspicious behavior. Lipton ended up as co-manager of the program, and over three years a team of 13 institutions developed a batch of algorithms that could distinguish and track unique objects using video. The project spawned the technology behind two of the leading automated surveillance companies in the United States, Sarnoff and ObjectVideo. Now a dozen companies and labs, in the United States and overseas, are racing to market video surveillance software based on similar technology.

The key to automated surveillance is what computer vision researchers call "blobology." First, the software takes pixels of information provided by the camera, which are represented by a series of numbers. Those numbers are used to create a statistical model, which over a period of time determines what is "normal" in the scene. "If you are looking at a forest," says Lipton, "and the trees blow in the wind, you can statistically speaking say that kind of motion is normal behavior." The system then begins to incorporate more features into its definition of normal: changes in light due to the rising and setting of the sun, say, or the rain from a passing storm. "We are basically understanding the scene," he says. "Not from a semantic sense. I don't know that it's rain. I don't know that it's wind." What the computer does know is what kinds of changes regularly occur in its data.

Once normality has been established, the software looks for blobs: objects in the scene that don't fit that pattern. A person walking through the frame, a bird flying over, or a vehicle driving past would all be identified as blobs. "Anything that happens at a faster time period than we are used to, that is not normally part of a scene, we can detect," says Lipton. Each blob is labeled with properties such as size, color, direction, and speed of movement. Over multiple frames, the software then tracks each of these blobs, using the same algorithms developed by NASA in the 1970s to track deep-space probes.

When all the elements of a blob are established, the program uses its characteristics to determine what kind of object it is. "We say, 'Here is something, we have a trajectory, we know what color it is, we know what size it is, we know how it moves from frame to frame,' " he says. "All that kind of stuff you can glean just from your blobs. That's when you can say, 'I reckon that's a human, because I reckon a human moves like this, and I reckon a human is about this big, and I reckon a human looks like that, with that kind of color distribution.' "

Once the objects are classified—a human, a car, a piece of luggage—the algorithms can turn to identifying aberrant behaviors. Take a hallway where all people are supposed to be traveling in the same direction. Automated surveillance software can reliably detect when a person is moving the opposite way, and sound an alarm for a human operator. Major airports in the United States, including Dallas/Fort Worth and Salt Lake City, already use software to detect these kinds of extremely simple anomalous behaviors. It can also set up "virtual barriers" in the camera's field of view—essentially invisible lines that cannot be crossed by a human or vehicle without alerting the program. The permutations are endless. Sergio Velastin, at Kingston University in London, has developed specialized algorithms—for CCTV cameras in the London Underground—that pinpoint passengers who loiter too long on the platform or leave a bag behind. (He founded a company, IPSOTEK, to market the technology.) Mohan Trivedi at the University of California at San Diego developed crowd density rules that alerted authorities when certain stadium areas became too crowded during the Super Bowl. ObjectVideo itself signed a multimillion-dollar deal with the Department of Homeland Security in 2003, to install the company's technology at vehicle checkpoints along the largely remote 5,000-mile United States–Canada border.

The common thread among all the programs is that they look for violations of predefined rules. "What we can do is take someone's policy and turn it into rules that the computer can understand," says Lipton. As in, "I care if someone parks across the road from my facility but nobody gets out. Maybe there is some hostile surveillance going on. We can even do it in a coordinated fashion, and say, 'If a car parks on my bridge, I don't care. But if 15 cars park simultaneously on 15 of my bridges, that's a coordinated attack.' "

Such cases provide opportunities in which technology, unhindered by the human limits of time and attention span, could potentially serve as a more perceptive watcher. Imagine, for instance, that the same white van appears on consecutive days at an airport, driving slowly through the passenger pickup area—but each day at a different time. (Video surveillance tapes at Logan, in fact, reportedly showed hijacker Mohamad Atta doing just that, driving through the airport five times prior to September 11.) To individual guards monitoring the area, the van would be a singular occurrence for their shift, not enough to represent an anomaly. But over multiple days, the reappearance of the van would constitute a *pattern* of unusual behavior—potentially someone surveilling the airport as a target. "That's where computers have the edge," says Lipton. "They don't blink, and they can remember forever. If an activity is occurring over a wide enough area over a long enough period of time, a human being isn't going to spot it. But a computer can."

The next step, he says, is to develop automated surveillance that doesn't need to be told in advance what constitutes a suspicious event, but that can learn this on its own, using its own observations. Because a camera can spend 24 hours a day, 365 days a year watching a scene, it could potentially be trained to catalog the range of normal behaviors in its field of vision—people usually walk a certain direction, vehicles typically stop for short periods of time. The system could then use that learning to develop its own rules for aberrant behaviors. "Then I can start to say, 'What are the outliers?'" says Lipton. "'What is the behavior that doesn't fit the pattern?' And maybe the system will be able to volunteer that something is amiss."

But an entirely software-based video system—especially one designated to detect any activity it generally deems to be suspicious behavior—risks being plagued by false positives. The human ability to detect anomalies is one that has not proven easy to replicate through technology—which is why instead of relying on technology alone, MassPort is building its own human-machine partnership. For starters, Logan officials gave out free cell phones to local clam diggers—who know better than anyone what constitutes an anomaly on the water—so they can report any suspicious activity.

Meanwhile, the agency is evaluating more than 20 potential automated surveillance systems—including ObjectVideo's technology—to install at Logan and other facilities. Besides training its officers in behavior pattern recognition, it is outfitting them with handheld computers that allow them to check the driver's licenses of the people they stop against criminal databases, similar to vehicle cops on the beat. Eventually, automated surveillance cameras will be linked to the handhelds, says Treece, so that when a virtual trip wire goes off, the nearest officer will be alerted and routed to the security breach. The camera, unblinking, would be busy tracking the intruder.

• • •

In December 2001, aboard American Airlines flight 63 from Paris to Miami, Richard Reid, the now infamous "shoe bomber," attempted to set off explosives contained in his sneakers. He was stopped only by an aware flight attendant, who smelled a lighted match and, with the help of passengers, was able to subdue him. The near-success of Reid is often cited by the Israeli experts as an example of the flawed American security system. Six months before, in what Israeli officials believe was a probe of their security, Reid attempted to board an El Al flight from Paris to Tel Aviv. The Israeli security personnel flagged him—based on his appearance and lack of luggage—subjected him to a complete search, and seated a sky marshal next to him on his flight. Reid later boarded an American Airlines flight unimpeded, and with plastic explosives hidden in the hollowed-out heel of his shoe. To Israelis such as Rafi Ron, Reid personifies the need for U.S. air carriers to target people, not weapons.

Reid's failure is an example of something else: the unique human ability to recognize dangerous behavior. The flight attendants and passengers on board flight 63 had no previous exposure to shoe bombs. They had no advance knowledge that Reid carried such a weapon. But they knew instantly that a man flicking a match around his shoes was not engaging in normal behavior, and they acted to stop him. No camera, heat sensor, or other piece of technology would have been able to spot the aberration that was Richard Reid.

As automated techniques for recognizing behavior advance, the central role of humans in seeing intentions will remain, enhanced by tools that extend our senses. But even with the most sophisticated tools, we still need a Diana Dean to connect the dots of suspicion and make the final choice about how much suspicion requires action. "I think there will always be a human decision-making element," says Ron. "The technology will never provide the total answer. If you do allow technology to provide the total answer, then what do you do next? You put a gun in the camera and let it shoot him? Of course this is nonsense. At the end of the process there must be a person."

A few months after his ObjectVideo demonstration along the shore of Logan, Alan Lipton was catching a flight out of Dulles International Airport in Washington, D.C. When his carry-on bag went through the x-ray, security personnel immediately grabbed him and escorted him to a nearby holding room. His bag, they said, was full of makeshift weapons—two butter knives, a plastic picnic knife, and a broken pair of scissors. Lipton told them the knives weren't his, and that he had no idea how they had gotten in his bag. The police charged him with an attempt to carry dangerous weapons onto a flight, and released him.

It was only later that Lipton recalled, as he explained to the judge who eventually dropped the charges, that during the Logan experiment, the MassPort police had loaned his team a collection of knives, impounded in the post-9/11 prohibition of sharp objects. Lipton's team had used the blades as survey markers in their system test— signaling the 50-, 75-, and 100-foot approach from the water. "They don't have any infrastructure at this point, on the perimeter," says Lipton. "So we had to bring our own and set it up. They used one of my travel bags to carry equipment out on the tarmac and back." After the experiment, Lipton had never unpacked.

In the diligent hunt for weapons, an x-ray machine and a rigid anti-terrorist policy had combined to nab not merely a person with no ill intentions, but someone who was working to combat the terrorist threat in the first place. It was an irony that Rafi Ron would appreciate. Yet in the end, the security officers were using their ingrained sense of suspicious behavior detection after all. When they confronted him about the knives, "I had a total mental blank," Lipton says. "I said, 'I'm sorry, I have no idea how they got in my bag.' And they didn't like that answer. They thought, 'Well, that's suspicious enough that it's worth charging him.' I imagine that if I had known what they were, admitted what they were, and said, 'Go ahead and confiscate them,' that would have been fine. But it was suspicious, because I simply had no idea."

THREE

Inside the Internet

W HEN DEPARTMENT OF JUSTICE prosecutor Mark Rasch tri-
umphed against Robert Tappan Morris in a crowded courtroom in Syracuse, New York, in 1990, he wanted to see the
24-year-old go to jail. Rasch didn't want the promising grad student—
whose father happened to be the top computer scientist at the National
Security Agency—to rot in the pen. But he believed that Morris
should serve some time. Rasch wanted to make an example of Morris,
for all those who might mimic what the hapless defendant had done,
and for those who would do worse. After all, this young man had
created a piece of dangerous software code, which in effect was a
weapon, and he had sent it across the nation, crashing thousands of
the Internet's computers. He had committed a crime.

The judge, however, had a more fatherly approach to the case and
sentenced Morris to a mere three years' probation. For Rasch, who recalls his reaction to the sentencing with still-fresh ambivalence, the
outcome was one of many early clues that jurisprudence and digital
technologies were on a collision course, and that the law would have to
leap or be run over.

Rasch's wide frame and fighter's posture more than make up for his
short stature. With a deep tan and a constant watchfulness, he at first

comes off as a man trying too hard to prove himself, but in fact is just a high achiever who is extremely smart. His tough-guy sound bites show up regularly in places such as *Information Week* and the *New York Times*.

A private consultant on legal issues related to computer security since leaving the Justice Department, Rasch is a political liberal with a cop's realism about human vice. While still in high school at Bronx Science, he was out in the streets, witnessing unedited reality: through a New York Police Department auxiliary program, the type-A teenager had trained as a paramedic and traveled around in a SWAT-like van to help in emergencies. He gave up on dreams of medical school the moment he hit undergraduate organic chemistry and made an about-face to political science with a minor in history. From law school, he went straight to the U.S. Department of Justice, where he worked on cases involving espionage, until he landed back out on the streets—this time in the nation's capital—spending six months prose-cuting city crime. "Fence jumpers" were a specialty, he says—usually guys "who'd gone off their meds" and climbed the barricades at the White House.

Back again at the Justice Department in the mid-1980s, Rasch was assigned to the department's fraud unit, where he went after the no-torious Gambino crime family and helped build the case against Lyndon LaRouche, accused of soliciting $34 million in loans, with phony assurance of repayment, to fund his flamboyant presidential bid. While investigating LaRouche, Rasch encountered numerous other fraud cases, some of which weren't the standard scam phone sales or mail solicitations; they were a novelty. These more creative misdeeds involved computers, and because they were new twists on old crookedness, the statutes then on the books weren't always a good fit. "There were gaps in the law," says Rasch, "with respect to things like trespassing and stealing." Intriguing questions had begun to sur-face, Rasch recalls, such as, "What constitutes 'trespassing' in cyber-space? What does it mean to 'steal' information, when the information you've stolen is still there? If you use someone else's account name, are you criminally impersonating them?" Rasch began to see that the legal tools then available weren't sharp enough for computer crimes.

There was the 1970s case of John Draper, for instance, who went by the name Captain Crunch, after the cereal that contained a toy whistle, which he blew into a phone receiver, imitating the tone AT&T's system used to announce an open long-distance line, thus allowing Draper to make free calls; stealing phone time was a new sort of fraud. And there was the 1982 case of Theodore Weg, a computer systems manager for the New York Board of Education, who used school computers to handicap horses at Yonkers Raceway; was this use of his employer's computer a "theft of services"? Eventually, attorneys such as Rasch, continuing to confront examples of new technology crammed into old law, would run up against ambiguously criminal acts such as spamming and music file-sharing and all manner of digital copyright infringement. In time, Hollywood would join in the huge, national legal row over the definition and punishment of computer crimes. But Rasch, a good 10 years before movie moguls even began to fret, recognized the plight of attorneys who were mired in old definitions of terms such as "theft" and "property."

Then in 1984, Congress passed its first attempt at a computer fraud law, which made it a crime, for instance, to access a government computer without authorization or to traffic in computer passwords—and at that moment Rasch's career began to crystallize. "I liked science and technology, so I read the statute," Rasch says—and apparently he was one of the few who did, because he quickly became the Department of Justice's de facto head of electronic crime. For the next two years, Rasch took the lead in advising Congress's judicial committees on how to improve the law. (It was amended in 1986, the first of at least four times it would be altered substantially, as the technology continued to shape-shift.) He worked on the Hanover hackers case, among others, which involved a group of jaded German youth delivering stolen computer code to the KGB. Then, in 1988, he embarked on *United States v. Morris,* which would be the first successful jury trial under the Computer Fraud and Abuse Act.

Robert Tappan Morris was born in 1965 to a quirky pair of intellectuals who had settled into an eighteenth-century farmhouse off the beaten path in New Jersey. His mother, Anne, devoted and rigorous in her attention to her kids' moral and mental development, worked for an environmental nonprofit. His father enjoyed a long career at the famously innovative Bell Labs and then became chief scientist at the

Computer Security Center of the National Security Agency (NSA), a code-breaking and eavesdropping outfit so secretive that its initials are often said to stand for No Such Agency.

Katie Hafner and John Markoff, reporters for the *New York Times,* captured the peculiar tics and convictions of the Morris family in *Cyberpunk: Outlaws and Hackers on the Computer Frontier.* Their book surveys the Morris lifestyle with gentle amusement: "The family had three large dogs, and at least a dozen cats. . . . When Meredith [the eldest child] asked for a horse one year, Bob [Sr.] compromised and got her a pig. . . . Every lamb they owned was named Lambchop, lest the children lose sight of its fate. . . . They played in orchestras together, sang in choirs together and took regular trips to Manhattan. . . . The family library included six thousand volumes, their subject matter ranging from theology to natural history to sailing and navigation."

The middle child, Robert, who enrolled at Harvard in 1983, was a computing prodigy. ("By age nine," according to Hafner and Markoff, "he was devouring back issues of *Scientific American.*") He was also painfully shy and awkward—a perfect instantiation of the programmer stereotype. Bored by schoolwork because it was too easy, and prone to spending hundreds of hours alone, he was susceptible to the lures of experimentation, the enticements of dreaming up challenges no one had thought of and then meeting them. Which is exactly what he was up to while in his first year of a Ph.D. program at Cornell. One exercise he invented for himself was to create a computer program that would propel itself across the Internet and inconspicuously establish itself on as many machines as possible; then it would "re-execute"— which is to say, scout out yet more machines and establish itself there as well. These types of programs were already known as worms or viruses, depending on their methods of infection. His version, soon to be known as the Morris Worm, was explicitly intended to avoid interrupting the workings of any machine; the point was only to make the thing propagate successfully, not to interfere. Morris was just keeping himself occupied.

Accidents can make good stories. In a last-minute hurry to get his experiment up and running, Morris made some sloppy mistakes, causing the worm to multiply out of control, its sheer volume overwhelming and paralyzing computers from Princeton to Stanford to NASA.

Systems managers all over the country were baffled. Some disconnected their computers from the network, "islanding" themselves for protection, with the unfortunate consequence that they couldn't receive the e-mail advice of their colleagues working on fixes. Some were intrigued, some impressed, but enough were angry and sleepless that the media smelled excitement.

Morris was easily apprehended, given a fearful confession to his father and a friend's excited leak to the *Times,* but for the DOJ's Mark Rasch, nabbing this culprit was a key step. For the first time in the short history of computer crime, he had a law to use that was written to fit the crime: not some clunky old tool from the shed, but one crafted for dissecting just this sort of case. Plus, although Morris came forward himself, Rasch had managed to track him via the electronic traces of his wrongdoing. The smidgeon of forensic expertise Rasch applied in the six-month process of investigating the case would develop into one of his more critical skills.

Probably no courtroom trial has ever hosted so many computing luminaries. Bob Morris Sr. spent the nerve-wracking days of testimony among his pocket-protected colleagues from Bell Labs, who flew in to show their support. The press had all the color it could use.

What wasn't so clear back then, however, was just how rare this episode would turn out to be. As a civil servant, traveling the world from courthouse to courthouse, and now as a security consultant in the private sector, piloting companies through the labyrinthine legalities of electronic privacy policy and evidence collection, Mark Rasch has learned as well as anyone just how hard it is to catch vandals, thieves, and trespassers who lurk on digital networks. With Morris's conviction, it looked as if ensuing legal thrusts into the hackers' world would yield triumphs. But what promised at the time to be the first of a series of prosecutorial victories turned out to be one of the few such snares of Rasch's career.

Back then, finding the perpetrator of a computer crime was a convoluted process; today it involves further contortions. As the Internet has sprouted millions more nodes and links, its vandals have grown more elusive. Morris is one of only about ten computer-virus and worm writers ever convicted. Of the tens of thousands of other types

of attacks perpetrated each year, only about one or two dozen are
ultimately prosecuted and punished. (Several attempts at offering
quarter-million-dollar bounties for information about perpetrators
haven't improved the odds.) Anyone who knows the basics of Internet
technology—including Rasch himself—will tell you that finding the
human source of an Internet disturbance is next to impossible. With
extraordinary measures, maybe; with major top-down technical changes,
perhaps; with enough hacker slipups, surely; but the profound reality
of an increasingly crowded Internet is that it's a near-perfect place to
hide. Which is one of the reasons why the notion of cyberterror is
both believable and unnerving. The Internet's anonymity is one of its
most fascinating and fertile traits. It's also one of its scariest. Power
with no face and at great distance is no match for a gun at your
temple, but it has its own particular grimness.

The Anonymity is key to what has been extraordinary growth in Internet crime. (In the inner circles of the computer world, "Internet
crime" and "hacking" are far from synonymous, the latter term applying not necessarily to wrongdoing but to any sort of clever programming; these distinctions are becoming more and more blurred.) The
Computer Emergency Response Team, or CERT, Coordination Center,
one of the most respected cyber security watchdog groups in the
United States, based at Carnegie Mellon, provides salient evidence of
that growth. In 1989, the group's second year of operation, it received
132 reports of network attacks. Ten years later, in 1999, that number
had reached 9,859. In 2003, the figure was 137,529. Given corporations' reluctance to acknowledge that they've succumbed to an assault
(no customer will relish buying products from a website that's been
breached) and given the weariness of systems administrators, managing hack after hack, experts suspect that only half of the number of
incidents are even reported. According to software maker Symantec,
nonprofit organizations and companies across the spectrum of industries on average experience between 20 and 40 attacks a week, about
one-quarter of which are targeted, while the rest are unleashed at
random.

The vast majority of these events are attempts to stir up trouble
that is no more diabolical than spray-painting subway cars or smoking
marijuana in the school basement. This kind of malfeasance finds its

roots in the theft of long-distance phone time, the favored crime of Captain Crunch in the 1970s, and remains almost entirely a pastime for teens.

Another sometimes overlapping group of computer attackers is made up of political crusaders or self-styled renegades of one form or another, whose exploits are also not hugely damaging. In his portrait of the culture, *The Hacker Crackdown,* science fiction writer and journalist Bruce Sterling spells out the ethos: "The most important self-perception of underground hackers . . . is that they are an elite. The day-to-day struggle in the underground is . . . for power, knowledge, and status among one's peers." Sterling plumbs the lawlessness: "When you are a hacker, it is your own inner conviction of your elite status that enables you to break, or let us say 'transcend,' the rules. It is not that all rules go by the board. The rules habitually broken by hackers are unimportant rules—the rules of dopey greedhead telco bureaucrats and pig-ignorant government pests."

Yet another subset of the category of hackers who break into computers is the community whose good-Samaritan mission is to strengthen the general security of the network. Often lawbreakers-gone-legitimate, these computer engineers are professionalized security advisers, usually authorized and under corporate contracts, who break into machines in order to find weak spots and inform the owners. In recent years, they've come to be known as "white hats," though frequently they're also called "black hats," because they're good guys posing as bad guys, breaking into company property in order to detect trouble and help out.

Since Sterling's account was published in 1992, the world of unauthorized electronic offenders has grown so large it seems unlikely its rank and file still perceive themselves as "elite." Because of the wealth of shared information online, learning to commit basic attacks has become as simple as learning to use the advanced features on a microwave. Software tools available around the Web automate attacks, so miscreants no longer even need to know how the ploys work. But the accessibility of weapons clearly hasn't dampened the thrill of transgression. According to a six-month Symantec study of Internet threats in the second half of 2002, large companies with more than 5,000 employees suffered on average 1,092 attacks.

This rebelliousness—or in some cases real malevolence—is of course stoked by the courage afforded an individual hidden away in a room where no one can find him. If 1 of every 10 hacks resulted in the shame and anxiety and family drama that Robert Tappan Morris suffered through, the increase in attacks each year, no doubt, would follow a much gentler curve. But hackers know that the chances of being caught are nearly zero. To actually sneak his way into the control center of a local phone company, a terrorist or vandal would risk ending up in handcuffs; get there via a cable modem, however, and the odds are excellent that the crime will never be reported.

A communication medium that hides its users' identities also provides a rich environment for the study and refinement of reviled ideas. For instance, an instigator trying test runs of new, potentially destructive hacks faces practically no danger of punishment. In security circles, the term "Day Zero" refers to the window of time computer managers spend analyzing and containing a novel form of weapon. Until security specialists understand the code that makes up a virus or the methods that produce a deluge of electronic messages suddenly crashing victims' machines, they—and the Internet—are defenseless. So Day Zero is when the real damage gets done. A terrorist, free to research and plan in obscurity, devising an unprecedented offensive, might defer Day Zero until the most auspicious time—the moment, for instance, when network engineers are frantically rescuing phone company facilities from a coordinated truck bombing or when fellow attackers have brought down some part of the nation's power grid.

Unlike the early phone networks, designed to distinguish individual customers in order to clock their usage and charge them fees, the Internet has no system of identification built in. Rasch likes to compare it instead to the postal system. "Somebody puts a bomb in a mailbox," he posits matter-of-factly, "a letter bomb—best you can hope to do is find out what post office it passed through first." But who carried it to which public box on what street corner or what other stopovers it's made? Rasch shrugs, a Bronx kind of shrug, and smiles amiably. "The post office is in the business of delivering letters, not preventing terrorism."

Yet no one out there—neither Rasch nor his colleagues in the burgeoning network security industry—wants to completely give up on the idea of online accountability; bad actors ought to be found and named. Even supremely jaded tech experts haven't given up on the idea of deterrence through law enforcement and by example; they just can't figure out how to do it without taking apart the Internet as we know it.

Online accountability for serious criminals requires climbing two Mount McKinleys: first, inventing the technology for tracing attacks to their sources; second, convincing the world to use that technology. The first is a formidable intellectual challenge, and the second is perhaps even tougher. Any means of identifying wrongdoers online threatens all users with a loss of privacy, and whether the benefits of locating hackers outweigh a potential assault on civil liberties is a potent question. The prospect of introducing tools for peeking behind people's digital curtains is chilling, regardless of its intended purposes.

Privacy battles aside, though, many computer security experts argue that the risks malicious hackers pose are simply not dire enough to prompt drastic technical measures anyway. To them, worries about terrorist attacks on the Internet have been exaggerated and are a distraction from more pressing if less sensational security problems. Bruce Schneier, a brilliant computer scientist whose books and newsletters on security have widely influenced the field, considers terror-scale exploits that "disable air traffic and emergency systems, open dams, or disrupt banking and communications" unlikely. In an attempt to put cyberterror in perspective, Schneier refers to the nuisance attacks we're used to as "cyberhooliganism," and he invokes a hypothetical scene: "Imagine for a minute the leadership of Al Qaeda sitting in a cave somewhere, plotting the next move in their jihad against the United States. One of the leaders jumps up and exclaims: 'I have an idea! We'll disable their e-mail.' "

Schneier argues that real cyberterror would be extremely difficult for even the most sophisticated hackers to pull off. "Conventional terrorism—driving a truckful of explosives into a nuclear power plant, for example—is still easier and much more effective," he points out in one of his newsletters. A report written for the Defense Intelligence Agency by officers at the Naval Postgraduate School concurs:

"The group that wishes to strike at large-scale command and control, industrial or infrastructure networks with surety must develop a complex-coordinated capability . . . but personnel at this skill level make up an extraordinarily small percentage of the hacker population." Even Mark Rasch speaks of electronic terror as a what if? rather than a pressing reality. "The worst cyber attack we've had," he says, "has been no more disruptive than a large-scale blizzard. Terrorism is terror—people afraid to leave their houses, people afraid to drink the water. How do you do that over the Internet?"

It is true, however, that increased connectivity has created new forms of aggression, the effects of which can seep out from the virtual world into the physical. As far back as the 1960s, technicians who managed basic infrastructures such as power plants and roadways began taking advantage of electronic technologies to automate and remotely control their systems. For instance, rather than trudge out to a tank on a hill to check its water level or adjust the action of a pump, engineers could sit in a control room, take readings, and activate machines over phone lines, via radio, or by satellite. Meanwhile, in the tidy offices where the business of running an infrastructure company gets done—where well-dressed receptionists greet visitors and executives manage billing and hiring—computers were becoming pervasive. By the 1990s the temptation for companies to connect their now-networked office computers with the gritty control centers of power and water systems, railroads, manufacturing, heating and food processing systems, was irresistible. What corporate leaders wouldn't be curious about the real-time operations of their assets? Why not send water-use figures directly to your billing department without any steps in between? The trouble was that those office networks were also connected to the Internet; hook in your control center and it too is now linked to that great, unregulated, public Web that spans the world. Once again, efficiency—in the form of directly and tightly connected procedures devoid of lag times or detours—spells risk.

In the peculiar chronology of Internet history, the issue of securing the data that ran between these various networks was ignored until long after the technology was incorporated into day-to-day business. Today, technicians dial into electronic control boxes that handle valves and switches and gates the same way we dial into AOL. They can pull data from equipment miles away, make operating decisions and change settings, without getting up from their chairs. Yet so might someone else, via the Internet, whose intentions are less banal. Old industries such as oil or railroad shipping have seized on these amazing efficiencies without applying the care and know-how to secure them. Workers may share a single account for convenience or use passwords set at the default—"password." Even if the control room itself—or the metal boxes that house mechanisms for, say, dispensing chemicals—is secured, when it comes to software for these basic devices, there are no commercially available applications called firewalls, which sift through data entering a network and block what's unauthorized or unwanted.

What worries the people at CERT is that we are becoming more vulnerable. CERT was born in 1988, in the weeks after Robert Tappan Morris's worm crippled computers from coast to coast, and it is home to some of the high priests and vaunted deacons of network security. The institute has several functions. A team of "incident handlers" take frantic calls and e-mail from owners of computers under siege, analyze the problems, and frequently devise a way to help, sometimes working into the night dissecting the code in a new virus or helping restore machines post-attack. In addition, the organization catalogs each incident for future study, functioning as the main security library for the Internet, and posts warnings of continuing strikes and new "holes" discovered in popular software.

CERT also supports several researchers who think about the big picture of safety in cyberspace. Howard Lipson, at the institute since 1992, spends a good deal of time contemplating the dangers of the new connectivity between both virtual and physical systems. "The world is actually doing serious things on the Internet now," he says, suggesting that the scientific research and fantasy games of the past were frivolous compared with the industrial processes and financial transactions that take place online today. For Lipson, the networked devices and software that are taking over the management of major systems, such as shipping or chemical production or power generation, weren't built with protection from hackers in mind. "We're going out on a limb and not looking back—without sound engineering practices." Moreover, he fears that the drive for efficiency in these newly networked environments will force more connectivity and less diversity: So far, he reasons, providers of the up-to-date, digital technologies that are changing yesterday's gritty industries have been small, niche companies selling one or two highly specialized chunks of electronics; but eventually, Lipson fears, like most maturing industries, this one too will consolidate. Those obscure companies, straining for profits, will make their products more uniform, sell them more widely, and ultimately merge with other small players. And homogeneity equals weakness: thousands of hackers focused on breaking into one operating system will have a lot better luck than the same thousands hammering at many different types of targets. "We've been saved so far," says Lipson, "by the arcane nature of these control systems."

Lipson has another worry as well. "Security degrades over time," he explains. "It's not that the bits rust out; it's that your adversaries get smarter." Combine this pretty convincing axiom with the prospect of today's relatively diverse network environment dissolving into uniformity, and it's hard to imagine that virtuoso coders won't eventually be clever enough to shut down parts of our infrastructure for significant lengths of time. There have already been previews: in March 1996, a hacker tied up 911 emergency dispatch lines in Florida; a year later, another managed to shut down the control tower at a Massachusetts airport; in April 2000, yet another got inside a waste management system, in Maroochy, Australia, and spilled gallons of sewage across the surrounding area. Unless a stunning, unexpected surge of innovation in network security transpires rather soon, these types of crimes will get easier.

Nevertheless, the destruction of infrastructure is still a step removed from the drama of real injury and death, making tampering with the Internet in order to damage other infrastructure a step further removed. To break into a computer with the intention of disrupting the running of railways in Madrid and scrambling a day's commute would have been immeasurably less devastating than planting 10 bombs in four trains, killing nearly 200 people and injuring nearly 2,000. And this distinction—along with the technical difficulty of getting inside control systems to begin with—explains the relatively sanguine approach to cyberterrorism by people such as Bruce Schneier and Mark Rasch. The real threat, at this point, is that villains with serious intent to do harm will use the Internet not to cause primary damage but to escalate disruptions caused by more lethal simultaneous attacks. Take the Code Red worm, which defaced a large number of U.S. government and military websites in July of 2001. Released and contained during the otherwise calm days of summer, it constituted an instance of what Schneier calls "cyberhooliganism," yet even so was unnervingly effective against Department of Defense computers. A worm such as that released in concert with more gruesome events would worsen public fears, interfere with emergency responses, slow down recovery, and generally exacerbate the crisis.

For a guy with such speedy synapses, Mark Rasch does a convincing send-up of Colombo: He puts his fingers lightly to his forehead, musters up a Peter Falk rasp, and says with studied innocence, "That Internet thing you mentioned earlier . . . how does that work again?"

Today, in a similar spirit of connecting-the-dots, a very different type of gumshoe is trying to use traces of data to find hidden perpetrators of one of the most common types of Internet mischief: denial-of-service attacks, in which thousands of automatically produced messages pummel a victim's computer with more data than it can handle, forcing it to crash. These present-day network detectives are searching for the digital footprints, the Hansel-and-Gretel breadcrumbs, that will locate criminals and put some muscle behind deterrence. Their effort and the computer science it represents is referred to as "traceback." With nothing but what's in a victim's computer to work from—with only the bits that flood a bunch of machines forcing them to slow down or stop functioning—how do you figure out whodunnit? That guy, that kid, that terrorist, who concocted this digital mess and dumped it onto the network the rest of the world relies on—where is the bastard? Traceback refers to the idea that you could burrow into the depths of the Internet, following the route of the damaging data backward, traveling from node to node, down into distant hubs and out through gateways and switches, finding signs of a nefarious presence, until you reached its source.

The idea is seductive, and it's also a nice means for grasping how the Internet actually functions, a tour through the great medium's in-

nards. Lieutenant Colombo's ability to calculate the trajectory of a
bullet by studying the state of the corpse requires some math, imagination, and lots of context. Likewise, traceback. To do it—or to watch
those who try to do it—you have to reckon with the fundamentals of
the environment itself. The Internet doesn't just hand over the evidence (as Rasch would say, it's not in the business of preventing terrorism). Researchers have to dream up the methods and formulas that
will determine the bullet's origins.

What the network does provide these researchers is a little bit of
structural help—along with some very pesky constraints. For instance, routers—the traffic sorters of the Internet—are busy. They receive and pass along more data in a second than you can wrap your
mind around, directing the Internet's oceanic flow of "packets," its
smallest units of transported data. Routers literally don't have time to
do much else. So recruiting one of these sorting devices to help reconstruct an attack is an enormous imposition on it. In addition, the originating address that gets stamped on not just every e-mail but every
request—every click—to see a web page is unreliable. It can be
forged, rewritten, and randomly generated. Furthermore, a large part
of the Internet's efficiency is its manic concentration on destinations
rather than the journey itself: other than the originating address of a
packet (which can easily be forged), the system keeps virtually no
records of what's happened in the past. Any traceback scheme has to
overcome all these inherent limitations.

Chris Morrow has a graceful solution. Morrow is an advanced
system security manager for phone giant MCI, which owns, operates
and leases many of the Internet's main arteries. Its customers are companies such as Earthlink, which in turn provide their own customers—homes and businesses—with connections. MCI owns one of
the largest and most expansive networks in the world, with more than
98,000 physical miles of terrestrial and submarine cables spanning 6
continents and reaching directly into 140 countries. It's a big job for a
guy of 31, who needs a shave and wears loose jeans as he ambles down
the third-of-a-mile-long hallway at corporate headquarters in Ash-
burn, Virginia. But in the world of electronic networks, especially in
these upper echelons of it, brains are still the ticket, and Morrow has
an enormous one.

Morrow's interest in traceback was driven by necessity. On a sprawling network that includes thousands of routers stationed around the globe, it's a challenge to trace a packet through your own system, never mind through other service providers' networks all the way to its source. A typical packet can easily visit 15 routers and sometimes up to 25 before it reaches its destination, and it contains no clues about which routers those might have been. So when desperate customers began calling Morrow more and more frequently to report that surges of useless packets were hitting their web servers—such a flood, in fact, that machines trying to read them all essentially jammed—it was no simple matter to trace the path of the bad packets and stop them. Some intermediary routers keep logs of bits that have flown by, but these records pile up so quickly it's nearly impossible to store them. Often logs are written over—erased to make room for new records—within minutes. What Morrow had to do, while the attack was under way, was first inspect the traffic that was passing through the closest router to the victim's machine; there he had to determine what types of packets were causing the crisis and ask the router to log them briefly. Next he had to determine from those logs the address of the previous machine those bad packets had come from. Then he'd repeat the same task, analyzing the traffic at each node and working backward through each previous router—a process that could take him eight hours.

Response time such as that might have been tolerable a few decades ago, but it's an eternity today. After all, downtime for a transaction-based business, such as a big online seller of books or stock trades or movie tickets, means that whatever revenues you could be earning in each minute of those eight hours are at best deferred and at worst lost forever. According to the *Economist,* a stock brokerage can lose $6 million (consider fees on lost trades, recovery costs, and losses in employee productivity), if its systems are down for 60 minutes.

So Morrow and his colleague Brian Gemberling found a way to reduce the eight hours spent manually blocking the attack to a mere five minutes. What it took was, in a sense, gaming the Internet, bending its features to their own purposes, "hacking" it in the best sense of the term.

To understand their approach, you need to know two more of the

Internet fundamentals that Morrow and Gemberling had to work with. First, of the billions of addresses the Internet has available for connected computers, several whole swaths are still unused. The Internet Address Naming Authority (not as Kafka-esque as it sounds, though its nonprofit, consensus-driven, coalition-style governance does give it a disembodied, ethereal quality) hasn't yet allocated them, so they belong to no one. Which means that if any of those unallocated addresses appear on a packet out there, someone has done some tampering—analogous to using a fake ID; they've trumped up a return address that doesn't exist. Second, routers can be instructed to sort packets according to rules devised by their human managers; that is, Morrow and Gemberling can command the thousands of MCI devices to answer, or reroute, or reject messages that have particular features, and they can do this within minutes.

Morrow has turned these traits into tools and created what's called "backscatter," which in principle isn't all that different from spaghetti thrown against the wall to see if it's cooked. When the deluged customer calls in for help, Morrow and Gemberling immediately announce to their routers, all around the globe, that this customer's computers no longer exist. This is a lie, of course, but it means the routers will treat the victim's address as a "destination unreachable" and do what they always do in such cases: generate a "return to sender" message, just like the e-mail reply you get if you, say, type ".cmo" instead of ".com" in the To: field. The destination of this reply will be the address of the original sender; the source address of this reply will be the address of the router that rejected the incoming packet. This latter point, that the return address on the bounceback packet is not the address of the victim anymore but of the MCI router that rejected it, is key.

Meanwhile, the two engineers give all their routers another task: to watch out for any messages with suspicious return addresses—such as, in many types of attacks, ones taken from among those known to be unallocated. Morrow and Gemberling instruct their routers to send these to a special "bit bucket" the team has established to catch junk. This means that packets with dubious origins—such as unallocated addresses—will be shunted off and saved, while packets with legitimate source addresses will go back to their probably-somewhat-confused senders (who will presumably try again a bit later). So immediately following the victim's complaint, the obedient electronic mail carriers begin rejecting the flood of incoming packets headed for the new "destination unreachable." And when the return message this produces appears to be destined for a suspicious address, they redirect it, as instructed, to the bit bucket.

Here's where Morrow and Gemberling can see if the spaghetti's done. Any "return to sender" messages dumped in the bucket are necessarily part of the attack, and now all have a new source address on them: the address of the first MCI router they hit, the first spot on the company's network to reject them, the first spot to note that bad address. Rather than check each router on the bad packets' route, Morrow and Gemberling have homed in on the "edge" router where the attack entered MCI. Now they can lift the "destination unreachable" rule from everywhere else, and continue to collect the cooked spaghetti on the wall. Whereas those bad packets would have poured into MCI, hopping perhaps 10 times through the thicket of MCI routers, using company resources from Singapore to Alaska, now they've been blocked at the first company router they hit, and Morrow and Gemberling can pinpoint it. This maneuver traces the attack packets to those routers that got the first handoffs from abutting networks such as Sprint's or AT&T's. The "backscatter" from those routers reveals their address.

All this cleverness, however, has not gotten Morrow to the real source of this nuisance. It may only have gotten him 10 hops in reverse—10 hops back, say, out of 20. From his perspective the job is done: a continuing filter on that edge router protects both his customer and his own network from excess traffic, all in a matter of minutes. But the identity of the actual troublemaker is no clearer than it ever was. The fact of the matter is that MCI's team gets 10 or 11 such attacks per day. Unlike perpetrators of attacks on other types of infrastructure, and unlike terrorists in general, electronic vandals significantly outnumber the forces that can fight back. Here, the asymmetry of warfare is reversed, with the good guys a small minority. The Internet's anarchic nature is thus in part a matter of limited resources—human resources: How can Chris Morrow keep 11 hackers accountable each day, when he's used so much creativity just to track them half the steps to their hideout?

•   •   •

The computer research community is a network in its own right. It's one subset of that greater web of engineers and scientists in general who know something about protecting the populace, but in itself this subnet exhibits copious links. In fact, in the surge of network science scholarship over the last several years, one physicist, M. E. J. Newman, actually conducted "A Study of Scientific Co-Authorship Networks," in which he examined the characteristics of a group of researchers defined by whether they'd written papers together. His study targeted fellow physicists, but it would have turned up the same results were it focused on computer research. Through cowriting or coteaching, meeting on sabbatical, reading journals, and applying for jobs, an ad hoc array of connections has evolved in computer science that rearranges itself over time. It brings people together to produce and disseminate new ideas and then sets them loose again to bump up against others. This happens in the subcommunity of computer scientists, and also in the sub-sub-community of networking engineers, and in turn in the smaller sub-sub-sub-community of people who've worked on traceback. Every person in this last set has heard of Chris Morrow and the magic of backscatter. They've also heard of Bill Cheswick, who came up with a scheme for tracing barrages of bad packets called "controlled flooding," and hardly one of this cluster has failed to include in his or her bibliography Steve Bellovin and his solution, called "itrace."

Some of the links between traceback researchers are in the form of mere acquaintanceship or shared ideas. Some represent attempts at improvement, as in the case of Stefan Savage, a grad student at the University of Washington in Seattle: his link to Chris Morrow was his wish to trace back further than the team at MCI had. Although Morrow had come up with a brilliant solution for his immediate problem, it was no help in establishing the accountability of the perpetrator. Once a system administrator has used backscatter to locate which router on his own network was the first one to pass the bad packets, he's stuck. He has followed the trail back to the last refrigerator-sized stack of electronics that he owns. And now he has a human problem. The adjacent Internet Service Provider (ISP) might not be inclined to help; after all, it's Morrow's customer who's getting bombarded, not his. Perhaps this next-door neighbor won't give up information without a subpoena. A subpoena takes a long time to get, and perhaps by the time the system manager has contacted the lawyer who will get a subpoena to serve, the attack has stopped. Now there's no more potential backscatter to work with. Every time a traceback effort hits another company's network, a host of new obstacles arises.

And so far, the telecommunications companies involved have shown strikingly little interest in cooperating. Howard Lipson from CERT has written a thorough overview of the best ideas for adding traceback capacity to the Internet as a whole (this part of his work is sponsored by the U.S. State Department, which has an obvious interest in methods for locating foes), and he believes encouraging better teamwork is one part of the solution. "Tracking beyond an administrative domain," Lipson writes, "depends on the trustworthiness, cooperation, and skill of upstream ISPs"—which, incidentally, may be foreign companies whose staffs don't speak English. "Having international agreements in place to facilitate the required cooperation will be an essential element of a traceback capability."

Stefan Savage, however, thought there might be an approach that didn't rely on good corporate citizenship. He thought, How can we get as close as possible to our menacing hacker without negotiating with neighboring ISPs? How can we automate traceback and build it into the basic mechanisms of the Internet? His answer is a trick called "packet marking," which requires enlisting the help of those hard-working routers—but only minimally. Just as the customs officer stamps a traveler's passport, providing proof of where the passenger has been, Savage's routers would stamp their addresses on packets—but not every packet, which would slow traffic down. Savage's scheme would put a stamp on only one packet, say, out of 20,000. While one in 20,000 may seem like a meager sampling, it's not: These sorts of flooding attacks easily pummel their quarry at a rate of 500 packets—and some 500,000 packets—*per second.*

In understanding packet marking, you run into another of the Internet's defining attributes, one that's essential to how the whole worldwide system functions. Each packet the network delivers, each individual stream of 1s and 0s, includes a preamble, called the "header," containing delivery instructions. Established by the creators of the Internet, this header includes details such as whether the packet is a routine one or high-priority; where the packet, which may be a fragment of a larger message, belongs in the sequence of parts; how much time the packet has to find its destination before it's considered "lost"; and most important, the address of the packet's sender and of its destination. These headers even contain information about themselves: their format, for instance, and their length (which indicates to the receiving computer when to expect the real meat of the message to begin). It's the information in these headers, along with the cooperation of routers, that conducts the glorious, clamorous din of the orchestra.

Savage's idea was to configure routers so that they'd borrow a tiny piece of space in the header and "stamp" it with proof that the packet had passed through. When the victim, overwhelmed by offending packets, gathered up a heap of them to analyze, he'd find that 1 in every 20,000 carried the name of a router along the attack path, and from that he could assemble a map. Here was the catch: Header real estate is crowded. The space Savage chose to camp out on could fit only about a quarter of the information he wanted it to hold: the address of the previous router, the address of the current one, and a number that each router would increase by 1 as a way of tabulating how many hops the packet was from its origins. So, to get all of it in, he performed some dazzling math. He compressed the two router addresses into one, chopped this amalgam into eight pieces each, and then inserted these smaller slices into the 1-in-20,000 sample of onrushing packets. At the destination computer, a program could collect these marked packets and, using a set of algorithms, reunite the eight fragments and decompress the two-in-one addresses. Presto! The series of routers that had facilitated the assault was revealed.

Savage's brainstorm was cunning, and along with three coauthors, he produced a paper about it that, in a fitting demonstration of the reach of the research network, would be cited by at least 51 other articles. Now the list of references at the end of every traceback article would be one citation longer, and Savage, nearing the completion of his Ph.D. and applying for jobs, could put another publication in his portfolio.

One of the places he went looking for work was the University of California at Berkeley, where, as part of the regimen candidates undergo, he gave a seminar to faculty and students. In it, he described his packet-marking invention, at which point, unbeknownst to him, a new link in the web of traceback enthusiasts was forged.

Dawn Song and Adrian Perrig are reed-thin and look about 19 years old. Song, who is actually in her late twenties and originally from China, has a steady gaze and a deep reservoir of self-respect. Perrig, in his early thirties and from Switzerland, has boyish-smooth skin and a frequently irrepressible zeal for his work. Both earned Ph.D.s in computer science at the University of California in Berkeley and both have become professors at Carnegie Mellon.

It was at Berkeley that the two began wrestling with the problem of traceback, when Song attended Stefan Savage's candidate-for-hire seminar and heard about his scheme for marking packets. "While I was listening," she recalls, "I realized it wouldn't work for large attacks." She conferred with Perrig, and within two weeks the duo had written a paper that repaired the flaw in Savage's thinking.

What Song objected to in Savage's scheme was that he didn't account for a type of assault that pummeled victims' computers from multiple places simultaneously. Rather than generating traffic from a single machine, hackers were fielding little programs (these are often referred to as "zombies") that would camp out on the far-flung computers of unwitting, ordinary users and when commanded, attack the enemy from all sides at once. These strikes had begun knocking out sites in 1999, and incident responders at CERT had quickly gathered a group of 30 experts from around the country to discuss how to cope. They realized the technique would radically amplify the damage of previous attacks. As one CERT staffer put it, "We knew this was something different. We'd crossed a line." But by February 2000, the new game was just getting started: Over the course of three days, Yahoo, eBay, CNN, Amazon, and Etrade, among others, were all knocked off line by these new, remotely controlled, multiply powerful assaults. Today, it's not uncommon for a thrill-seeking hacker to send out thousands of zombies, each producing ruinous outpourings, as ISPs use im-

proved firewalls and a variety of evolving tactics to defend themselves, including Morrow's backscatter technique.

With so many widely spread sources of bad packets, Song thought,
how could you possibly do all the sorting that Savage's traceback technique demanded? Whereas 10 routers—10 hops—might have been
used for a simpler, old-fashioned attack coming from one computer,
suddenly 10,000 routers might be involved in an attack coming from
1,000 machines. Slicing up the addresses for those routers, writing
them intermittently into packets, and reconstituting them on the
other end would, at that scale, take days of computer time.

So Song and Perrig took on traceback, too. They liked Savage's use
of the 1-in-20,000 (or 1-in-40,000, or even 1 in 1 million) sampling to
reduce the burden on the system, and they liked his choice of fields in
the header in which to stamp the IDs of routers along the attack path,
but they dispensed with the idea of using eight slices of the routers'
addresses which later had to be matched back up again. They relied
more heavily on another approach for shrinking the routers' ID to fit
the space in the header. This was a mathematical technique that re-
duces numbers of any length into numbers of a consistent, shorter
length, creating an abridged version or a digest. The procedure is
called a "hash function," and coders reach for it frequently, the way
cooks might reach for a baster, as a familiar tool for getting certain
tasks done. One task hash functions can accomplish is compression,
condensing lots of numeric data into a little. You could, of course,
shorten numbers by simply lopping off the last several digits, but if
you drop the last three digits, say, of the numbers 1,898,300 and
1,898,276, you'll have lost the initial figures' individuality. What hash
functions do is formulate digests for a series of numbers, each digest
corresponding to its original and as few others in the series as possible.
A weak hash function will spit out repeats: the same compressed
figure representing lots of originals, so that it's impossible to tell them
apart. But a good hash function produces such a range of digests, they
are almost unique to their originals.

Not only are these procedures common, they're also illustrative of
how programmers tend to think: recursively. (Think of the "begats" of
the Bible, which require a similar, iterative kind of discernment, or
the mounting layers of harmony produced by singing a round.) In

order to assign an unwieldy number its own compressed identifier and one it will share with the fewest others possible, hash functions perform operations reflexively and repeatedly on the original. Here's a dramatically simplified example: rather than lop off the last three digits of your numbers, you could add all of them up—1+8+9+8+3+0, etc.—and then divide the result by the last two digits of your new figure; then you could add up these resulting digits all over again. In other words, you'd reuse the elements of your initial figure to preserve its specificity. Do like a programmer: repeat a self-reflexive operation many times over.

Song and Perrig replaced each router's address with a hash, thereby squeezing the lengthy identifier into the skimpy header field. They applied several more mathematical routines, cramming the digest with enough information that the 1-in-20,000 marked packets, put together, revealed a map of the routers leading back from the victim to the source of the attack. In 2001, they presented their paper at the respected IEEE Infocom conference, where for 20 years a network of computer networking researchers have met to circulate their ideas.

Yet, the traceback problem still lingers, and law enforcers on the trail of a hacker have no automatic switch they can flip to find their suspect. The SoBig worm clogs computers for days and nobody knows who unleashed it. An attacker inundates a website's servers with garbage for an hour, preventing other users from getting in, and the event causes sweat, fear—and no arrests.

Still, the traceback researchers are, if not gaining ground, at least learning by doing. They've released their propositions onto the network of ideas, which weaves together the programmers and professors, consultants and system managers who think about security. Their concepts are moving among the research community's nodes. Chris Morrow downloads Song and Perrig's paper, and says, "No way—no way anyone will do that." For one thing, he believes intervening in even 1 in 20,000 transmissions is far too much work for his routers; for another, he argues, the header field they're borrowing may be rarely used, but it is used. "You can't write off all the packets that rely on that field just because they're a small percentage!" Meanwhile at CERT, Howard Lipson mulls each solution and what it would mean for privacy. At BBN Technologies, a Boston company venerated for building the original Arpanet and inventing the router, Craig Partridge dispatches another scheme into the whirl, an ingenious variation on how routers keep logs, which uses hash functions and other maneuvers to address the problem of storage. Steve Bellovin's itrace is discussed among key figures who help guide improvements to the Internet.

The question is at what point, if any, will one or several of these schemes be implemented. Morrow's backscatter technique is used daily, but the others are mere suggestions. Without any one of them in place, the job of law enforcement is monumental; and even with one of these accountability schemes adopted, nabbing computer crime suspects would hardly be guaranteed: tracing attacks to their electronic source might get investigators no farther than an Internet café in New Haven or a breached computer in a Cypriot university. As author and security luminary Bruce Schneier puts it, "The hardest part of computer forensics is knowing who is sitting in front of a particular computer at any time." Even Stefan Savage has grown discouraged about the prospects for traceback since the publication of his packet-marking paper: "I'll be honest—I'm a bit of a cynic on this stuff. Traceback requires you to change a lot of infrastructure and doesn't give enough of a win. Plus, none of these schemes is perfect." Capabilities added to a system that's already been built and is already in use globally, he says, are never just what you wanted. "They're fit in sideways as hacks on the side of the protocol."

In part, the impediment to widespread adoption of any change in the Internet's architecture—like many difficulties that arise when loosely governed cooperative communities attempt change—is a lack of shared incentives. Routers won't mark packets unless their manufacturers design them to, and manufacturers won't change their products unless customers demand it. But what benefits would providers such as Earthlink or Comcast get out of routers that allow for traceback? None. Like Chris Morrow at MCI, these companies are not focused on capturing criminals and terrorists; their highest priority is keeping their networks running.

Craig Partridge, at BBN, has a different perspective. Since its founding in the 1940s, BBN has worked for some influential organizations—such as the U.S. and British navies, the U.S. Department of Defense, the Marine Corps Central Command, the Air Force Research Laboratory, and the Joint Chiefs of Staff. "My best guess," says Partridge, who calls himself BBN's General Networking Mr. Fixit, "is that the military will eventually sign up for this." He's straightforward about why. "They need to know." And if the military wants manufacturers to add traceback to their routers (especially Partridge's form of traceback, which doesn't sacrifice user's privacy), manufacturers will have their incentive. "At that point," Partridge speculates, "the vendors might just ship it for free."

Trace back through the history of the Internet all the way to its beginnings and you'll find an intimate, collegial, comfortable world. Today, we travel through a vast and indifferent online space, where hundreds of millions of strangers log on with countless different goals. But trace back even a single decade to the early 1990s, and the online population was more sheltered, limited to computer professionals and hobbyists, graduate students, faculty, and a slew of verbose journalists logging on to the predecessors of AOL. Back a decade further, in the 1980s, when Robert Tappan Morris unleashed his worm, the Internet was even more cloistered, still chiefly an appendage of the academy. Go back all the way to its birth, and the Arpanet connected a mere four nodes: UCLA, UC Santa Barbara, the Stanford Research Institute, and the University of Utah.

Howard Lipson likes to say, "The Internet is not rated for its current use," by which he means, we're not just dialing up to chat rooms anymore, swapping tips on how to build home computers. The great flowering of telecommunications has meant steadily shedding the familiarity and trust of the early days. Today's Internet is so porous and inclusive that any sense of asylum there has vanished—unless you're a cautious malefactor.

Or, put another way, at the starting end of history's traceback path, the present, we're in a position unnervingly analogous to the victim's.

But we haven't succumbed. And luckily the present is a very active and industrious place, where lots of ingenious thinkers are firing ideas back and forth, searching for breakthroughs and solutions, connecting. Which is good, since the Internet is infinitely complicated, and to tame an environment so boundless—or to quiet it enough, at least, that the work of terror there can't go unexposed—you need an even better network, not bigger, not faster, not busier, but human.

FOUR

Mortal Buildings

W HEN EVE HINMAN talks about the Oklahoma City bombing, she does so slowly and deliberately, as if calling it up frame by frame in her mind. She visualizes the scene and then searches for a comparison, landing on the most natural one, that of a war zone. "It was ghoulish, hellish," she finally says. When she arrived, two weeks after the bombing, workers had already lost any hope of finding survivors. The operation had shifted from a rescue to a recovery. But tensions remained high among emergency personnel and engineers working with frayed nerves around the clock to recover the victims' bodies.

"And of course, after 16 days, there was this smell in the air," she says. "They were spraying stuff on it to try to keep it down."

By the time Hinman showed up, the basic facts of what had happened at the Alfred P. Murrah Federal Building were already established. On the morning of April 19, 1996, a truck packed with a large explosive device had been parked next to the north side of the building and detonated. One hundred and sixty-eight people were dead—19 of them children—and 782 wounded, in what was at the time the worst terrorist attack ever executed on American soil. That, essentially, was all the country would need to know about the Murrah

Building itself. Americans moved on to aiding the living, mourning the dead, and finding and punishing the perpetrators.

Hinman, though, was interested in something else. As an engineer working for Exponent (then called Failure Analysis, Inc.), a company that specializes in understanding why systems fail, she wanted to know how the Murrah Building itself had responded to the attack. Why had so many people died? It was apparent from the beginning that most of the casualties had come not from the blast itself but from the collapse of almost one-half of the building. Yet the Murrah Building, constructed of steel-reinforced concrete, was not in any obvious sense a flimsy structure. Hinman's goal, as she later described it in the book she coauthored with David Hammond, the chief structural engineer at the disaster site for the first 12 days of the recovery operation, was to understand "the mechanics of structural failure because of explosions and what can be done to reduce their consequences."

Hinman had come to Exponent after spending 11 years at Weidlinger Associates, a New York–based structural engineering firm that specializes in the blast-resistant design of buildings. She had worked on designs for U.S. embassies around the world, "hardening" them against possible terrorist attacks. But she had never before seen the effects of a bombing firsthand. She arrived in Oklahoma City with free rein to investigate the cause of the building collapse, a brief given to her by her boss, John Osteraas, the chief engineer who relieved Hammond on the site. Like a physician conducting an autopsy, Hinman probed the building for the source of its demise. She scrambled over and around the rubble, a hard hat covering her shoulder-length red hair, photographing and measuring the remains, seeking clues to explain the catastrophic scale of the destruction.

In their book, *Lessons from the Oklahoma City Bombing: Defensive Design Techniques,* Hinman and Hammond methodically and dispassionately analyze the sequence of the Murrah Building disaster. The bomb, consisting of approximately 4,800 pounds of ammonium nitrate fertilizer and fuel oil, was detonated about 10 feet from the building. The explosion generated a roughly spherical shock wave that shattered windows as much as two miles away and created a crater 30 feet wide and 8 feet deep. The blast shattered the supporting concrete column closest to the point of detonation, designated as column G20. The pressure wave coursed through the building, punching floor slabs upward and sending debris and shards of glass flying, before it dissipated on the far side. All of this happened within the span of several hundred milliseconds, about the duration of an eye blink. Then gravity took over. Slabs that had been forced upward by the blast now came crashing down, and, with column G20 missing, other columns began collapsing under the additional weight. Nearly half of the building imploded.

From Hinman's analysis—and another conducted by the Federal Emergency Management Agency—it was clear that the design of the Murrah Building had contributed significantly to the large number of dead and wounded. In particular, the inability of other columns to carry the load displaced by G20 precipitated an almost inevitable failure. Flying debris, typically the second greatest cause of casualties after collapse, was the other culprit.

But a harder question was whether or not the building's structural engineer had been at fault. After all, no one had anticipated that the removal of a column could actually occur; withstanding a truck bomb hadn't been a requirement in the original plans. The only thing that was certain was that the Murrah Building design had failed to do so.

Today Hinman runs her own engineering firm, Hinman Consulting Engineers, specializing in the design of blast-resistant buildings in the United States and overseas. She's designed everything from the United States Courthouse in Las Vegas to the embassy in Uganda, and been hired to conduct post-9/11 vulnerability assessments of the New York Metropolitan Transit Authority, the New York State Power Authority, and the Newark Airport. Sitting at an outdoor table a few blocks from her San Francisco office, she recalls traveling to Oklahoma City for months after the disaster to meet with survivors and the relatives of the dead, who wanted to push the government for new building-design regulations. She would spend hours with them, listening to their stories of horror and loss. She found herself trying to help them work through their sense of helplessness and confusion by explaining what had happened to the building. "I think it was some part of the healing process," she says. "They did want to know." But in the end, the technical facts were of little solace. The trauma was too great, and eventually she stopped going. After a certain point, she says, "There was nothing we could do."

But the Oklahoma City experience had a lasting effect on Hinman. "It was important in my career, because it drove home that it's about protecting the *people*," she says. Doing so, she realized, meant recognizing that for buildings at risk of attack, protective design should not be optional. "I like to think of the protective measures that we recommend as being a natural part of a building," she says. "A building needs to protect itself. Our bodies are designed to protect themselves—we have eyelashes to keep dust from getting in our eyes, and our ears are shaped to prevent things from getting in our ears. To have a building that's totally open and vulnerable is inconsistent with nature."

We tend to think of our buildings as empty, static vessels, shells brought to life only by the people who inhabit them. But in fact buildings—and any other man-made structures, whether bridges, dams, skyscrapers, or highways—are themselves animate systems. They are the product of complex interconnections of materials, designed to constantly shift and transfer the forces of gravity, wind, and water out of themselves and harmlessly into the ground or air. They are built to bend without breaking.

As the engineers Matthys Levy and Mario Salvadori wrote in their 1992 history of building collapse, *Why Buildings Fall Down*, "A building is conceived when designed, born when built, alive while standing, dead from old age or unexpected accident." Each structure contains its own maze of invisible, internal networks—ventilation ducts, electricity grids, water pipes, stairways, and elevators—the bodily systems that activate it. "Like most human bodies," they wrote, "most buildings have full lives, and then they die."

What does it take to send a building or a bridge to an early death? Structural engineers, whose job it is to translate an architect's plan into a standing edifice, have long considered that question. They have devised structures to resist the forces of earthquakes, wind, fatigue, and decay. They have developed their own language to discuss the challenges of keeping a building aloft: The forces applied to a building are "loads." Abnormal loads are "insults." When an insult destroys (or renders useless) a part of a building, it is said to "fail" that component.

Computer simulation programs that test designs before they are built are "codes."

Knowledge of the effects of one specific type of force—explosions—has long been confined to a small cabal of engineers, most of whom cut their teeth examining the effects of nuclear weapons during the cold war. For 40 years they studied the impacts of atomic weapons on buildings, first with live tests and then with computer simulations designed to model the impacts. Because of their sheer size and power, however, nuclear weapons offered a different problem than that posed by modern terrorist bombs. The blast pressure—on a scale of milliseconds—lasts much longer, and the effect on the building is uniformly devastating. A localized bomb's complex effects, such as the single shattered column and displaced floor slabs of the Murrah Building, required entirely new research. As the nuclear threat subsided, structural engineers turned their attention to understanding the fine-grained effects of smaller blasts. Emerging from the military world of unlimited resources and total control of design, they encountered a daunting cost-benefit balance in protecting civilian buildings. Scattered among a handful of specialized structural engineering firms, academic labs, and government agencies, they have spent the last two decades developing design techniques to counter the impacts of terrorist attacks on the structures around us.

Now the work of this quirky elite is suddenly in the spotlight, as terrorists at home and abroad have taken their own violent interest in structural collapse. Despite our growing fears about weapons of mass destruction, explosives remain far and away the weapon of choice for terrorist attacks. Bombs are simple to design, easy to construct from basic materials, safer to handle than gas or germs, and easily transported and concealed in ordinary vehicles. The catalog of carnage from explosive attacks on buildings is extensive, stretching from the United States Embassy and Marine barracks in Beirut in 1983, to the first World Trade Center bombing, to Oklahoma City, to Khobar Towers in Saudi Arabia, to the U.S. embassies in Kenya and Tanzania, to the nightclub bombing in Bali and dozens more. Insurgents in Iraq have used car bombs to terrorize nonmilitary targets such as the United Nations compound and the Red Cross headquarters in Baghdad. The attacks forced the U.S.-led occupying authority to erect thousands of 12-foot-tall concrete barriers around soft targets such as police stations and government administration buildings. Within the United States, bomb attacks on buildings remain a risk. In August of 2004, the *New York Times* reported that the Department of Homeland Security distributed a memo to state security agencies titled "Potential Threat to Homeland Using Heavy Vehicles." A terrorist alert the same month, issued to financial institutions in New York City, focused on the threat of truck bomb attacks.

A common feature of such attacks is that their goal has been not only to kill—something just as easily accomplished in a crowded bus or marketplace—but, by assaulting the structures, to make them accomplices in the killing. Terrorists have realized what blast engineers have long recognized: in an explosion, more people will die from a building's damage or collapse than from the blast itself. And, as the September 11 attack on the World Trade Center demonstrated, collapse has a symbolic value all its own.

Like other ordered systems—such as power, water, and transportation networks—structures are vulnerable to cascading events, disruptions that start small and spiral into out-of-proportion consequences. In structural engineering, the phenomenon is called "progressive collapse": a relatively minor insult leads to a chain reaction and a total or near-total crumbling of the structure. If critical components are suddenly altered or removed from a building—whether through an explosion, an earthquake, or a flaw in the original design—it can become a house of cards. As the Murrah Building showed, progressive collapse is a terrorist's best friend. "Generally speaking, what the terrorist is trying to do is induce massive failure," says Ted Krauthammer, director of the Protective Technology Center at Penn State University. "This is what has to be prevented. As a society, we should be willing to accept local failure, but not massive failure."

There is, of course, one easy way to fortify buildings against explosive attack, and that is to turn them into bunkers: no windows, no public or vehicle access, no vaulted lobbies, no glass of any kind. This was the obvious answer of structural engineers to the nuclear weapons threat during the cold war. But short of filling our landscape with steel-and-concrete monoliths, the solution to the problem of bomb-proof design remains elusive. As a former undersecretary of state for political affairs, Thomas Pickering, said about the vulnerability of the Kenyan and Tanzanian embassies after both were bombed in 1998, "Bomb-proof is a relative term. Tell me what size bomb, and I'll tell you what the proof is."

The question that bedevils blast engineers is how to address the risks of progressive collapse and dangerous debris without knowing where the explosion will come from or how large it will be. A wide-open research vista confronts these engineers—they're trying to understand and apply technologies even as the problem they're working on remains hazily defined. "We're making stuff up all the time," says Hinman, "because there is nothing to go on. From an engineering standpoint, that's what makes it exciting. From a reality standpoint, it's nerve-racking."

Ted Krauthammer of Penn State has spent four decades studying explosions and their effects on structures, publishing more than 300 papers on such topics as "Structural Steel Connections for Blast Loads" and "Assessing Negative Phase Blast Effects on Glass Panels." On a warm afternoon in early September, he is teaching the first session of what is, in the view of most blast engineers, the only academic course in the world on blast-resistant design. Krauthammer, who is a small, slender man of 58, has an archetypal engineer look—parted hair and a short-sleeve button-up shirt tucked into khakis, with two pens and a spiral memo pad jammed into a pocket. His voice is tinged with his native Israeli accent, and he has a deeply serious, almost sad, demeanor. "People are talking about terrorism quite emotionally, which is understandable," he tells the 10 graduate engineering students seated before him at a conference table. "It is very frightening. It is very difficult to protect against. But I think we need to look at the problem much more objectively, and much more professionally."

After reviewing the major terrorist bombings of the last decade, he clicks his slide presentation forward to a photo of the Khobar Towers military housing complex near Dhahran, Saudi Arabia. Terrorists detonated a truck bomb outside the building in June of 1996, killing 19 United States servicemen and injuring more than 300 people. While at the time considered the most devastating terrorist attack against the United States on foreign soil since Beirut, Khobar is cited by blast engineers as an example of successful building design. Subjected to an estimated 3,000 to 5,000 pounds of plastic explosives (the equivalent of 20,000 pounds of TNT) loaded in the back of a tanker truck, the complex did not collapse. The fact that the building remained standing after the explosion, Krauthammer tells the students, saved countless lives. "The number one thing" that they must remember on every project, he says, is to design against progressive collapse. "I do not want to have a pile of rubble at the end. What you are saying between the lines is, I am willing to accept localized damage. If there is an explosion, the offices close to it will be destroyed, and the unlucky people in them will be injured or killed. But you will save hundreds or thousands of other people."

When an explosion such as the one at Khobar Towers occurs, it creates a short, high-energy blast of pressure. The pressure can damage or destroy parts of the structure—a bridge column, say, or the concrete block wall of a building—and that destruction forces the network of connections to respond to a new distribution of gravity loads. "Now the building is trying to find a new state of equilibrium," says Krauthammer. "The loads are moving, trying to find a way back to the foundation. As they are moving, they can overwhelm elements that were not designed for these changes." The loss of one key supporting member can create a cascade effect. "If you start losing columns, the load transmitted through the column has to go somewhere. So it will go to another column. If that other column doesn't have enough reserve capacity, that other column is going to fail and now you start getting a progressive-collapse scenario."

The property that distinguishes buildings that fail, such as the Murrah Building, from those that remain standing, such as the Khobar Towers, is called "ductility." In structural engineering parlance, ductility is the ability of materials to absorb energy without having their integrity compromised. "Many of these design philosophies come right out of the seismic community," says John Crawford, president of the Los Angeles–based blast engineering firm Karagozian & Case. "You are not designing for strength. You are designing for strength plus ductility. You want motion."

Indeed, many of the philosophical principles of earthquake-resistant building design can be applied to blast resistance. Ductile buildings, because they can absorb energy, respond better to forces of any kind than so-called brittle structures and materials. But when it comes to precise specifications, blast and seismic design don't always coincide. Think of an earthquake as a long, slow push against the bottom of a building. The localized force of an explosion, which arrives and departs an order of magnitude faster, is more like a quick punch in the stomach. "If I punch you," Crawford says, "you rock backward and absorb it to stay on your feet. If you stand there like a rock, it's going to hurt a lot more. The best way to design a building is design it to take a punch."

Whereas earthquake motion primarily has an impact on the connections between vertical members and horizontal structures (since entire floors are moving in relation to one another), the more directed blast from an explosion affects columns, walls, and floors at different angles and pressures. Both seismic and blast design require ductility of the connections between columns and floors, the places where components of the building are tied together. But in an earthquake, the building needs only to move with the "push." Blast hardening requires that the columns, walls, windows, and floors themselves—because they have to withstand the direct pressure of a blast—can bend without breaking.

It was in part Khobar Tower's ductile connections, in the form of steel cables literally tied together within the concrete frame, that kept it intact. In class, Krauthammer turns to the Saudi bombing to illustrate another lesson: it matters as much where the attack is located as it does what steps have been taken to harden the building. The attackers tried to back the tanker through a hedge toward the complex, he explains, but then fled, leaving the truck in a perpendicular stance to the building. Since the blast wave from the tanker bomb was cylindrical, much of the force dissipated harmlessly to the side, instead of striking the building. The terrorists didn't understand the physics of the bomb they had built. "Had they taken this course," he says, matter-of-factly, "they would have known to park parallel to the building."

After class, sitting in his office, Krauthammer talks about his fascination with explosive effects, which began as a boy in Israel. "Seeing weapons systems and incidents was almost a daily occurrence," he says. "It was part of growing up." After serving in the Israeli army, including a stint in the combat engineers, he enrolled in Ben Gurion University, where he earned a master's in mechanical engineering. He studied under an American named Eugene Sevin, the visiting head of the department and a luminary in the world of blast hardening. "He exposed me to the discipline side of this field," says Krauthammer. "Before that, it was just how to blow things up and how they function. There was no theoretical foundation to this problem. He introduced me to the computational side."

Krauthammer came to the United States to pursue a Ph.D. in civil engineering, and decided to stay. After working for a consulting firm and studying nuclear blasts for the military in the 1980s, he returned to academia—to study the structural implications of conventional bombings, model them mathematically, and teach the next generation of designers how to defend against them.

At the end of his class, Krauthammer looks up and announces that next week will be the first two-hour session of the course. "We just did our first double session," one student says, pointing out that the class has gone an hour beyond schedule. "You should have told me," says Krauthammer, looking disbelievingly at his watch. "When I get into this stuff I can go on for days." Afterward, another student approaches and asks the question that was likely on everyone's mind: How does the professor think society should defend buildings against airplane attacks? "You stop the planes from being hijacked," Krauthammer replies without hesitation. Explosion-resistant design, he says, is a last-ditch defense. Without other layers in front of it, it's futile. "If you let the bad guys do whatever they want, then you will lose every time."

Both blast-resistant design specialists and the tools they use to understand explosions are cold war legacies, born out of the hefty budgets and grim requirements of mutually assured destruction. In the 1950s and 1960s, with the emergence of two nuclear-armed superpowers and the adoption of a strategy guaranteeing each side's annihilation, the United States military set to work on structural designs that could withstand nuclear blasts. The job fell to structural engineers within the military and among commercial contractors, who found a brisk business in creating hardened missile silos, command bunkers, and airplane hangars. But the defense industry has a way of finding new niches as the security landscape changes. For blast engineers, that meant evolving with the end of the cold war—much like the rest of society—from the strangely simpler world of nuclear explosions to the myriad threats of terrorism. "This is not an unusual story," says Jeremy Isenberg, the president and CEO of Weidlinger Associates, the dominant world player in the field of blast-resistant design. "The military-industrial complex has a lot of stories like this." Developing the science of civilian building protection is a story that blast engineers are still writing.

Isenberg is sitting at his desk on the twelfth floor of Weidlinger's headquarters, a glass-encased building on the west side of Manhattan. His office faces away from the World Trade Center site; behind him are uninterrupted vistas of the Hudson and midtown New York City, a landscape dotted with apartment and office towers. With 300 employees in nine offices around the world, Weidlinger has provided the engineering brains behind such high-profile structures as the new $200 million international terminal at Logan Airport and the eastern span of the San Francisco/Oakland Bay Bridge, now under construction. Between 35 and 40 percent of its $40 million business derives from protective engineering. Isenberg and his engineers have designed dozens of U.S. embassies, federal office buildings, and courthouses to stand up in the face of explosive attacks. They have constructed buildings to resist accidental explosions for companies such as Chevron and Union Oil, and are the lead experts advising the lessee of the World Trade Center in litigation over 9/11. Using their computer simulation software, they have precisely modeled every detail of the towers' collapse. The company is also consulting on the World Trade Center's replacement, the Freedom Tower, helping to design it against a variety of terrorist events.

The story of the firm's founder, Paul Weidlinger, exemplifies the origins of blast-resistant engineering in the military age. A pioneer in the field, Weidlinger educated a generation of engineers such as Isenberg and Eve Hinman, who graduated from the engineering program at Columbia in 1982 and got her start at the firm. He also elaborated his ideas about explosive effects as a visiting professor at both Harvard and MIT.

Born in Budapest in 1914, Weidlinger studied engineering in Hungary and Switzerland before emigrating to Bolivia in 1938 and then the United States in 1943. Six years later he opened Weidlinger Associates in New York City, consulting for architects and builders in the post-war boom of city construction. In 1953, Weidlinger met a young and brilliant Columbia University engineering graduate by the name of Melvin Baron, who had just coauthored the textbook *Numerical Methods in Engineering* with legendary Columbia professor Mario Salvadori. Weidlinger recruited them both to join his firm, and Baron and Weidlinger discovered that they had a common interest in blast effects, an area the military was just beginning to investigate in earnest. By the late 1950s, the firm had carved out a business in the military-industrial complex of the cold war. "In another life, they might have been called defense intellectuals," says Isenberg. "They had connections with government officials, and they were very smart people. They found niches—only niches, not General Electric, not Lockheed, not Boeing—but they found niches where very smart people could prosper."

Using first slide rule calculations and then the growing power of computers, Weidlinger and Baron worked out the effects of shock and vibration on submarines to help the Navy design quieter and more resilient ones. On land, they built structures at the government's nuclear test site in Nevada, using them to gather data on the effects of nuclear weapons—usually simulated using large-scale chemical explosions rather than atomic bombs—on buildings and silos. Baron was an early pioneer in applying computational techniques—in the form of complex mathematical models—to engineering questions, and began to develop computational methods to simulate the effects of explosive events and then predict the impact. Starting with the test data and some mathematical physics, he created simulation programs that took the size of the bomb and the distance from ground zero, and returned information on how the command bunker or missile silo was affected. As atmospheric and eventually underground nuclear testing were halted, the computer simulations allowed engineers to create military structure designs without having to actually set off a bomb to test every concept.

In designing facilities against nuclear attack for the military, the primary goal was what blast engineers call "continuity of operation." The question was whether the structure would be able to perform its function, namely, ensuring the enemy's destruction. "The idea was that it could be in the crater of a nuclear weapon hit and still survive to launch," says Hinman, who worked on superhardened missile silos while she was at Weidlinger.

In the calculus of nuclear deterrence, learning how to protect civilian buildings from bomb attacks simply wasn't part of the equation. "In the bad old days of the cold war, nobody ever designed tall buildings to resist nuclear explosions," says Isenberg, who has an undergraduate degree from Stanford and a Ph.D. from Cambridge and joined the firm in 1973. "The only buildings that were designed to resist nuclear attack were military structures." When it came to designing those structures, no amount of funds or materials was too great to accomplish the objective of keeping it functional. "We were interested in structural survivability," he says. "And if you are interested in structural survivability, all you have to do is keep adding metal until the structure survives."

That philosophy changed on April 18, 1983, when a van carrying 400 pounds of explosives detonated outside the United States embassy in Beirut, killing 63 people. The following October, a suicide bomber crashed a truck through a security barrier outside the U.S. Marine Corps barracks in Beirut, detonating an estimated 12,000- to 20,000-pound TNT bomb that turned the building to rubble, killing another 241 people. Truck bombs as a weapon of international terrorism had arrived, and like other attacks that followed, the casualties in Beirut came not from the bombs themselves but from the destruction and collapse of the buildings, generating a flurry of discussion about protecting U.S. facilities overseas.

It was at this point that Paul Weidlinger positioned his firm to take a central role, using their military expertise and government contacts to win the job of writing the State Department guidelines for the design of embassies. The cold war was winding down, and military-related engineering firms were going out of business in droves. Before long, Hinman, Isenberg, and other Weidlinger engineers had simply shifted from designing bunkers and silos to designing embassies.

As the blast experts at Weidlinger began to look at civilian structures, however, they quickly realized that the knowledge they had amassed in their military work often wasn't relevant. They were now confronting entirely different types of buildings, ones with budgetary and aesthetic restrictions that were never meant to accommodate blast hardening. Defending against truck bomb attacks required understanding not only how these buildings survive but how they fail, so that failure could be prevented. Suddenly the engineers had to address questions about how windows break, how columns shatter, how floor slabs hold and shift weight. The computational models built for nuclear effects weren't set up for the localized effects of a bomb, and there was little experimental data to feed into them.

"Nuclear weapons—it's a big load, big bomb," says Isenberg. "And the blast loading lasts a long time. Terrorist explosions—these are small bombs on the scale of megatons, and the load lasts a short time. And this caused a lot of problems and required an adaptation of the technology."

Mainly, it required examining the rapid, localized effects of smaller explosions, instead of the more uniform effect of a nuclear blast. With a truck bomb, the blast will strongly affect a limited area of the building—potentially crippling it. Blast specialists such as those at Weidlinger needed computer models that could simulate how a high intensity blast that lasted only a few milliseconds could lead to complete collapse—even hours later.

They began generating the data to create those models through small-scale experiments—a process still going on today. At test fields in New Mexico, the engineers build floor-to-ceiling windows (called "curtain walls"), masonry, and occasionally full-sized building mock-ups outfitted with accelerometers, high-speed video cameras, and strain gauges. Then they select and position the size of the bomb to test them. "The first time you put your design in front of a 1,000-pound bomb, and then you go see what happens to it and compare it to what you predicted—that makes an impression you will never forget," says Joseph Smith, a blast expert and vice president at Applied Research Associates, a Mississippi-based engineering firm specializing in protective design. "When you are standing a half-mile or a mile away and you hear it and you feel it and this thing pushes against you, and then you are looking at test specimens that are only 20 to 50 feet away, you think, God, how could anything survive? It gives you a much better feel for the forces you are dealing with."

As the engineers have experimented, the simulation codes have evolved to reflect the basics of what was happening in a building. Using the programs, blast engineers in the government and at firms such as Weidlinger can, for example, digitally simulate the removal of one column and observe how the building responds. They can ensure that there are alternate paths and structural redundancies to assume whatever weight the column had been carrying. But even with more sophisticated computer models today, getting accurate results requires engineers who have years of first-hand experience using the simulations. "The building broke," says Smith. "What we are asking is, How broken is it, and is that okay? That is a very hard question to ask. And to be honest, there is some art mixed in with the science. It does take a feel for how a facility or a thing will respond dynamically. After doing hundreds of them you end up developing the feel." An engineer could input a certain size bomb exploding at a certain distance, for example, and the model would return numbers on how far the closest walls and columns had been deformed, whether they had exceeded their "elastic limit." But it wouldn't be able to tell if people in the rooms behind them had survived. "Translating those numbers into real damage states and real impacts on people," he says, "is the art."

Beyond the basics of structural damage the problem gets considerably more difficult, and even the most experienced structural engineers still do not comprehend all the ways that explosive shock waves affect buildings. "The problem that we don't understand is the physics of how the loads are flowing in a complicated structure," says Krauthammer. The effects of an explosion on a building are what engineers refer to as "nonlinear"—meaning that changes in the size of the explosion may not follow a one to one correlation with changes in the amount of destruction. The gravity loads in the building can shift in chaotic and highly unpredictable ways. "Every time that you change an assumption," says Krauthammer, "you may end up with totally different consequences—simply because it is a nonlinear effect that you cannot predict ahead of time." Many models can make a column disappear from the structure, for instance, and then simulate the effect on the building. But they cannot take into account the impact that a falling column might have when it slams into the column next to it. Krauthammer and other researchers at the Protective Technology Center—and engineers at firms like Weidlinger—are working on computer simulations that will try to replicate those impacts.

Even with advanced codes such as those at Weidlinger, many of the assumptions aren't fully tested by explosions in the field. "It's a question in my mind whether we have moved very far toward certifying computational models for progressive collapse," says Isenberg. "It's not an easy thing to do experimentally. There have been some experiments conducted. They are unsatisfactory. They are smaller-scale experiments within reasonable budgets trying to give computational modelers a target, something to chew on. And it's in its infancy."

John Crawford has blown up his share of buildings. On his business card is a photograph of a fiery explosion, below which are listed his services: "blast consultants, structural engineering, explosive safety, blast resistant design." With his heavy frame, gray Jerry Garcia mane, thick beard, and round metal glasses, Crawford's look and off-the-cuff demeanor hide decades of experience with blast design, first in the Navy and now at Karagozian & Case. His office is located across the street from the Burbank Airport, and Crawford is discursive on the topic of how his building would react if the airport were blown up. "Almost any blast at that airport," he observes, "is going to shatter these windows." Like most blast engineers, his office walls are lined with books with titles such as *Failures of Materials in Mechanical Design* and *Design of Structures to Resist Nuclear Weapons Effects,* and his desk is stacked with brown file folders and project reports. Since September 11, 2001, his phone has rung nonstop with potential consulting gigs. He picks up a bound project proposal—an outline of possible options for retrofitting the Los Angeles International Airport—and flips through it, then picks up the one below it. "This is another airport. I've got similar reports for buildings. I just did three bridges."

Blast engineers such as Crawford have a myriad of tools at their disposal when it comes to building or retrofitting structures to resist collapse. But it's a delicate and complex process to figure out which ones are useful—and feasible. After 30 years in the business, Crawford is cynical about the building community's understanding of blast effects, and his words are weighted with the weariness of a man who feels his good advice has often been ignored. "There are lots of design concepts that will protect people," he says. "You can design for very large bombs that are very close to you. One of the things that bothered me a lot, around the Murrah Building time, was that a lot of people thought, 'Well, there is nothing you can do. It's an act of God, practically.' But you know the military, they've been designing structures for much higher loads for a long, long time." The real question isn't if a building can be protected, says Crawford, it's at what cost. "It really isn't so much whether you can design the structure that will resist the bomb. It's a question of whether you want to live in that, because it looks so ugly or massive that you are not going to build it."

In the balance between aesthetics, cost, and protection, Crawford and others have developed some clever techniques—starting with trying to keep the blast as far away from the building as possible. Blast engineers call it increasing the "standoff," a technique that involves setting the building back from the street and erecting barriers— walls, water features, or bollards, often disguised as benches or other landscaping to increase their aesthetic appeal—to prevent vehicles from getting close to a building. The absolute minimum standoff for protection, Crawford says, "is somewhere between 5 and 10 feet. Then you can start to design structures that resist those loads. When you have nothing, it makes it difficult."

For substantial protection, new U.S. embassies overseas are required to have 100 feet or more of standoff (a fact which has delayed the construction of a new United States Embassy in Berlin, where the site lacks the necessary setback, for five years). Such standoff can provide both protection and deterrence. According to the *9/11 Commission Report,* Al Qaeda operatives declared the United States Embassy in Kenya "an easy target because a car bomb could be parked so close by." Often, the hardening of a new structure also includes ideas as simple as moving the mail room to an off-site location, not allowing a loading dock to be included under a building, or restricting parking around the perimeter.

In urban areas, however, standoff is often not an option. Other times, architects or building owners refuse to alter their plans, and blast resistance has to be built into the structure itself. After engineers reach the standoff limit, that's when complex questions of simulation and progressive collapse arise. Crawford—who along with Weidlinger Associates has what are considered the most comprehensive simulation codes in the business—uses models to try to determine the most vulnerable areas of a structure and the ones most susceptible to progressive collapse. Certain areas of a building, mail rooms or loading docks, might be hardened with special materials. The idea is not to stop the blast, but to diffuse its energy to parts of the building that are less important—away from load-bearing parts.

One way to do that is to create ductile connections that hold components of the building together. "Many people are using conventional connections. They are not good," says Krauthammer. "You have to use special connections so that the connection does not fail. You want to 'fail' specific types of members, and take out energy this way, rather than other members that could trigger the collapse." In reinforced concrete that means tying together the steel rebar inside columns and beams. For steel structures, it means employing redundant steel-plate attachments such as those created by a Laguna Hills, California, company called SidePlate Systems.

Another method is to address the building's basic materials—the concrete, steel, and glass that make up columns, walls, and windows. On the topic of retrofit materials and minutia, Crawford is a walking encyclopedia. To wit: The simplest and often most effective defense is to wrap columns in carbon fiber or steel jackets, a technique used in California for seismic building and highway retrofits. Because concrete crumbles easily when bent, blasts and earthquakes that wrench columns can quickly rubbleize them. Adding the wrap serves to "confine" the concrete, allowing it to bend and still stand up under the load of gravity. "The wrapping changes the mechanics of the response," he says, "and puts you from a very brittle response in a column to a very ductile response." The same is true for floor slabs, which are not designed for the upward pressure that comes with a blast. A carpet-like carbon fiber covering can help resist the shattering from a pressure wave underneath.

Some of the techniques used against explosions can provide additional defense against chemical and biological attacks. A common blast requirement, says Hinman, involves constructing all walls that connect public and private parts of a building—those in the lobby, for instance—as structural walls that attach to the floor and ceiling slabs. The alternative, partition walls that stop short of the ceiling, could allow gas or germs to seep through. Hinman also advises builders at risk to place the air intakes for ventilation systems high up on the building or roof.

But the blast engineer's most important task, after defending against collapse, is finding ways to reduce the casualties created by flying debris. Crawford rummages through a closet in his office, pulling out samples of laminated glass, the type used in car windows, one of many shatterproof designs that can be used to replace normal windows. When the glass breaks, an interior layer of plastic contains the fragments. Other types of glass—which structural engineers refer to as "glazing"—are manufactured to shatter into tiny, harmless pieces instead of dangerous shards. Eve Hinman is known in the industry for her development of curtain walls that minimize the hazard of flying glass.

Crawford also describes methods that engineers use to contain the fragments and energy of a blast—in the same way a real-world firewall compartmentalizes a blaze. For some airport or building entryways, for example, he proposes installing large fish tanks—made with the special glass—to help dissipate an explosive force. "Even though you knock the tank out, the water can't move fast enough to get out of the way," he says. "Water tends to absorb energy; a foot of water is as good as a foot of concrete, as far as mass is concerned." Engineers can also add Kevlar or steel fabric catchers to masonry walls, and even apply a polyurethane spray—formerly used to protect truck beds—that helps hold the brick or cement blocks together.

"All of these things are very embryonic," Crawford says. "Either they haven't taken off, or they are actually used in a very few places so far." One of those places is the Pentagon, where blast-resistant windows reinforced with steel—among other advanced designs—were responsible for saving lives on September 11. The engineers rebuilding the Pentagon after the attack are using even more advanced materials such as Kevlar, as are an increasing number of embassies and federally owned buildings. But when it comes to the wider construction world, he says, "basically you are sitting on the ground floor with most of this stuff."

Crawford cues up a video from a test explosion in New Mexico, in which dummies were placed in offices to show the varying effects of flying glass. The slow-motion clip shows a dummy sitting with its back to the window. The bomb goes off, and the dummy's head is shredded by the shattered glass. It is a rare study on the effects of blast on the occupants of a building. "If you had a building with a thousand people in it," he says, "you could reduce casualties by some large number by just laying the office out more effectively. Just by putting people off-axis from the window."

In determining what buildings are at risk of attack, and what kind of attacks they might face, there is little to go by but history. As Ted Krauthammer notes, "Terrorists do not leave you a manual that says, 'This is what we are planning. This is the size of the attack.'" But when structural engineers talk about progressive collapse, they rarely point to the disintegration of the twin towers on September 11. This is because many of them don't believe that the World Trade Center was an example of progressive collapse at all. A progressive collapse is a situation where "the damage that occurs is disproportionate to the mechanism that initiated it," explains Eve Hinman. "If you have a house of cards, and you pull out a card and the whole thing falls down—that is a progressive collapse. With the World Trade Center, the buildings were assaulted in a very gross way, where many, many structural elements failed. The fact that the building collapsed is kind of consistent with the level of damage that occurred." To most blast engineers, in fact, few if any applicable lessons specific to explosion-resistant design can be drawn from the attack. "With the World Trade Center—it's not clear that that is an example of anything," says Crawford. "You can probably argue, 'Well, we should have had better exit strategies.' But you could probably also argue, 'Well, we don't design buildings for 767s striking them, especially 100-story buildings. What are the lessons from WTC? I don't think much of anything. I think there will be a few fire-related issues that come out of it. I'm not so sure that it's a blast-effects issue at all."

As it happens, the first explosion-related collapse that people in the blast business point to has nothing to do with terrorism. It happened in London on May 16, 1968. That morning, Ivy Hodge, a resident of an eighteenth-floor apartment in the 22-story Ronan Point Tower in Canning Town, on the east side of the city, climbed out of bed and went into the kitchen to make tea. Unbeknownst to her, a leak in her stove had filled the apartment with gas. When she struck a match to light the burner, it set off a relatively small explosion, which blew out what was later revealed to be flimsily designed kitchen and living room walls. Levy and Salvadori, in *Why Buildings Fall Down,* describe what happened next:

> Floor by floor, in a domino fashion, the entire corner of the tower collapsed. Miss Hodge was lying, dazed, in a pool of water spilled from the overturned kettle, looking out at the morning sky all around her. The living room of her flat and half her bedroom had vanished. She suffered second-degree burns on her face and arms but was otherwise unhurt. Four sleeping residents died on the lower floors, crushed in their bedrooms when the giant concrete panels making up the structure of the building came tumbling down. . . . Once one wall panel blew out, the wall panels above it were left unsupported and fell. The floor panels that were consequently left virtually unsupported then crashed down on the floor below, overloading it and causing a progressive collapse of all the walls and the floors below.

While the tower itself remained standing, the end result was that one corner of the building had essentially been sheared clean off. The collapse at Ronan Point fueled a reevaluation of concrete-based building methods around the world and led to a revision of building codes in the United Kingdom, mandating redundancy in all buildings of a certain size. The Khobar Towers complex in Saudi Arabia, in fact, was built to the revised British standards. The lesson of Ronan Point for many engineers is that there is the potential for general standards to address the risk of progressive collapse.

If the United States had its own Ronan Point, it was the Oklahoma City bombing. The toppling of the Murrah Building spawned a new set of regulations, promulgated by the General Services Administration, which mandate that some 8,000 federally owned structures—from courthouses to office buildings—conform to certain blast-related and progressive-collapse standards. Such standards produce the vast majority of business for blast engineers like Hinman, Crawford, and Isenberg. Embassies have even more stringent rules, but even those are not always followed. Former national counterterrorism czar Richard Clarke, in his book *Against All Enemies*, recalls the bureaucratic resistance to spending on protective design for embassies. "If there were new funds," he writes, "the Department of State had many things it wanted to do with the money other than build more fortresses."

Without a perception that buildings are vulnerable to attack, blast protection has yet to filter into commercial building codes. Krauthammer compares the current state of blast design to seismic engineering in the early part of the twentieth century. "In the 1930s, people said, 'There is absolutely nothing that can be done to resist earthquakes,'" he notes. In the 1950s, structural engineers began focusing research efforts on earthquake effects, and by the 1980s, seismic designs and building codes were more established. With the further push after the 1989 Loma Prieta and 1994 Northridge quakes, seismic engineering has become a mature field. In the same sense, "We know that there is a problem," adds Jeremy Isenberg. "And we probably can do a pretty good job of addressing the technical problems on a case-by-case basis. What we don't know how to do is to address a global problem of making technology the servant of public policy and sound economic decisions."

Meanwhile, building codes in other countries encourage the use of innovations. Some skyscrapers in Japan, for example, contain sensors that detect an earthquake and then trigger the movement of a large mass of material in the building, to help give it ductility and mitigate the damage. The technology, called "active control," was actually invented in the United States, but liability issues have so far kept it from being used here. In Israel, new blast-resistant designs and materials are rushed into service as soon as they are tested. "They get an idea, they go blow it up. If it works, they stick it in a building," says Joseph Smith. "They don't really know what they are getting, but they know that it's better than what they had before."

Buildings designed to resist blasts would, in general, be more resilient against earthquakes, and vice versa. Better standards can also help prevent collapse from accidental explosions. (The worst blast-related collapse in the history of the United States occurred in 1937 in New London, Texas, when a gas explosion below the steel-frame high school killed 298 people.) "Eventually," says Krauthammer, "there will have to be a unified building code that addresses all abnormal loading conditions. In other words, there will be no attempt to separate wind from blast from earthquakes." A multidisciplinary design approach would enable engineers to take advantage of points where the threats to buildings overlap. "Then when you add up the total cost for all these things," he says, "it will be far less than if you had done each one separately and superimposed them."

Eugene Sevin, Krauthammer's mentor, is one of the last surviving members of the first generation of blast engineers. Melvin Baron died in 1997, followed two years later by his friend Paul Weidlinger. At 75, Sevin is still called in to consult on blast projects and sit on government committees. After five decades of studying explosion-resistant design, he knows that engineers can't control which buildings are protected. "The trick is to figure out how you can protect buildings and people inside buildings without having quite all the wherewithal that the military has," he says. "It's a much tougher problem, but it's not a mysterious problem. The biggest difficulty lies less in the engineering and more in convincing the developers and the building owners to invest that kind of money."

Of course, not all buildings need the same defenses; in fact, many may not merit any at all. For now, it is up to individual developers to decide how much they are willing to spend on protection. Should an office building in Omaha have the same hardening as one in Manhattan? Should skyscrapers have greater protection because of their symbolic value? (For the Freedom Tower, the answer is clearly yes. The architects and structural engineers are creating a redundant structure that will retain its integrity even in the face of powerful attacks from the air and ground.) Is it worth the cost to retrofit urban buildings? Those are questions just barely being asked, on an ad hoc basis. And unlike in Baghdad, the government can't simply try to surround all buildings deemed as potential targets with concrete barriers.

In the face of so much uncertainty, some building owners simply don't want to hear about their vulnerabilities, for fear of opening themselves to liability if they fail to correct them. Others are acting on their own, hiring blast specialists to assess their vulnerability and propose solutions. "The client—often with our help and advice—decides that he wants to consider a threat of a certain size," says Isenberg. "There is a range of responses: standoff, harden the structures, or you can do nothing. Our responsibility to the client is to advise on that range of options." With no clear sense of the threat of attack, however, and no building codes to go by, blast-resistance engineers have no way of guaranteeing their work against an actual event.

Even more complicating is the fact that if one building owner decides to change setback or harden a building, that will alter the threat level for other less-protected targets. "The harder you make your target, the greater the chances that the terrorists will choose another target," says Rafi Ron, the airport security consultant. "So even if you are not 100 percent [hardened], if you are 20 percent better than other targets, you are probably almost as safe as 100 percent. Because you are better off than other targets, and terrorists like to succeed." Ron likens the situation to an old joke about two guys trying to outrun a charging bear. "One says to the other, 'We can't outrun the bear, he's too strong for us,' " says Ron. "The other guy says, 'I'm not trying to outrun the bear, I'm trying to outrun you!' The fight against terrorism has that nature. We are all fighting back. The one who puts the stronger fight back avoids the results. It's the weaker one who pays the price."

In a 2002 paper on risk and building protection, Richard Little, director of the Board on Infrastructure and the Constructed Environment at the nonprofit National Research Council, proposes a "probabilistic approach" that uses decision matrices and probability formulas to determine whether a building should be protected. Risk, as security experts define it, is the consequence of an attack multiplied by the probability of that attack. Blast engineers can enumerate different bombing threats and then calculate vulnerabilities and potential numbers of casualties. But when it comes to figuring out the likelihood of an attack, the formula still has one blank to fill. "I don't know what the hell the probability element is," Little says, "and nobody else does either. Don't rely on the Department of Homeland Security to tell you what the answer to that is."

Ultimately, choosing the acceptable level of risk will become a societal question: How much are we willing to pay to avoid unlikely but devastating events? "It really calls for a significant cultural change," says Sevin. "Unfortunately, that probably comes about by more World Trade Centers, more disasters. We have a tendency to sort of find the edge of the cliff by falling off it."

From a practical standpoint, says Krauthammer, owners of buildings deemed at risk will have to combine some level of redundant design with more comprehensive defenses that include surveillance and access control. Applying explosion-resistant design concepts smartly, he says, will require more forethought than just adding layers of concrete and steel. "Unlike seismic engineering, where seismicity is a natural hazard," he says, "this thing is an intelligent hazard. It evolves. The bad guys are not stupid. When they see that we can do something, they will come up with a better contraption to try and defeat what we have come up with. You always have to be one step ahead. Unfortunately, until recently we have always been one step behind."

FIVE

Biology Lessons

G EORGE POSTE—tanned, fit, and affluent, British-born but now forcefully American—is sitting on the terrace of the Hotel Majestic in Cannes having lunch and holding forth. In fact, he's boasting a little. He has a fair amount to boast about. He's respected enough by his scientific peers in the United Kingdom to have been elected to the Royal Society—the rough equivalent of the National Academy of Sciences—an honor that goes to very few. As chief science and technology officer of the Anglo-American pharmaceutical giant SmithKline Beecham, from 1992 to 1999, he presided over the successful introduction of 26 vaccines and drug treatments. In 2003, he was wooed by Arizona State University to run the Arizona Biodesign Institute, an ambitious new research establishment with a budget of $400 million over five years. He's over in Cannes to dispense wisdom and advice to a select gathering of venture capitalists and fledgling biotech companies. Gifted, respected, intellectually combative—all in all, George Poste is a very impressive guy.

But Poste isn't boasting about himself—he's talking up the talents of his grandson. He's describing how this 12-year-old boy has been sitting in a school biology lab, cutting up genes and moving them from one organism to another. The child is in sixth grade, and he's conducting experiments worthy of a Nobel Prize. Not, admittedly, this year's prize; but in 1980 a Nobel was awarded to Paul Berg, a Stanford professor, for just this sort of DNA cutting and splicing. The pace of progress in molecular biology over the last couple of decades has been so fast that such Nobel-level work is now gifted child's play.

And that's what George Poste is really going on about. The sheer dynamism of modern molecular biology has few, if any, precedents in scientific history. Unlike some scientific revolutions, moreover, the spate of knowledge being produced today is remarkably applicable in the real world; already the impacts are being felt in medicine, with the development of new medicines based on proteins copied from nature, and in agriculture, with the introduction of pest-resistant soybeans and the like. For the first time ever, we can take biological systems that have evolved over billions of years and induce them to remake themselves. Mice have been made to "bulk up" like body builders; fish have been made to glow; bacteria and free-floating cells have been made to produce drugs that save thousands of lives a year. The molecular biology revolution offers remarkable possibilities for new orders of human development and happiness. But it also brings with it quite spectacular dangers.

One of Poste's further claims to fame is that he is a member of the Defense Science Board, a group of experts appointed to advise the secretary of defense. And in his role as head of the Board's bioterrorism task force, the tanned, relaxed mover and shaker is an anxious man. In a world where gene splicing is now part of the grade-school curriculum, literally hundreds of thousands of people around the world have the know-how to move genes—including those that can cause illness—from one organism to another. This worries Poste deeply. He knows that the Soviet Union was making bioweapons until the 1990s and that other countries and groups may already possess or intend to produce some of them. He knows that stranger, subtler, and quite possibly deadlier weapons may be designed and built in the future by unfriendly states and terrorists abroad—or by the disaffected at home.

George Poste is afraid that for all his acumen, insight, and connections, he may not be able to keep his grandson safe.

•　　•　　•

The molecular biology revolution erupted in 1953, when Francis Crick and James Watson discovered the structure of DNA, the molecule in which biological systems store their genetic information. Unzip that famous double helix, and each of two strands consists of a string of chemical "letters" that shape and make possible every form of life, from bacteria to humans. By the mid-1960s, a mixture of inspired hypothesis and meticulous lab work had led biologists to the next critical breakthrough: this sequence of letters, this information, is used to make proteins, the molecules that do most of the work in any organism. Proteins catalyze the reactions needed to create everything else, they pass signals around, they join some molecules together and pull others apart. They come in a bewildering array of shapes and sizes. But each protein starts off as a string of components called amino acids. Sequences of DNA are the code that determines the order of the amino acids in those strings. Cracking the DNA code in the 1960s made it clear that, to a rough approximation, a gene—that basic unit capable of transmitting traits from one generation to another— was a piece of DNA that described a particular protein. If you changed a gene's DNA sequence, you could change the protein. If you deleted the gene, there would be no way to make the protein anymore. And if you picked up the gene and put it someplace else, you could produce that protein anywhere you wanted.

Soon tools for editing and copying genetic sequences became available. By the mid-1970s, these advances had given birth to the biotechnology industry, which took genes that describe medically useful proteins—insulin was the first commercial prospect—and put them into bacterial cells that could then be induced to churn out the desired protein in bulk. By the 1980s, "recombinant" insulin—so called because it had been made by recombining insulin genes from humans with the genomes of bacteria—was available to diabetics, and more and more proteins were being studied as treatments for diseases ranging from cystic fibrosis to cancer to AIDS. Between 1980 and 2001, the revenues of the top five biotech companies—Genentech, Amgen, Biogen, Chiron, and Genzyme—rose from $9 million to $9 billion, mostly thanks to the use of proteins as medical treatments.

As successful as commercial biotechnology has been, it's the pace of the science itself that is truly impressive. Molecular biology is a discipline unique in its ability to feed off its own fruits. It sometimes seems that every revelation about how nature works—about how genes are expressed as proteins, about how the information in DNA is managed, about how the immune system functions—leads almost immediately to new methods for finding out yet more.

The most spectacular example of this was the discovery, in the mid-1970s, of how to "read" the sequences of letters in DNA molecules. At roughly the same time that the first biotech companies were being dreamed up, a biologist at the University of Cambridge named Fred Sanger devised a technique for ascertaining the order of the letters in a DNA sequence. In 1977, Sanger achieved yet another breakthrough, deciphering the entire 5,386-letter sequence of a tiny virus that preys on bacteria. For the first time in the earth's history, an entire genome was laid bare for inspection.

Since then, our ability to abstract the information stored in a genome has grown exponentially, largely because the process is highly amenable to automation. In 1995, the Institute for Genomic Research in Rockville, Maryland, a nonprofit founded by an ambitious genetic pioneer named Craig Venter, published the genome of the bacterium *Haemophilus influenzae,* roughly 300 times longer than that of Sanger's virus. (Viruses can have very small genomes because they are parasites, hacking into the genomes of their hosts in order to reproduce; bacteria such as *H. influenzae* are more self-reliant, carrying all the genes they need to get by in the world on their own.) In 2002, after a race between a public consortium and a private effort headed by Venter, the 3 billion letters of the human genome were published. By the middle of 2004, more than 200 genomes had been sequenced. They included almost all the bacteria that cause major human diseases; the most widely used laboratory species, such as the mouse, the fruitfly, and brewer's yeast; the mosquito *Anopheles gambiae,* which spreads most of the world's malaria; and rice, the staple crop that feeds more people than any other. Hundreds more were underway, including those of the chimpanzee (very nearly finished) and the honeybee, wheat, maize, chickens, silkworms, the cat, horse, rat, hosts of relatively obscure algae and bacteria, and legions of parasites.

This progress has been possible because, for the past decade, the machines that actually produce the DNA sequences have been getting better at an incredible rate. The information revolution has been driven by Moore's law, named after Intel founder Gordon Moore, which says that computer chips double in power every 18 months. The genomics revolution has depended on sequencing machines improving at roughly the same rate, and it shows no sign of stopping. Indeed, newer gene-reading technologies capable of working with smaller amounts of material and reading more letters at a time are actually *outpacing* Moore's law—the capacity of these machines for reading a DNA sequence is doubling in just a year or so.

The same is happening with technologies that "write" DNA molecules—that create whole genes to order from simple molecular building blocks. This gene-synthesizing technology is already capable of creating short sets of genes. In fact, it's possible, using a gene synthesizer, to write a whole viral genome from scratch. If current trends were to continue, by 2010 it would take a single person a single day to read a human genome—or to write one. If the relevant technology fails to maintain its current pace, such tasks should still be well established in grade-school curricula by the time the next generation of George Poste's family enrolls.

Even if you ignore the accelerando of wonders-yet-to-be, the first half-century of molecular biology has already seen enough progress to rank it among the great intellectual and technological revolutions of the modern age. But there is a striking difference, in the West, between biology's remarkable ascent and the historic rise of other sciences. In the growth of synthetic chemistry, of aviation, of electronics, of nuclear physics, the potential for military applications was readily apparent early on. Within a few decades of the founding of the modern chemical industry in the 1860s, the military uses of poison gas had been foreseen and (ineffectually) banned. Or look at aviation: the Italians dropped bombs on Tripoli eight years after the Wright Brothers first took off at Kitty Hawk. In 1932, James Chadwick discovered the neutron in Cambridge University's Cavendish Laboratory (the same institution that would later be a home to Watson and Crick); 13 years later, countless billions of the tiny particles were being put to work splitting atoms in the first atomic bombs. In the 1940s, the earliest computers were swiftly charged with deciphering enemy code.

Although molecular biology also has clear military implications, for most of its history, researchers and policy makers in the West have either ignored or denied this reality. In 1972, the United States, the United Kingdom, and the Soviet Union together sponsored the Biological and Toxin Weapons Convention, a UN treaty banning biological weapons that has now been ratified by 148 countries. While the Soviets ignored the convention completely, secretly expanding their large and inventive biological warfare program, called Biopreparat, the United States had by that stage already unilaterally renounced the weapons. Its renunciation was in part the result of a belief that bioweapons simply weren't practical—and this may indeed have been true. The life sciences weren't robust and predictable, two qualities weaponeers clearly value.

But there are other reasons, too, for the historical lack of military interest in biology. First, there's the bomb. Once the nuclear age got into its swing, it offered ways of killing people in untold numbers, ways of wiping out cities and countries, that put plagues to shame. When you already have one way of waging annihilatory war on a global scale, do you really need another?

The second reason, clearly linked to the first, was a difference in the standing of physics and biology. After the development of the nuclear bomb, governments around the world saw physics as the supreme science. Test sites, photographs, and the scorched remains of Hiroshima showed them what it was capable of. Looking back on the Manhattan Project as early as 1947, Robert Oppenheimer told an audience at MIT that "in some sort of crude sense, which no vulgarity, no humor, no overstatement can quite extinguish, the physicists have known sin." They also knew power. Although Oppenheimer himself ran afoul of the U.S. government, the physicists as a group—Edward Teller, E. O. Lawrence, and Hans Bethe among them—enjoyed enduring fame and access at the highest levels. They served both as weapons makers and advisers—wizards and chamberlains at the court. One of the results was that physics grew into the archetypal "big science." A system of national laboratories was created both to make nuclear weapons and their components and to further all sorts of other physics research. The big three—the labs that actually made the bombs—were Los Alamos and Sandia in New Mexico and Lawrence

Livermore in California. The labs in the system geared toward non-military research included Argonne and Fermilab, both outside Chicago; Oak Ridge in Tennessee; and Brookhaven on Long Island. By the 1980s the physicists' atom smashers had come to be measured in miles of circumference and billions of budget dollars.

While the leaders in Washington, D.C., lavished such attention on physics after the war, they never really "got" biology. The federal government invests a great deal of money in the field through the National Institutes of Health (NIH), but it does so in a piecemeal and unstructured way. The goal is relatively vague—that somewhere down the road, we'll end up with better medicine: more drugs, faster surgeries, cheaper care, longer lives. The government, in other words, supports biology without knowing much about it. While the United States' physics-based national laboratory system knows how to make nuclear weapons, knock the quarks out of atoms, send robot explorers into the depths of space, and ignite miniature suns with mile-long lasers, its biological capabilities are a lot more modest. As Roger Brent of the Molecular Sciences Institute in Berkeley, California, points out, not a single one of the national laboratories has the technological and procedural know-how to design and develop a new drug of the type that is the pharmaceutical industry's bread and butter.

Last but not least, bioweapons have gone un-discussed until very recently because most biologists just *wouldn't* discuss them, or even think very seriously about them. Biologists like the idea that their work will create a better world; they are less attentive—you might say even unreflective—when it comes to the reverse: spawning a dystopia. On this point, George Poste offers a warning echo of Oppenheimer: "There is a profound naivety on the part of the life scientists," Poste says. "Biology is poised to lose its innocence, and they have no understanding of that."

The dilemma, in a nutshell, is this: there may be good reasons why the United States and its allies have deemed molecular biology unsuitable for weapons making, but these rationales haven't held everywhere, or at all times. Biological weapons are not only feasible; they already exist. While the United States was building its biotech industry, the Soviet Union was producing biological weapons by the tankload, in some cases using quite similar techniques. The West knew

almost nothing about this parallel universe until two senior Soviet officials, Vladimir Pasechnik in 1989 and Kanatjan Alibekov in 1992, de-
fected and brought to Britain and the United States their stories of the
Biopreparat. They stunned national security circles with revelations
of a program that had employed tens of thousands of individuals and
produced a wide range of virulent pathogens in large quantities.

As described by Alibekov—who was Biopreparat's second-in-
command, and who now, under the name Ken Alibek, has remade
himself as a biodefense expert in the United States—the application of
molecular biology in the Soviet Union was intimately linked to the re-
newal of its weapons program in the 1970s. Soviet academic biology
was incapable of understanding the potential of the new technology;
biologists keen to explore its possibilities found that the best way forward was to tie their research to the military and promise that new
weapons could be made with it. The Soviet bioweapons program gathered pathogens of all sorts from around the world, sometimes through
the KGB, sometimes by simply asking foreign academics for them (the
work was largely carried out under the cover of pharmaceutical research). Soviet biologists developed the first human-engineered
bioweapons, cultivating traits such as antibiotic resistance in bacteria.
The main focus, though—and the focus of all the major bioweapons
programs of the twentieth century—was taking dangerous microorganisms out of the lab and "weaponizing" them.

Weaponization is a term for the morally ugly and, for the most
part, dull and inelegant process of turning a natural pathogen into an
instrument of warfare, crime, or terror. Anything done to turn a disease into a weapon of war can be seen as weaponization. (In the 2003
SARS epidemic, the Taiwanese, imaginatively but not unreasonably,
described China's refusal to allow Taiwan access to the World Health
Organization's resources as the "weaponization of SARS.") In the
Soviet Union's case, weaponization meant the manufacturing of selected pathogens in forms that were reliably virulent, that could be
stored and transported relatively easily or produced on demand, and
that could travel efficiently from a weapons delivery system into the
bodies of its targets. Factories across the Soviet Union produced
weaponized pathogens literally by the ton.

The public health experts at the U.S. Centers for Disease Control

(CDC), the country's primary public health agency, list six pathogens as category A potential bioweapons: smallpox, anthrax, plague, tularemia (a particularly infectious bacterium), viral hemorrhagic fevers (Ebola and some other viruses with similar effects), and botulism (the toxin made by *Clostridium botulinum,* which at very low doses is used in Botox treatments but at higher levels is a fatal poison). In the early 1990s, Ken Alibek revealed that Biopreparat worked on weaponizing all of them, plus half a dozen more.

To those in national security circles who thought about such things, the idea of these newly revealed weapons getting into the hands of rogue states or ambitious terrorists was becoming a waking nightmare. The Soviet Union might have vanished, but no one could be sure that all its bioweapons stockpiles had vanished or that none of the expertise developed by Alibek and his colleagues would surface in former Soviet client states, or in states that had the money to buy the know-how, or in the hands of terrorists. It may still be true that bioweapons are less than ideal for standard military deployment, but they are beyond a doubt well-suited to catastrophic terrorism. A bioweapon used against civilians does not need to create casualties either quickly or predictably, as a weapon of war is meant to: the amount of panic that even a poor biological weapon might produce would amplify its effects many times over. The immediate threats are daunting enough—with access to the tools that molecular biology has already developed, with access to some inexpensive facilities, some (currently) graduate-level expertise, some imagination and a bit of luck, bad actors could do great harm. But using the academic knowledge now emerging, these same people could make weapons that are even more varied and deadly. "The creative intellect of a group of Dr. Evils could find many more ways to usurp what's going on in biology than they could in any other discipline," says George Poste. He continues with distinctive techno-grandiloquence, "Nature has only sampled a very small fraction of biospace. The total combinatorial assembly of every life-form on the planet is still only a very small element of the combinatorial space that *could* be sampled." In other words, the creatures that now grace the earth are a mere subset of what's possible based on all the same building blocks. "Biologically inspired design can take the elegance and simplicity of nature's design and start to harness it to entirely new frameworks of engineered materials and engineered processes." At this point, a mighty gust of wind blows over the parasols on the Hotel Majestic's terrace. Poste laughs: "God got a little pissed off that we were about to duplicate his activities."

Many biologists, including Poste, see a future in which scientists can look at the molecular basics of a biological system and predict in detail how it will respond—what its own internal dynamics of growth and change will be—if someone tinkers with it. Among the fruits of this new perspicacity, he goes on matter-of-factly, "will be the knowledge of how to screw up every biological system in the body." By this, Poste means not only that the new science will reveal more ways to kill, though it will offer plenty of those. He is also referring to other sorts of interventions—possibilities arising from new biological insights in precisely the same way that modern pharmaceuticals have. He lists a few examples that come to mind: "Modulation of memory; induction of lethargy; induction of vindictive behavior; triggering of violence; hallucinations. You name which part of the motor, sensory, emotional, or intellectual repertoire you want to play with, and you can begin to find a way to perturb it with small molecules."

Sitting in an office suite looking out over Baltimore's Inner Harbor, Tara O'Toole is worried about the future Poste imagines. But she's also worried about the present—and blunt in her assessment of it: "The advantage is now firmly with those who would seek to deploy offensive bioweapons; the state of biodefense is relatively weak." O'Toole is a forthright woman with a dry sense of humor; her rectangular face, softly creased, has the slightly embattled look of someone who has been running into barriers for a bit too long. She was the assistant secretary for environment, safety, and health in the Department of Energy during the Clinton administration, and now she's the chief executive officer of the Center for Biosecurity at the University of Pittsburgh Medical Center. (Until being signed up by the University of Pittsburgh en masse in the fall of 2003, she and most of her 20-odd colleagues made up the Johns Hopkins University Center for Civilian Biodefense Strategies, and their offices are still in Baltimore.) Her job, as she sees it, is to get the United States and other governments to adopt sensible strategies for new biodefense systems, which may account for that running-into-barriers demeanor.

The most famous—O'Toole calls it "notorious"—of her efforts to date was a war game called Dark Winter, which was staged in the summer of 2001. A group of senior political and government figures—former Georgia senator Sam Nunn, former presidential counselor David Gergen, and former director of central intelligence James Woolsey among them—played the roles of the president, members of his National Security Council, and various others. They gathered at Andrews Air Force Base on the outskirts of Washington—which, as Air Force One's home, is well-prepared for looking after visiting dignitaries. For two days, they sat around a table in the middle of a dimly lit amphitheater pretending they were in meetings of the National Security Council, as O'Toole and her colleagues in the shadows fed them information from a previously prepared bioterrorism scenario and recorded their reactions.

In the Dark Winter scenario, an airborne virus was released in three shopping malls—one in Oklahoma City, one in Atlanta, and one in Philadelphia—at the beginning of December. By the time the NSC members heard about the attacks, CDC was confirming 20 cases of disease in Oklahoma and there were disturbing reports from other states. It was at this point in the unfolding of this hypothetical event that O'Toole's role players got into the game. The deadly weapon was not particularly sophisticated—in fact, quite the opposite. It was smallpox, a scourge that has been around for millennia, quite possibly killing more humans than any other single infectious disease in history.

For history's most dramatic example of what the smallpox virus can do, look to the conquest of the Americas. Like many infectious diseases, smallpox evolved through the close proximity of large numbers of people to large amounts of livestock, circumstances that occurred mostly in the Old World. Over millennia, natural selection equipped Europeans and Asians with immune systems able to cope with the disease to some extent, as did childhood exposure; but Native Americans had developed no defenses at all. In 1520, an epidemic of smallpox traceable to an infected slave from Spanish Cuba killed off perhaps as much as half the Aztec empire, including the emperor himself. In the century after Cortez arrived, the population of Mexico shrank from 20 million to about 1.5 million under an unremitting onslaught of disease and deprivation, beginning with smallpox and followed by measles, typhus, and the starvation that results when there is no one left to farm the land.

The pattern of massive mortality was repeated throughout the New World, a mostly unintentional genocide that invaders found militarily useful if economically damaging (with so much of the indigenous population dead, colonizers had to import slaves rather than simply round them up from among the local nations). On occasion, though, the disease was spread deliberately. In the war of 1763, British officers attempted to introduce smallpox to the Indian tribes allied with the French. In 1777, concerned that something similar might be perpetrated against his own army, George Washington ordered that all his soldiers be inoculated against the disease. Inoculations, in which material from someone else's pustule would be placed on cuts or scrapes in the skin to provoke immunity in the way that a vaccine does, may have originally been tried in China as early as the first century B.C.; records of the practice date back to the Sung dynasty. The practice was brought to the West from Turkey in the eighteenth century by Lady Mary Montagu, wife of the British consul in Constantinople. In the 1790s, the British physician Edward Jenner discovered that the cowpox virus, which he called vaccinia, would do as good a job as inoculations of genuine smallpox, with considerably less risk of disease; the modern era of vaccination was begun.

It was more than two centuries later, in 1958, that the Soviet Union proposed a concerted, worldwide effort to eradicate smallpox. At that time, the disease was still killing 2 million people a year, almost all in Africa, Asia, and Latin America. But deadly as it was, the sickness had various characteristics that made it vulnerable to public health campaigns. Many viruses can flit back and forth between humans and animals; the animals can thus serve as a reservoir from which new cases of disease can flow. Though smallpox undoubtedly evolved from an animal virus at some point, in its modern form it infects only people, so if expunged from the human population, it has nowhere to hide. It is also easily diagnosed by the distinctive pattern of "pox" marks all over the skin. And by 1958, a good vaccine was widely available—the benign vaccinia virus.

So if outbreaks could be spotted and all people exposed were vaccinated, it would be possible to break the chains of infection and quash the disease. A decade-long program introduced just after the death of Lenin in 1924 had used these same principles to eradicate smallpox from the Soviet Union in the mid-1930s, a source of considerable pride, and now the Soviets were suggesting that the World Health Organization (WHO) commit to achieving the same victory on a global scale.

Which it did. In the 1960s, the WHO sent a crack corps of young, idealistic individuals from Europe and North America, including a large number from U.S. public health headquarters, the CDC, into the field to get the eradication going. Their procedure was not unlike that of the CDC's Epidemic Intelligence Service, a system set up in the early 1950s specifically to address cold war worries about biological attacks, which involved sending surveillance squads to investigate outbreaks of disease around the United States. With bravery and not a little ruthlessness, the WHO teams tracked down every smallpox outbreak they could and would stop at nothing to vaccinate everyone at risk. They went into war zones and faced down armed gunmen with their medical kits. Daniel Tarantola, a tough and resourceful French doctor who went on to play an important role in the WHO's fight against AIDS, took his life in his hands to vaccinate a band of brigands in Bangladesh. By 1980, the WHO was able to announce that there were no longer any humans infected with the disease—it had been eradicated from the wild and lived on only in a few secure laboratories. For the first time, one of nature's greatest killers had been eliminated.

As a choice for Tara O'Toole's Dark Winter scenario, smallpox made sense for several reasons. First, it is a highly dangerous disease—without good medical care, it might kill as many as a third of those infected. Second, O'Toole's partner at the Center for Civilian Biodefense Strategies was a Johns Hopkins public health legend named D. A. Henderson, who had led the WHO's smallpox eradication campaign and had a serious personal interest in warning that his work might be undone. And third, there was no longer any dispute, given the revelations of the two defectors, that smallpox could be produced for use as a weapon; the Soviet Union had continued to do so for years after the virus was eradicated in the wild. The Kremlin had grasped the tragic unintended consequence of eradication: vanquishing the disease had made it possible to halt smallpox vaccinations in the early 1970s, and had thus produced a world in which ever fewer people were immune. The extent to which those vaccinated in the past maintain a residual immunity is not clear. What is clear is that the unvaccinated population is entirely vulnerable.

What an outbreak of smallpox would look like in today's United States and, in particular, how quickly it would spread are hotly debated. Estimates of how many people each infected person would go on to infect vary from 1 or 2—in which case the disease would probably be very easily controlled—to 12 or more—in which case it could quite easily get completely out of control. In Dark Winter, O'Toole, Henderson, and their colleagues used a high number, and the results were correspondingly grim. From the moment the airborne virus was released into the mall in Oklahoma City, it began infecting Americans in large numbers. When the National Security Council was alerted, the president, played by Senator Nunn, ordered the vaccination of anyone who might have come in contact with those who were sick. This containment effort, called "ring vaccination," works in the same way as cutting troubled "islands" out of the energy or communication infrastructures. But six days later, as the exercise progressed, there were 2,000 confirmed cases in 15 states, and others in Canada, Mexico, and the United Kingdom, all of which could apparently be tracked back to the three malls. Three hundred people were dead, vaccine stockpiles were running down, the public was panicking, and international borders were closed. A crash vaccine manufacturing program was set in motion.

By the thirteenth day, there were 16,000 confirmed cases and 1,000 people were dead. The epidemic had spread to 10 more countries. The aggressor was still unknown, but threats of further attacks with different bioweapons had been received in Washington, D.C. The situation was becoming uncontrollable. The worst-case scenario, at this point, was that within six weeks the disease could be afflicting 3 million people, and that eventually 1 million might die.

•   •   •

A weapon such as smallpox plundered from some far-flung Soviet lab—or engineered by a disgruntled researcher at home—is a horrifying prospect. But it's only one part of a larger arc in the changing concerns of the infectious-disease community. Beyond bioweapons, researchers in the West have begun to rethink our fundamental, age-old conflict with germs. This was a war that, having been fiercely fought by humans for almost a century, recently seemed nearly won. Yet in the 1990s, the fighting flared up again on unexpected fronts.

The war between humankind and microbes began in earnest when Robert Koch, Louis Pasteur, and a few lesser-known pioneers identified the existence of germs in the late nineteenth century. Infectious disease, they showed, was not just a natural scourge or an act of God; it was the coordinated assault of subtle enemies. And science could make these enemies visible for the first time, give them names, assess their allegiances, and learn their behaviors. Most important, there were ways to fight back. The fight was obviously an unequal one—the enemy was savage in its attacks, legion in its number, elemental, and subhuman—yet to the nineteenth-century mind, especially the nineteenth-century colonial mind, such inequality was common in conflict. Military metaphors—"surveillance" and "campaigns"—became inseparable from the language of public health. (This governing metaphor—the war against the microbes—may explain the particular and pervasive disgust attached to the idea of biological weapons. To use them is not just to add to the arsenal a new way of killing soldiers and civilians; it is to side with the greatest of enemies against the whole human race.)

For much of the twentieth century, the war went well for humans. Cleanliness and sanitation removed the enemy's ambush points and supply lines, and then specific treatments started to appear. In less than a century, the enemy we had discovered in nature seemed to be in the final stages of retreat. By the 1970s, antibiotics were routinely curing infections that would otherwise have been fatal, while vaccines were protecting the young from broad swaths of disease. The sanatoria were emptying for want of new tubercular inmates; the iron lungs of the polio wards were gathering rust. And the greatest of all the killers, smallpox, was on the brink of surrender.

When President Nixon declared his "war on cancer" in 1972, it was in part because the war begun by Pasteur a century earlier seemed over. At a time when biology in general was rushing forward, basic scientific research into infectious disease increasingly came to be seen as a backwater. At the beginning of the 1970s, Macfarlane Burnet, a pioneering virologist who had won the Nobel Prize for his work on the immune system, was able to write, "The most likely forecast about the future of infectious disease is that it will be very dull."

By the mid-1990s, however, things looked rather different. A new and incurable (though eventually treatable) infection called AIDS had come out of nowhere to kill millions around the world. In the developed world, lesser but still frightening infections—Legionnaire's disease, Lyme disease, mad cow disease—had arrived. In some developing regions, especially in countries in Africa, progress made against infectious diseases in the 1960s and 1970s seemed to be stalled, or slipping away, even if you ignored the new ravages of AIDS. Malaria was making a comeback. And harrowing stories of new diseases were appearing in magazines and television shows, diseases such as Ebola and Marburg that seemed to dissolve their victims into blood and goo. Compared with the upper-respiratory-tract infections and waterborne diarrheal diseases that kill millions of children each year these new killers were exotic rarities, but they were spectacularly unsettling, and they were only a plane ticket away.

Meanwhile, throughout the world, diseases not previously associated with infections, such as stomach ulcers, turned out to be caused by bacteria that no one had even known were there. Tuberculosis cases, including some that were drug-resistant, were on the rise—especially in places, such as New York City, where systems for reporting and tracking cases had been dismantled. And hardy strains of bacteria in hospitals were developing resistance to more and more antibiotics. Humanity had fallen into what arms control experts call the "fallacy of the last move"—the belief that a good-enough technology will solve the problem once and for all.

But nature does not believe in last moves. The malleability and relentless adaptability of life-forms have not been trumped by the great powers of medicine, despite what we may have wanted to believe. Even if the terrorist attack enacted by Dark Winter never comes to pass, emergent diseases and resistant strains are already with us, a daunting challenge to science.

The reversals in medicine's great war on germs, along with Ken Alibek's shocking revelations about the Soviet pathogen industry, provided a galvanizing context for the reassessment of bioterrorism and biodefense. Then the fall of 2001 provided near simultaneous evidence both of bioterrorism's reality as a threat—in the anthrax attacks— and, in 9/11, of the sheer scale of the destruction some terrorists were willing to unleash. Suddenly, response and defense initiatives moved into high gear. By 2003, the NIH's biodefense budget had leapt to $1.5 billion from a couple of hundred million the year before; according to Anthony Fauci, head of the National Institute for Allergy and Infectious Disease, where most of that biodefense money goes, it was "the largest single increase of any discipline, in any institute, for any disease in the history of the NIH." The Department of Health and Human Services started stockpiling drugs and vaccines. Production of smallpox vaccine was restarted, and new forms of the vaccine went into development at a number of different companies.

All this activity might make you think that Tara O'Toole and her team at the Center for Biosecurity are sitting pretty, but they're not. Having drugs and vaccines isn't enough—you need the means to use them properly and quickly during an attack. You need techniques and incentives to encourage people to develop new drugs and vaccines, and to be able to do it in haste. You need a research establishment that's focused on the most basic mechanisms by which microbes cause disease, which can serve as the intellectual basis for new defense technologies. In short, you need a high-level strategic approach to the problem, not just a grab bag of tactical responses.

It's the systems needed to create and deliver remedies, not the specific remedies themselves, that O'Toole cares about—and in her view, because we don't have the strategy, we don't have the systems we need. True, we have some isolated systems, but that's just what they are—isolated; they aren't aligned with, and don't communicate with, each other. O'Toole likes to quote Kevin Kelly, the former executive editor of *Wired* magazine, saying that a system is just "something that talks to itself." So far, she warns, no biodefense program has figured out how to create the conversations, the feedback loops, that are needed to connect up government, public health workers, the military, the medics, and, crucially, basic research.

If the problem were just one of known weapons to which there were known responses, this deficit might not be so bad. But unfortunately, that's not the problem. To O'Toole and her colleagues, pathogens such as smallpox are twentieth-century bioweapons—yesterday's news. They are great for grabbing attention and they do need to be dealt with, but in the very near future they will no longer be the only game in town. It's the twenty-first-century bioweapons that she's worried about: the pathogens that have been reengineered with today's molecular biology, or tomorrow's.

At the first and most trivial level, almost all the CDC's category A threats could easily be altered to make them more deadly or harder to treat—most obviously by making the bacteria resistant to all common antibiotics, something the Soviet program had already accomplished for some of its biological weapons. There is a whole range of ways of doing this. One is to splice in the relevant genes; another is to do it the evolutionary way, growing the bugs in the presence of the antibiotic and picking out the ones that cope best in each generation. Much the same applies to a far broader range of more obscure pathogens. The Australia Group, an intergovernmental organization that monitors pathogens, keeps a list of those that its experts consider potential weapons. In 2003 it boasted 32 viruses, 15 types of bacteria, and four rickettsiae (peculiar parasites that live inside cells), as well as all the genetic sequences that make these various bugs pathogenic in the first place.

Would it be possible to simply develop vaccines and treatments specific to all these ever more obscure potential pathogens, as is being done for the CDC's Big Six? Perhaps, but O'Toole insists that this would be missing the point: "You can't just say, 'We'll do the class A list and the class B list and the class C list.' You're going to spend a lot of time and money on things that will never be used and you may miss the main threat." Roger Brent, a researcher at the Molecular Sciences Institute in Berkeley, likens such an approach to building fixed defenses; even a really impressive line of fortifications, such as the Maginot Line that France erected along its border with Germany in the 1930s, is of little use if the enemy can just go around it. And in biology, there's always a way around.

The main threat, in O'Toole's analysis, is a weapon based on one of these pathogens—or on some genes from one or more of these pathogens—but genetically altered in some entirely new way. The raw material with which to plan such modifications is freely available; of the 15 bacteria on the Australia Group list, the genomes for 11 had already been published by the middle of 2004, revealing their genetic makeup for all to see, and the genomes of the rest were soon to follow. Couple that information with increasingly sophisticated insights into how biology actually works and the options for the bad guys become more or less limitless.

Here are a few possibilities the early twenty-first-century bioweaponeer might play with: Pathogens that normally infect one species tweaked to infect others; a mouse virus, for instance, might be made to infect humans. Genes inserted into a virus or a bacteria, producing a protein that subverts the human immune system; our resistances to pox viruses, for example, could be overridden in this way. Changing how the Ebola virus expresses a single protein and thus producing a super-Ebola, even more devastating than the one we know.

The startling thing all these possibilities have in common is that they have already been demonstrated. Not as part of biological weapons programs—as far as we know—but as straightforward research with benign, even laudable, goals. Redirecting a virus, for instance, so that it infects a type of cell it normally wouldn't, or carries a gene it would never be found with in nature, is a routine occurrence in labs. It's a way of putting genes where the biologist wants them for purposes of well-intentioned research. Some proponents of "gene therapy"—in which a genetic disease would be remedied by transplanting healthy genes into the sickly cells that lack them—expect viruses to become vital medical tools, since putting genes into targeted cells is exactly what viruses do for their sort-of-living.

Nor is the super-Ebola merely speculative. Its creation was reported in 2001 by German scientists researching the disease. Their intention was to better understand normal Ebola by genetically altering the bug slightly; a minor change, they found, could make the bug more damaging to the cells it infects. What interested them was the likely converse: that certain Ebola genes must function not to make the virus more aggressive, but rather to rein in the nastiness. (This is actually a common trait of pathogens, because killing the host too quickly weakens the chances for the organism to get to its next host.) If researchers wishing to conquer Ebola could understand this genetic accelerator and brake, they might be able to keep the disease in check. Of course, a weapons designer would be more interested in flooring it: eliminate those more temperate genes and make a virus that is ever more destructive.

The super-Ebola, however, was not the most famous of accidental bioweapon developments. More chilling was the creation of the invincible mousepox. In 2000, Ronald Jackson, a researcher in Australia, was looking into the possibility of engineering the mousepox virus—a relative of smallpox and cowpox that infects mice—in order to create a kind of humane pest control. The idea was to trick female mice into a deep immunological dislike for their own eggs. A mousepox was engineered so that it carried the same proteins on its outer coatings as the eggs of mice do. If a mouse were infected with the virus, the antibodies it made in response would attack its own eggs, too, and thus make the animal sterile. The virus would be a sort of contraceptive vaccine passed from animal to animal.

The first version of Jackson's mousepox virus didn't succeed in turning the females' immune systems against their own eggs, so Jackson decided to beef it up. He and his colleagues added a gene for an extremely potent small protein that regulates the immune system. They expected this souped-up virus to instruct part of the immune system—the part that goes after viruses—to back off a bit, allowing the virus to infect more cells and thus provoke a greater flood of antibodies. It didn't work out that way. Instead of ultimately generating more antibodies to take aim at the eggs, this engineered mousepox virus just killed the mice outright. It also killed mice that had been bred to be immune to mousepox. It also killed mice that had been vaccinated against mousepox.

It is not immediately clear what the same techniques would do to the powers of smallpox, which is related to mousepox but not identical. But were such a reengineered virus possible, a smallpox with the power to trample over human immunities and ignore vaccinations, it would be an already horrendous disease rendered unstoppable.

The mousepox experiments, along with two other causes célèbres— the creation, from scratch, of a polio virus at the State University of New York at Stonybrook and a paper by University of Pennsylvania researchers describing how vaccinia, the smallpox vaccine discovered back in the 1790s by Edward Jenner, might be made to behave a lot more like its deadly relative, smallpox—have became the focus of intense debate. Should such experiments be carried out? And if so, should the results—including the standard descriptions of methods and materials needed to reproduce them—be published openly?

In the old days of national security—the days of nuclear weaponry—this sort of problem didn't often come up. The techniques needed for making bombs were not of much interest to non-bombmakers, and information on them was reasonably easy to control. Arms control officials expert in technologies and tools kept lists of socalled dual-use technologies—technologies that had innocent uses but were also vital to a bomb program. The lists were not exactly short, but they weren't endless, either. In the world of biology, though, whole tranches of the tools and techniques needed for peaceful civil research and for military research—whether defensive or offensive— are pretty much identical.

Tara O'Toole recalls going to a meeting on biodefense that the National Science Foundation organized in 2003 at the behest of the CIA and the Defense Intelligence Agency: "The DIA put up a list of all the bioscience advances that had caught its attention over the past year as being possible ways of enhancing bioweapons, or making new weapons, or disseminating weapons in new ways. And it was like everything neat and cool in biology. It was RNAi [a recently discovered mechanism by which cells control gene expression], it was stealth viruses [viruses that escape the immune system], it was the new DNA synthesis machines [such as those used to build a polio virus from scratch]—I mean, it was biology."

Biologists don't normally see things in this distressing light. The normal view is that they study life, and life is a good thing, and that makes their studies a good thing too. While most biologists don't work directly on curing disease, they feel the possibility that their work might have applications in medicine validates its moral worth (as well as justifying the huge public investment in basic biology). The idea that their work might be restricted because some people—some biologists, even—might use it to do harm is one they have tended to resist, often angrily. The dual-use issue came to a head at a National Academy of Sciences meeting, held at the organization's headquarters on the mall in Washington early in 2003. This summit produced a document outlining a class of "experiments of concern" that should be reviewed by biosafety committees before being conducted. This category includes experiments that make vaccines, diagnostics, or drugs against known pathogens ineffective. Experiments that would lead to pathogens that could spread more easily (both between humans and between species) or become more virulent were also singled out for review.

The NAS called for self-regulation on the part of scientists and the journals that publish their work, under the guidance of a new advisory board. The advantage of this system is that it preserves autonomy and freedom of inquiry, which scientists value beyond rubies. To critics such as George Poste, however, this system relies on scientists thinking through the implications of their work with a thoroughness they haven't exhibited in the past. As Poste said at the NAS summit, "I do not wish to see the coffins of my family, my children, and my grandchildren created as a consequence of the utter naiveté, arrogance, and hubris of people who cannot see that there is a problem." He gives short shrift to the widely held argument that people will be deterred from working in biodefense by more regulations and restrictions, along with the risk of going to prison if they break the rules. "The scientific community is sufficiently prostitutional that it will do research on anything that people are prepared to pay money for," Poste says. "We've seen it with cancer, we've seen it with AIDS, and now we see it as they run toward the trough of biodefense." (Poste has little patience for what he sees as academic preciousness; he uses the term "prostitutional" with uncommon frequency and some relish.)

More temperately, Gerald Epstein, a researcher at the Center for Strategic and International Studies in Washington, asked the scientists at the NAS how they would feel if the methods-and-materials portion of a paper on some "experiment of concern" were "found in a cave in Afghanistan with sections highlighted in yellow." This is not an idle point. In 2001, at an Al Qaeda camp in Afghanistan, the CIA found a number of papers that dealt with isolating, culturing, purifying, and identifying different strains of bacteria. They were papers from the 1950s and 1960s, but the methods they described were no less workable for that—and the equipment needed was far from the cutting edge. Agents also found popular books on biowarfare and letters to researchers in other countries. And nearby, they found a laboratory with equipment that could have been used to produce dangerous pathogens.

That discovery highlights the risks that may be lurking in the methods section of a scientific paper. More broadly, it underscores the reality that biological knowledge is not something that is basically good with a few dangers thrown in. Biological knowledge is knowledge of how the world around us and within us works; it can be used for good or ill, for making those worlds function better or cease to function at all.

Tara O'Toole remembers a meeting with Harvard faculty in 1999, when her Center for Civilian Biodefense Strategies was still young and unknown. "We went down and met with six researchers who had no idea what they were doing talking to us about bioterrorism. I gave them a 15-minute rap about why we have a center for biodefense, while they looked at me like I had three heads and had fallen off a cliff on one of them. Then somebody else in the room said, 'Could any of you build a biological weapon that could kill tens of thousands of people?' And that engaged them. They started thinking, and they said, 'Well, yeah, I could do that.'" Well-taught, well-read, experienced biologists with a background in infectious disease already know how to make weapons. It's conceivable that a tiny minority of these people are a greater threat than the people in that camp in Afghanistan. There's no reason to believe that all biologists are well adjusted, and some surely have irrational, even dangerous beliefs. Theodore Kaczynski, the Unabomber, was a mathematician. Next time, he could be a biologist. Indeed, given what is so far known about the 2001 anthrax attacks, it seems quite possible that such a homegrown bioterrorist has already attacked.

But the problem is not just that the potentials for both good and evil are woven into the very fabric of biology. It's that, at the moment, the two strands are not evenly matched. Evil has a significant advantage, at least at the conceptual level. Though there are far more people using biology for good than for ill, knowledge of how to mess things up is easier to grasp than knowledge of how to put things right, giving the small number of would-be weaponeers a serious head start. The would-be bomber of a suspension bridge, for example, needs some understanding of civil engineering—but not nearly as much as a suspension-bridge builder. It is far easier to make a bacterium antibiotic resistant than to develop a new antibiotic. And the know-how is piling up remorselessly. The long-term goal of understanding how biological systems work provides an awful lot of short-term information about ways to disable them.

One reading of Tolstoy's claim that, while all happy families resemble one another, all unhappy families are unhappy in different ways, is that he hit on an asymmetry that goes far beyond parents and their children. A happy family is a complicated organism, and complex systems run properly only when they're able to assess their position and adapt to their environments—when the components are able to talk to each other. Building up that ability is hard work; for a ruthless outsider, breaking it down is not. The hard work that turns the tiny clockwork of our molecular components into living, breathing, feeling people is mostly hidden from us. But thanks to molecular biology, the many, many means by which that happy hidden system can be shattered into a multiplicity of miseries are increasingly obvious to those who wish to look for them.

SIX

Being Bioprepared

KEN ALIBEK DIDN'T GO to medical school while serving in the Soviet army intending to learn about biological warfare. In fact he wanted to be a psychiatrist. But when he studied the facts about the battle of Stalingrad as part of an assignment in one of his courses, he started to think he was seeing the workings of an efficient and powerful biological weapon. He was fascinated by what he read in the relevant history books and journals, and particularly in one epidemiological study. According to this study, 10,000 cases of a bacterial infection called tularemia, which can spread from animals to humans, were reported in the Soviet Union in 1941. In 1942, as the German military was encroaching upon Stalingrad, the number of cases shot up to more than 100,000, with soldiers from both camps incapacitated by pneumonia-like symptoms; in 1943, the figure settled back down again.

These records, along with the fact that the disease had disproportionately affected German troops, suggested to Alibek that the Soviet Union, in a just and valiant effort to protect itself, had released the bacteria in the area of the enemy forces. Although mainstream historians maintained (as they still do today) that the outbreak occurred naturally, Alibek thought otherwise. The emphatic way in which he was advised to shut up when he expressed his surmise persuaded him further. He abandoned his plan to study psychiatry and took up infectious diseases.

A few years later, in 1976, Alibek was shipped off to a secret facility where he was to learn a great deal about how to grow bacteria. Officials informed him circumspectly that his interest in infectious diseases could be of service in the defense of his country. Though the purpose of this stealth training was rarely, if ever, explicitly discussed, it was clear to Alibek and his classmates what they were being schooled for. Some had qualms, some didn't.

Alibek had a few. He asked his father, a much-decorated veteran of the Great Patriotic War, to help get him transferred to another assignment, but to no avail. And in a short time he became reconciled to his new calling. Though initially a klutz in the lab, he developed a certain flair for making bacteria do his bidding. Within a few years he was in charge of a project that successfully turned tularemia, the disease he thought had been deployed at Stalingrad, into a state-of-the-art bioweapon.

In the early 1980s, Alibek, by then a senior lieutenant, was intimately involved in the production of vats of *Francisella tularensis* at a weapons plant in Omutninsk, between Moscow and the Urals—more intimately than he would have liked. One night, called into the lab because of an accident, he found himself standing in a puddle of concentrated bacterial culture the color of milky tea. Despite the protective gear he wore, and despite the fact that he sprayed his boots with peroxide repeatedly as he inched back toward safety, he ended up infected. It was only through massive doses of tetracycline, procured from a friend of his wife's, that he kept himself safe when the fever hit—and covered up the severity of the incident from his superiors.

By the late 1980s, Colonel Alibek had not only weaponized tularemia, he had also overseen the construction of a fearsome anthrax production facility and talked through plans for putting smallpox on ICBMs. He had worked on weapons capable of killing millions. And if the Soviet Union hadn't fallen, he says, he might be doing much the same thing today: "Maybe I'd be a major general, enjoying Soviet privileges to have cheap sausages and red caviar every other Wednesday." But it did fall. Alibek, a native of Kazakhstan, saw no need to stay in Moscow once the empire he had served no longer existed. He knew his knowledge of the full extent of the Soviet Union's bio-weapons program would be welcomed in the West—and he knew that a good scientist could earn a lot more in the United States than in Kazhakstan or the new Russia.

And so on November 25, 2003, rather than enjoying his sausages and caviar, Alibek is sitting in the bar of a hotel in Washington, D.C.'s corporate suburbs with the theme music to *Dawson's Creek* trilling away in the background, taking his pleasure from one of the strong-smelling cigarettes his new country will not allow him to smoke in the scientific conference going on next door. The subject of the confer-ence—which Alibek organized—is how to protect the United States from weapons such as those that he and his colleagues developed in the last decades of the cold war.

In part, Alibek's interest in biodefense is a propitiation for his ear-lier activities, born of concern that, as he wrote in *Biohazard,* his 1999 memoir, "even today, in Iraq or China, another father of three may be sitting down at a conference table to plot the murder of millions." And in part it's good commercial sense. While it's not unheard of for a post-Soviet immigrant to become a professor, or an entrepreneur, or a sought-after consultant, to manage all three, as Alibek has, is something of an achievement. Add on a line of "immune-system-enhancing" vitamins sold under his name and the achievement looks unique. But then, Alibek has a unique selling point: an understanding of how the largest bioweapons program in history did its stuff—knowledge that goes all the way from the puddles on the floor to the strategies at the top.

Today's conference is an Alibek showcase. Some of the presenta-tions are made by former colleagues at the Soviet Biopreparat, and a large number are made by coworkers at either George Mason Univer-sity, where he is now a professor, or Applied BioSystems, the biomed-ical company at which he is the chief scientific officer. (Indeed it's quite hard to make out where the university ends and the company begins. As Alibek puts it, "Everything is, let me say, in one basket.") Shuttling from one huddled conversation to another, the 53-year-old scientist displays his charms, reaching to touch his interlocutor's elbow with used-car-lot ease. Heavy-set and slightly stooped, with jet-black hair and a complexion prone to allergic reactions—the result, he says, of a lifetime of getting vaccinated on a very regular basis—he moves through the crowd from contact to contact. Fresh-faced young officers tout the abilities of the U.S. Army's medical research facilities. Less fresh-faced corporate executives troll for business opportunities in biodefense.

Although Alibek was a tremendously powerful officer in the Soviet Union—and although he has undeniable historical importance in the West as the man who did more than anyone else to reveal the scale of the Soviet bioweapons program—it remains to be seen if he has a lot to contribute to American biodefense. The work of defense is considerably different from that of aggression (the element of surprise exploited in the latter being the most obvious contrast), and the conditions in his new home—systems of finance and policy, scientific priorities and institutions—bear little resemblance to the world of the Biopreparat. So far, Alibek and his colleagues have performed a range of interesting experiments on various pathogens, and some of the approaches described at the conference—mostly centered on the immune system—have had success in small-scale animal studies. But whether that success can be translated into effective treatments for humans infected by terrorists is not clear.

Several obstacles define the nature of biodefense development in the United States. One is the role of the commercial sector: the West's pharmaceutical industry is a massive, mature set of interests, not likely to change except under great duress. The molecular biology revolution has provided a startling range of new approaches to infectious disease, new ways of making vaccines and targeting drugs, new insight into how pathogens attack and how bodies defend themselves. But it hasn't provided the economic incentive for pharmaceutical companies to turn those leads into real clinical progress. For decades infectious disease has been a low priority for drug companies because the most lucrative markets for drugs and vaccines targeted at infections are thought to be pretty well served by existing products. Untreatable infectious diseases are a relatively minor problem in the rich economies where drug companies amass their profits, so policy makers in those countries have done little to interfere with this rational allocation of resources.

In other, poorer parts of the world, infectious disease is still the biggest of deals. It kills millions of people every year, many of them children. This is not, for the most part, because the drug companies don't make the right drugs. It's because the people who need the drugs can't afford them and their governments can't or won't provide them. We don't need new technologies to prevent most of those deaths. We need new priorities. The death and sorrow caused by treatable infectious disease today vastly outweighs the amount of damage any plausible bioterror attack might bring with it.

The fact that global health could be greatly improved without much by way of innovation doesn't mean that innovations wouldn't make the prospects brighter. While saving lives in developing countries is, in absolute terms, quite cheap, in practical terms it seems to be too expensive, in that the countries which could afford to help save those lives choose to spend the money to other ends. One answer is to press for more money to be spent. Another is to develop technologies that make saving lives even cheaper and easier. The two are far from mutually exclusive. For example, it is worth making insecticide-treated netting for children to sleep under more widely available, but it is also worth working on a malaria vaccine. It is worth spreading the use of proven treament regimes for tuberculosis, but it is also worth developing better drugs that can treat the disease more quickly. It is worth investing in AIDS prevention programs, but it is also worth investing in an AIDS vaccine.

The technologies needed for better drugs and vaccines are very closely related to biodefense technologies—improvements in one will allow improvements in the other. Here, again, we see the possibility of dual use—but of a potentially benign sort. Well-planned biodefense research might help alleviate today's suffering, as well as providing the basis for protection against emerging diseases that are no less deadly for being natural, unexpected surprises, such as SARS or West Nile virus.

And though that humanitarian objective may seem a long way from biodefense, the two are linked by more than just technology. In much of the world, disease and poverty form a vicious circle, each causing the other. Disease saps the power of the poor to improve their lot; poverty in turn breeds more disease. Breaking that cycle with effective health interventions would be a huge step forward in human development, and the subsequent empowerment of the poor could do much to spread democratic governance. Making the world healthier will not in itself make it as safe a place as we might wish it—but it would make many countries more affluent and more stable.

The complacent laissez-faire attitude toward new treatments for infectious disease is in urgent need of change. And it's possible that, catalyzed by worries about biodefense, change may now be set in motion. Biodefense thinkers such as Tara O'Toole at the University of Pittsburgh argue that it's time to look for dramatic new innovations that would capitalize more quickly on the knowledge biological research is churning out. Which points to another major obstacle in the path of biodefense: the creeping pace of safe drug development. Biodefense requires speed. And yet the development of a new drug typically takes a decade. Every stage takes time—finding the disease-specific proteins, receptors, enzymes, or genes in the body that you want to target; finding a small molecule that will act against that target; tweaking the molecule so that it can survive within the body and get to its target; testing the product for safety in animals, then in humans, and then testing to see if it actually works, and if it works better than alternative treatments. At each stage things can go wrong, sending you back to the beginning. Even a crash program to develop a new vaccine that was merely a modification of an existing one would take more than a year, according to Thomas Monath, chief scientific officer of the British biotech company Acambis. (Designing the vaccine and conducting initial tests would take six months. Further tests and the development of properly controlled manufacturing procedures, undertaken simultaneously, would take another six months. Final safety testing and review by regulators would then take three more months.) Five quarters is incredibly fast in business-as-usual terms, and if an emerging disease such as SARS can be kept in some sort of check for a year or so, then it might be enough. But if the problem were a designer plague introduced into the population in many different locations at once, 15 months would be catastrophically slow.

So how do you make the "bug-to-drug" process more agile? How do you respond to a devastating outbreak, provide a new medicine for a new disease, in 10 months or 8 months or 5 months instead? Several approaches might help. Some of them are quite specific: the idea, for instance, that you could accelerate the painstaking process of testing a drug in animal and human trials by developing better animal models through genetic engineering. If mice were provided with human immune systems, for example, they would be much more accurate predictors of which medicines were really worth pursuing; some mice with partially "humanized" immune systems have already been developed. (Such developments themselves pose dual-use worries; better animal models make it easier to develop and test weapons, too.) Other research focuses on the actual creation of antibiotics and vaccines, which has often been a matter of repeated trial and error; new molecular techniques could bring rational design—and thus new speed—to the process. And then there are ideas about ameliorating the way intellectual property concerns constrain research. In all these areas, though the challenges are rife, scientists are already working to make twenty-first-century biodefense, if not brisk, at least conceivable.

The tools of biodefense are medicines. Here's a brief guide to how they work, tricked out in appropriately military metaphor. A bacterial infection can be seen as a war between two types of cells—the bacterial cells invading the body and the immune system cells defending it. The bacterial army has an awesome ability to resupply itself (it can reproduce quickly) and some powerful weapons (toxins) at its disposal. It may also have the advantage of surprise, going unrecognized during the attack's early stages and thus setting up strong-points over wide swaths of the battlefield before the fighting gets underway. The immune system, on the other hand, has highly specialized troops, a more sophisticated chain of command, and a considerably wider range of weaponry. Its biggest problem is getting enough ordinance to the battlefield quickly enough—and not letting loose too freely with its full firepower for fear of collateral damage.

The most obvious way of intervening on the body's side in such cases is the use of antibiotics, which do vastly more harm to the bacteria than to the immune system. What makes antibiotics possible is that bacterial cells and the cells of the human body have very different metabolisms; bacteria contain vital molecular machinery that is both essential to their lives and pretty much unlike anything used in human cells. Antibiotics are small molecules that gum up those bacteria-only machines but have fairly negligible effects on the body's own cells. By killing bacterial cells but not body cells, antibiotics function as outside intervention, tipping the balance of the battle.

Because the drawbacks of antibiotics—most notably the risk of resistance—have recently become more obvious, it's easy to forget how amazing this ability to wreak selective havoc on bacterial metabolisms really is. In military terms, it allows the intervening power to ignore all of war's normal requirements for well-thought-through strategies and tactics and just get on with the simple task of annihilating the enemy. It's like flying at 30,000 feet, dropping swarms of idealized smart bombs that really do kill only the bad guys. Kill enough of them and mopping-up operations can then be left to the immune system troops on the ground, who will go about the job with gusto, pausing only to gaze up at the antibiotics' high-tech contrails with moist-eyed gratitude.

But not all wars can be fought this way. For a start, viruses can't be targeted by antibiotics because they take over the workings of the body's own cells. They don't have an independent metabolism for the antibiotics to disrupt. Instead they take over the body's own machinery, which you really don't want to attack, to make ever more copies of themselves. The takeover process itself can be targeted; there are antiviral drugs that seek to sabotage the way viruses enter cells or the way body cells help in the production of new virus particles. But there are still very few such antivirals; those that exist are mostly specific to a single virus or family of viruses (unlike antibiotics, which can attack wide ranges of bacteria) and viruses are pretty quick to evolve resistance.

Bacteria—especially those that have taken up residence in hospitals—evolve resistance to drugs as well. So some researchers are attempting to develop more antibiotics, ideally drugs that work in new ways, thus rendering existing forms of resistance futile. The most radical of these approaches would be to use a completely different class of bacteria-specific weapon—a type of virus called a bacteriophage, or simply phage, that infects only bacteria. The term builds on the Greek word *phagos*, "to eat," and translates roughly as "bacteria eater." The unlikely sounding idea of fighting bacterial disease by giving the bacteria a disease of their own is undeniably neat, and there is a tradition of using phages as medicines in some parts of the former Soviet Union. But as yet no phage has pushed its way into the armamentarium of mainstream medicine.

Another way of intervening is to provide the immune system with some military intelligence and thus give it the ability to prepare for likely attacks. This is what a vaccine does. The immune system has all sorts of wonderful weapons for fighting invaders, including antibodies that recognize telltale molecules on the attacker's surface; cells that engulf the enemy whole; or, in the case of viruses, a system for killing the very cells under attack before more harm is caused. Vaccines brief the immune system on the traits of a potential attacker by presenting it with a simulacrum which, at a molecular level, looks like an enemy yet to be faced, thus encouraging the immune system to prepare a response. In this case, forewarned really is forearmed. After vaccination, the immune system will be able to mount a full response as soon as it spots the enemy it has been primed to recognize. But a vaccine can help only if you know what diseases to vaccinate against, and have developed a vaccine that can do the job. New diseases, natural or otherwise, will require whole new vaccines.

A messier way of intervening is to actually get in the trenches with the immune system. Antibodies can be made outside the body and then shipped in like Stinger missiles. Crude antibody preparations have been used this way for a long time—that's what the gamma globulin shots for mumps and hepatitis are. Now, genetically engineered and precisely targeted antibodies are becoming available, though they are as yet fearsomely expensive to produce. Cheaper and in some ways more powerful—though less precise—resources are cytokines, proteins that communicate with and regulate the immune system (the gene that souped up Ron Jackson's killer mousepox virus so alarmingly was a gene that produced a cytokine).

Many talks at Alibek's conference focused on cytokines—they're an enthusiasm of his. He argues that if you could interrupt and rewrite communications carried by cytokines, you could reinforce immune reactions to a wide range of infections with variations on a single technique. A one-size-fits-all solution would be a critical step forward, but even if cytokines can't be harnessed as a force for good, just stopping them from causing harm would be a huge help: many infections do damage by overstimulating the immune system and setting off "cytokine storms," the cause of various extreme immune reactions, such as those in toxic shock syndrome. Cytokine therapy is the equivalent of getting in among the immune system's outnumbered but heavily armed foot soldiers, grabbing a megaphone and broadcasting orders to them in a language that you barely know. You have to send the same orders to everyone—you have no way to make use of the customary chain of command that turns high-level strategy into local tactics. And your very presence overrides the troops' normal ability to respond to changing circumstances with a bit of initiative, and thus moderate the amount of damage they inflict on their environment. Moderating environmental damage hasn't ranked high on lists of war goals in the real world's history; when the environment in question is your own body, though, it starts to matter a lot. An immune system left to celebrate its victory in a body it has reduced to scorched earth is not a success.

These destroying-the-village-in-order-to-save-it risks make some people understandably wary. But if there's a threat from viruses and bacteria against which a population has not been vaccinated and for which there are no effective antibiotics and antivirals, shouted orders in a foreign tongue may be an option you have to explore. As yet it has shown few successes.

There are still more options further down the road. While the immune system gets all the limelight, the body's other cells have various methods of their own for fighting infections. These systems lack the precise targeting that the immune system's cells are capable of, but if they could be kicked into action effectively early on in the course of an infection, they might easily turn the tide. Consider this approach as a Second Amendment form of intervention—arming the untrained citizenry. Another possibility might be to use nonviolent conflict resolution, persuading the bacteria that launching a war isn't really in their interests anyway. Most of the bacteria in the human body are inoffensive, and many are positively benign. If incoming antagonist bacteria could be chemically persuaded to lay down their arms and join these happy "commensals"—from the Latin meaning "to dine at the same table"—the body's wars could be won before they began. That option might be ideal for staving off some infections, once more is known about how bacteria process information and choose their lifestyles.

These creative schemes are all well and good, but medical treatment that can stand up to twenty-first-century bioweapons will require not just new versions of antibiotics and vaccines, not just research into phages and cytokine therapies, but also faster ways of designing and producing the drugs as needed. Compare the speed of a trip to the emergency room for extracting a bullet lodged in your side with the wait for a cure for a mysterious disease; the latter can take, quite literally, a lifetime.

The development of vaccines is a highly current case in point. While vaccines have absorbed a great deal of the first rush of biodefense spending in the United States that followed Dark Winter and the September 11 attacks, most of that money has gone toward stocking up on tried and tested vaccines against known threats. Biodefense experts understand, however, that depending entirely on vaccines we already know about would be foolhardy, leaving us wide open to newly engineered biological agents. So leaders in biodefense are encouraging researchers to think about new vaccines and, most crucially, faster methods of designing them.

At the biotech company Acambis, Thomas Monath is working both beats: his company has benefited greatly from orders for old-fashioned smallpox vaccines from the United States and other nations around the world; meanwhile, it's also developing a novel means of producing vaccines—in double-quick time—against diseases for which there is currently no protection.

Monath looks like the sort of man you would want as your family physician. He's courteous, clear-eyed, and undramatically authoritative. In the 1970s and 1980s, he worked at the CDC, first studying yellow fever and Lassa fever in West Africa, then directing the Colorado-based division of the CDC, which is devoted to the study of diseases spread by insects and other creepy-crawlies. In the late 1980s, though, he transferred to the military biodefense center at Fort Detrick, Maryland, where he became chief of virology. At the CDC, he says, "I was always frustrated that we weren't making products and developing things. We had programs on developing vaccines, but we didn't really know what we were doing. The military was much more engaged in that, which attracted me."

This desire to actually make things soon drove Monath from the Army to industry—specifically, to a young biotech firm called Oravax in Cambridge, Massachusetts. Oravax had lots of good ideas for new vaccination techniques, but for most of the 1990s they failed to pan out. The company was close to bankruptcy in 1997, when Acambis, based in the other Cambridge, bought it.

By then Monath had already attempted to land a contract with the Pentagon to provide smallpox vaccine, by doing some research on new methods of producing the vaccine in the lab. (Many vaccines, especially for viruses, are made in oddly old-fashioned ways; the vaccine for smallpox was, until recently, harvested from the skin of infected calves; flu vaccines are grown in chicken eggs.) The Pentagon, though, had stayed with its old supplier. Then, in 1998, Monath was invited to the White House as one of a small group of virologists and biodefense experts asked to brief President Clinton. The briefing is sometimes said to have been a response to Clinton's reading of *The Cobra Event,* a thriller by Richard Preston about the release of a bio-engineered smallpox. Monath insists the meeting was part of a less precipitous policy-making process. He also says that, before advising the government to build up on vaccine stockpiles, he revealed his financial stake in the proposition; according to the *New York Times,* other participants think he didn't. Either way, he definitely spoke of the need to have more vaccines on hand in case of an attack, and when the administration subsequently decided to build up stockpiles, Acambis got the work. In 2000, the Department of Health and Human Services ordered 40 million doses of smallpox vaccine. After 9/11, the total number of doses ordered shot to more than 200 million. Monath says that eight other countries in Europe and Asia that he's not at liberty to name have also placed orders. As a result, Acambis is currently a business rarity: a profitable biotech company.

Monath's greatest enthusiasm, though, is for a fast-track vaccine-development program at Acambis called ChimeriVax. The often quoted but little read Canadian academic Marshall McLuhan used to teach that the medium is the message—that it is impossible to divorce what is being communicated from the way in which it is being communicated. ChimeriVax is an attempt to show that, at least for some diseases some of the time, this doesn't hold true.

A vaccine is a way of sending a message to the immune system, telling it to be on the lookout for something. Normally, the medium the vaccine uses looks, in some way or other, like the pathogen the message is warning against. This medium may be a component of the pathogen: a characteristic protein, say, as in the case of the hepatitis B vaccine—or a killed version of the pathogen, or a live strain of the pathogen that's been "attenuated" so that it can't cause disease, or a stand-in microbe that's closely enough related to the pathogen to provoke the desired immune reaction, as is the case with vaccinia and smallpox. There is one type of vaccine that breaks these rules; in DNA vaccines, a piece of DNA that describes a part of a pathogen is used as a vaccine. The idea is that this DNA will be translated into protein by the body's cells, and that protein will then prime the immune system. In effect, the DNA tells the body to make a more traditional sort of vaccine. This is potentially very promising, but as yet there are no DNA vaccines approved for use in humans.

For non-DNA vaccines, the message—the sounding of a particular alarm—and the medium—the stuff that's actually put into the human body—are obviously closely linked; vaccines, and the ways in which they are made, are currently specific to specific diseases. But genetic engineering holds out the possibility that a known medium, something safe and trusted and effective, might be altered so as to be able to carry different messages in different situations. If this were possible, developing new vaccines might become a lot quicker and easier.

Monath's choice of mediums for this foray takes him back to his work in Nigeria in the early 1970s, where he studied yellow fever. The vaccine against yellow fever is a strain of virus that has been successfully "attenuated"; by growing the virus for generation after generation in the laboratory, scientists have produced a strain that doesn't cause disease. The yellow fever vaccine has been in use for decades, and its side effects are mostly mild. (There have recently been a very small number of severe side effects, which Monath and others are studying with some urgency.) While the protection offered by some vaccines needs repeated doses to reach an effective level and can fade over time, the protection offered by just a single dose of yellow fever vaccine is sure and long-lasting.

ChimeriVax technology uses this effective and well-behaved vaccine as the medium for warnings sent to the immune system about other, related diseases. This is how it works: in the current yellow fever vaccine, the medium is the "attenuated" virus, and the message is a protein on the surface of the virus particles called the envelope protein. ChimeriVax changes the genes in the virus so that instead of presenting the immune system with yellow fever's envelope protein, it offers the envelope protein from some other, related virus. This should allow the chassis of the old vaccine to be used to create vaccines against any one of a family of viruses called "flaviviruses," which are put together like the one that causes yellow fever. The grouping includes dengue fever, a mosquito-borne disease that is second only to malaria in its toll, sickening millions of people and killing thousands of them every year. "It's a huge problem," Monath says of dengue fever, in the compassionate but enterprising manner of drug developers. And it was the first target for the ChimeriVax experiment. Acambis has since licensed the dengue vaccine, now in clinical trials, to the pharmaceutical company Aventis. "Unfortunately," Monath laments, "we gave that golden egg away. We were out of money."

There are a couple of shortcomings to this new-drillbits-for-an-old-drill approach, which has been tried before and will surely be tried again. One is that the immune system will often fail to make a clear distinction between the medium and the message, developing immunity to both. This makes reusing the medium difficult, if not impossible. As Monath explains it, "If you take the smallpox vaccine, vaccinia, and put in a foreign gene, most of the vaccinia virus is still there. So if you're immune to smallpox, you have an immune barrier to the vaccine." (In the context of bioterrorism scares, it's worth pointing out that this problem travels the other way, too: if a genetically engineered vaccinia-based vaccine was in use against some other disease, it might no longer be possible to use vaccinia against smallpox proper.)

Acambis's ChimeriVax products seem to sidestep this immune-barrier problem, which is one of the reasons Monath is so taken with them. The only thing the immune system seems to recognize in each of the vaccines is the pattern of envelope proteins that covers its surface, the one thing that changes completely from vaccine to vaccine. The medium really does seem to be invisible. But while sidestepping one of the problems with the technique, ChimeriVax walks into the other one—limited range of use. ChimeriVax vaccines can only be made using envelope proteins that behave very much like the yellow fever envelope protein. That pretty much restricts the technique's applicability to diseases closely related to yellow fever; it turns out that the proteins from other viruses can't be packed together on the surface of the viral particles in the right way. No magical medium has yet been discovered that could be used to send the immune system all the messages corresponding to all the world's infectious diseases.

Nevertheless, ChimeriVax does have some reach. There are two more ChimeriVax vaccines in clinical trials—one for Japanese encephalitis and another for West Nile fever, the most recent disease to emerge in the United States. If the West Nile vaccine works, it could be a big new product for Acambis, and Monath, an expert on insect-borne disease, thinks West Nile is not going away. Plus, there's a direct biodefense link to the technology: both dengue and Japanese encephalitis are on the Australia Group's list of biological agents to worry about. Monath lists tick-borne encephalitis, Kyasanur Forest disease, and Omsk hemorrhagic fever as further diseases that could be weaponized, and against which ChimeriVax vaccines might work.

Clearly, this approach of exploiting an existing vaccine to multiply your defenses is promising. Within less than a year of the first SARS outbreak in 2003, there were a number of labs and companies trying it. Acambis is using a form of the smallpox vaccine with added SARS proteins, and GenVec, in Gaithersburg, Maryland, is using a cleaned-up adenovirus, one of a family of viruses that cause sore throats and laryngitis. A similar adenovirus approach has already been used to make an experimental Ebola vaccine. As yet, the effectiveness of these vaccines is not known, but the speed of their conception has been impressive.

Some of the vital tools needed to speed up the bug-to-drug process and protect us from surprise biological terror attack will undoubtedly come from genomics, the revolutionary science that allows us to isolate, sequence, analyze, and synthesize all the genes in a given organism. At the Institute for Genomic Research, the nonprofit genome-sequencing outfit founded by Craig Venter, researchers collaborating with vaccine makers at the biotech company Chiron have ripped open bacterial genomes to discover scores of previously unstudied proteins that might form the basis of new vaccines. Others are applying similar techniques to finding new targets for antibiotics—targets, in this sense, being proteins that bacteria need and people don't. A striking example of what's possible in this vein is offered by Arrow Therapeutics in London.

Arrow is not a traditional pharmaceutical company. Big pharmaceutical companies have sprawling research campuses on the edges of towns they often all but own. Arrow's headquarters and research labs are to be found in the Borough, a bustling if slightly down-at-the-heels area just over the river from the City, London's financial center. The Borough still has a somewhat Dickensian air—it's where Mr. Micawber was jailed for debt. It's an unlikely setting, but Ken Powell, the virologist and pharmaceutical-industry veteran who started the company in 2000, reckoned he'd attract more of the people he needed to a site in central London than to some science-oriented business park. Filling a refurbished 1960s office building with chemistry labs and biology workbenches is an eccentric but pragmatic move that seems typical of the soft-spoken but clearly determined CEO. It was cheap, too.

Although Powell's background is in virology, Arrow's technology is largely geared to working with bacteria and finding completely new types of antiobiotics with which to kill them. Powell knows from experience that antibiotics are an area Big Pharma neglects—and he thinks that gives an opening to small, agile companies such as his.

As Powell explains, the industry's stance is, not surprisingly, purely economic. Unless it's very special, a new antibiotic is unlikely to be used as an initial treatment—doctors will prefer something tried and true, which may well be out of patent. The new drug will only be used on the cases where the old drugs fail, and though resistance means that such cases are increasing, they are not yet a vast market of the sort Big Pharma likes. Overall, the world pharmaceutical market is climbing toward $500 billion. Anti-infectives make up just 5 percent of that, with antibiotics accounting for less that $20 billion. Occasion-

ally, a giant pharmaceutical concern will still make a new antibiotic—
Cipro, much touted and stockpiled at the time of the anthrax attacks in
the fall of 2001, is one of the few introduced in the last two decades—
but they are all commercially attractive variations on well-established
themes, not radically new products. Cipro is particularly good at getting to bacteria lurking in the urinary tract, even at low doses, which
gives it a market, and the fact that it was still under patent meant the
drug brought its maker, Bayer, over $1 billion in 2002. But its patents
are running out, and generic forms of the drug will eat into those
sales. Moreover, though $1 billion a year is not chicken feed, there are
about 60 drugs in the United States that make as much or more. Lipi-
tor, a cholesterol-lowering medication, had sales of $10.3 billion in
2003. A small company such as Arrow, though, doesn't need a blockbuster to make its mark. And the increasing resistance of various
deadly diseases to the antibiotics we have today makes the idea of entirely new antibiotics ever more timely. As Howard Lipson, the computer security specialist, put it when talking about Internet hackers,
"Security degrades over time," not because safety mechanisms give
out but simply because "your adversaries get smarter." Exactly the
same sort of arms race goes on in biology. Arrow's main business target
is a bacterium, methicillin-resistant *Staphylococcus aureus* (MRSA),
that has evolved into a dangerous threat in hospitals as a direct result
of attempts to kill its ancestors off.

Arrow's strategy is to pinpoint the proteins in dangerous bacteria
that are essential to the bug's survival; logically, those are the only
proteins an antibiotic should bother to attack. The company also has a
few other criteria for its targets: the proteins must be vital not just in a
few sorts of bacteria but in almost every sort, no matter how distantly
related; they must be markedly dissimilar from human proteins, so attacking them doesn't mess up the host's metabolism; and they must be
"tractable" in terms of drug development, meaning they should look as
though they could be stopped in their tracks by a small molecule of
the sort that medical chemists know how to work with.

Arrow's way of tracking down these targets shows how today's
revolution in biology blurs the line between life and machinery—
between the organism and the computer. First, Arrow's technology
harnesses a peculiar genetic phenomenon that occurs in many natural

genomes and turns it into a precision instrument—a genomic scalpel. The phenomenon is a type of short DNA sequence that's commonly called a "jumping gene." These little sequences can insert themselves all over a genome with gay abandon, and as they do so they will frequently interrupt the DNA sequence of a less flighty gene—one that actually works for a living. When a jumping gene inserts itself into a working gene, that gene will no longer work. Arrow infests the genomes of bacteria with these jumping genes, allowing them to "jump" into as many positions as they can. Eventually, each gene in the genome will be disabled in at least one of the bacteria. By growing colonies of these mutants in the laboratory, Arrow can produce a sample of thousands of bacteria in one or other of which every gene will be disabled.

As these bacteria grow in the lab, therefore, some will survive and others won't. Those that succumb have clearly lost a gene they couldn't live without. And those are the genes Arrow would like to know about—they're the potential targets for new antibiotics.

But how do you figure out which mutants have made it and which perished—which genes are vital and which are not? For this sort of careful comparison of thousands of genetic sequences, you need some of the most powerful technology the biological revolution has spawned. Arrow's version of this technology is a set of extremely expensive machines sitting in a well-isolated room on the first floor of its London offices. At heart, these machines are simply very good ink-jet printers—but rather than ink, their tiny nozzles produce precisely metered sprays of customized DNA sequences. DNA synthesizers—tiny automated labs that can assemble any sequence you tap out on a keyboard—feed the printers with DNA components that are then sprayed in a pattern of tiny dots onto glass slides. The slides are called "microarrays," and the pattern of sequences sprayed on each is a strategically stylized representation of the entire bacterial genome. When the colony of mutants is matched against this design, the machine picks out the jumping genes, and thus reveals the important genes. If a gene has been interrupted by a jumping gene, then it wasn't really imperative to the bug's survival. The mutation of *that* particular gene, evidently, hasn't killed the bacteria. Where there are no jumping genes—in parts of the genome that actually show up as dark areas on the microarray—their absence indicates a gene that, if disabled, leads to the bacteria's demise.

In a single reading, with a single slide, Arrow can thus pinpoint a pathogen's critical, indispensable genes and the proteins they describe, precisely those proteins that a clever antibiotic might target. Variations of the technique can look for more subtle effects. For instance, it can be used to find the genes that make a pathogen pathogenic. This use of the technology might be very helpful to vaccine makers, since it offers a way to design "attenuated" strains of a bacteria for use as a vaccine without the trial and error that normally goes into such efforts. Acambis has already licensed Arrow's technology for exactly this sort of use.

But antibiotics are the main focus. As of the end of 2003, says CEO Powell, Arrow had come up with 17 proteins that appear to meet the criteria it set for potential targets. In the big-numbers world of genomics, where DNA sequences can be millions or billions of letters long and genes are turned out by the hundreds and thousands, 17 targets may not sound like very many. But, as Powell points out, the antibiotics in service today are aimed at just "half a dozen" targets. If only half of Arrow's possibilities bear fruit, the company will have roughly doubled the number of targets available. And if the company finds good ways of targeting just one of the 17 possibilities, it will have done more to increase the range of antibiotic therapies than the whole pharmaceutical industry has done in decades.

Possibly the most drastic way to accelerate the bug-to-drug journey is through the development of what's called "systems biology," or even "predictive biology." Systems biology is about producing computer models that, when fed the right data, can make precise predictions of what a biological system will do. Models like this could, in principle, provide drug researchers and other biological researchers with the sort of design tools that other types of engineers take for granted—ways of saying "if this, then that" without actually having to try it out. Engineers who design microcircuits have complex software to answer such questions; so do the engineers who design buildings and bridges. By comparison, people trying to design drugs that affect the metabolism do so more or less in the dark. Systems biology is meant to turn on the lights and open up the possibility of biological innovation that's as reliable as that associated with traditional forms of engineering.

Roger Brent is an engaging systems biologist with a big-picture vision of what he's up to and a keen sense of how it might aid biodefense. His great gift is a mastery of technique—of finding new sorts of experiments to do, new ways to interrogate the genomes of the creatures studied in molecular biology labs. While he was at Harvard in the 1980s, Brent became one of the editors of *Current Protocols in Molecular Biology*—a how-to manual for moving genes around that now takes up 10 volumes. It describes in gory and replicable detail pretty much every technique of any importance that molecular biologists have ever had cause to commit to print. Brent now runs the Molecular Sciences Institute (MSI) in Berkeley, an independent laboratory that he set up with Nobel Prize winner Sydney Brenner to stand "about five years in front of the cutting edge" of molecular biology. A slightly awkward man suffused with intellectual excitement, Brent talks with relish of the possibility of doing "world-historical science."

The MSI's first lunge for greatness is as the central node in a network of labs studying how to get yeast cells turned on. The idea is to produce models that can predict how different "mating types" communicate and decide whether it's time to grab a willing partner and recombine some genes. The path that carries these signals from the proteins that receive them as pheromones on the cell's surface to the proteins that eventually respond by turning genes on and off in the cell's nucleus is, both literally and figuratively, about as sexy as yeast biology gets.

A huge amount is already known about this communications pathway. But Brent wants to go beyond these basically qualitative accounts of how molecule A nudges molecule B, which shuttles off to molecule C and capture the system's workings with mathematical rigor. He wants to know how the system processes information—to follow the way the chemistry of a pheromone turns into a genetic command in the same way as a computer engineer might follow an instruction through a set of electronic circuits. He wants to know how the system talks to itself, and to listen in. Once he's built models that can make quantitative, testable predictions about the workings of this best-studied of pathways, he thinks, he will be able to follow many more, and more complex, systems. Within a few decades Brent expects nature's basic workings to be commonly thought of in terms of the information that they process—and routinely reproduced in software and silicon.

To start developing such models, MSI brings computer people and experimental biologists into close contact, with ideas constantly flowing back and forth between the "dry" side of the building—screens on desktops—and the "wet" side—experiments on benchtops. The models demand new data and extremely precise types of measurements; the biologists try to provide them.

It's not clear that such computer models will ever actually work. Evolution's systems may be just too byzantine for number crunching to crack their secrets. Some researchers believe that although biological chains of cause and effect can be understood in a qualitative way, reliable quantitative predictions of their future behavior will be impossible for decades or longer. But to Brent and his colleagues, there's no clear proof that what they're trying is impossible. Strategists sometimes say wars are only fought when there's genuine disagreement about the likely outcome; scientific breakthroughs tend to be made in fields beset by similar uncertainty.

Meanwhile, MSI is producing usable genetic tools in the present. One of these is a laboratory technique for counting molecules in tiny samples of cells. Brent calls it the "tadpole," and it could radically improve the detection of pathogens in laboratory samples, in terrorist casualties—or in raw hamburger, for that matter.

The tadpoles are designed to detect and measure ludicrously scarce amounts of almost any molecule. Picking up tiny traces of DNA—or its cousin, RNA—is a run-of-the-mill task these days thanks to a technology called polymerase chain reaction (PCR). PCR is a wonderfully sensitive technique that uses the molecular structure of DNA to detect specific sequences. Every DNA sequence has an "opposite" that it can stick to—the messages on the two strands of the canonical double helix are stuck together this way. PCR uses little pieces of DNA to pick up complementary sequences, and then uses proteins that copy DNA sequences to make lots of duplicates of what's been found. This duplication is PCR's "chain reaction," and it acts as an amplifier of the tiniest samples. But PCR can only work on DNA and RNA. Finding other large molecules, proteins in particular, is the job of specially created antibodies that recognize some aspect of their target protein's shape and stick to it. Making an antibody that recognizes a particular protein, however, is time-consuming; the traditional way is to first get a live animal to make the antibody by injecting the protein into it, then fish out the genes that describe that particular antibody, then transfer those genes into a cell line that makes antibodies well. Plus, antibodies can't amplify the signals they pick up; to recognize a protein with an antibody there has to be quite a lot of it around.

The tadpoles, developed by MSI's Rob Carlson, who's now at the University of Washington, Seattle, and Ian Burbulis, are a way of getting the best of both worlds. They are probes that recognize molecules the way antibodies do, but amplify the results the way PCR does. The trick lies in sticking a protein "head," which acts like a small antibody, on to a DNA "tail" that can be amplified by PCR. The details of sticking the head and tail together involve co-opting a weird self-mutilating property of a family of proteins discovered in the 1990s. Add this property to the protein heads and at a simple chemical command they'll dismember themselves, leaving a neat, DNA-friendly stump on which to stick the tail. Thus yesterday's discoveries become today's techniques.

Watching his colleagues come up with the tadpole idea, Brent immediately saw its potential for biodefense. Until the mid-1990s, he wouldn't have noticed: Brent, despite family connections with the CIA and a distinctly boyish interest in military hardware, had little interest in biological weapons. Then, while searching for new sources of funding, he met a remarkable man named Shaun Jones. Jones is a former Navy SEAL with a resumé that includes competitive harpsichord playing and neck and head reconstructive surgery, not to mention a college major in poetry. When Brent met him, Jones was running a project on "unconventional pathogen countermeasures" for DARPA. Jones was funding some very speculative research on ways that biological systems could be engineered to resist diseases, and he recruited Brent to the advisory panel for his program. Since then, Brent has taken a keen interest in the national security implications of biological research. When his young colleagues came up with the tadpoles, he knew they might have applications well beyond academic systems biology.

Before the tadpoles could be sent out into the world, though, there was an obstacle: intellectual property. There are lots of different ways of making small antibody-like proteins that could act as tadpole heads, and like many interesting bits of molecular biology, these technologies are buried in a thicket of potentially conflicting patents. Brent's response to this obstacle was simple: he decided to make useful tadpoles, give them away, and ignore any intellectual property violations. Says Brent: "Let's see anybody say, 'You, the Molecular Sciences Institute, are giving anti-smallpox detector reagents to the Centers for Disease Control, and we think that's bad.' Let's just see someone try that one."

In late 2003, MSI sent a multimillion-dollar grant application to the National Institutes of Health for funds to make the tadpole factory a reality. The idea was to streamline the making of detector heads and to develop some better-engineered processes for manufacturing whole tadpoles in large amounts. The goal is to make tadpoles capable of recognizing key proteins in a previously unanalyzed germ within a couple of weeks and get them shipped off to the people who need them: the U.S. Army lab at Fort Detrick, Maryland, which deals with biological defense; state health departments; the CDC in Atlanta; and any other labs dealing with an outbreak. Contrast this goal with what happened during the SARS events of 2003, where simple molecular tests weren't developed until six months later. According to Brent, the containment of SARS, even after it had shown up around the world, was "a great triumph of twentieth-century, even nineteenth-century, public health practices"—the use of quarantine and the tracking of contacts—"and a great failure of twenty-first-century technology."

Since Big Pharma's lack of interest in anti-infectives is clearly bad news for biodefense—no matter how many small companies such as Arrow and Acambis forge ahead—the U.S. government is attempting to provide incentives to encourage research that will fill the gap. This initiative, called BioShield, was announced by President Bush in his 2003 state of the union address and after some delays the relevant legislation made it through Congress in the summer of 2004. While the initiative offers the incentive of new product sales to the government, those lures can't really compete with incentives the market provides.

Nor did BioShield provide full indemnification for companies if a drug that couldn't go through normal clinical trials—such as a drug for smallpox, of which there are normally no cases—were to have rare side effects.

The difficulty of involving the pharmaceutical sector is one of the reasons why biodefense activists such as Tara O'Toole at the University of Pittsburgh see a need to bypass the industry at least partly by establishing a large publicly funded program of directed research. In a 2003 paper, O'Toole and her colleagues appealed for not only better development of specific vaccines and drugs but also a research enterprise that stretches all the way back to basic science, a biodefense capacity that encompasses research, development, and, crucially, research *into* development—the search for new ways of turning insights from biology into tools for defense.

Like O'Toole, Roger Brent has a vision for future biology that goes beyond new feats in the lab or on the computer. In fact, his disdain for intellectual property is a piece of it. (He says he has made about $250,000 from patents of his own, but that lawyers acting for him, for his employers, and for various companies have made "10 times that" arguing over them.) In the long run, Brent would like to see biology unshackled from the ties of intellectual property as a matter of practicality. He, Rob Carlson, and another MSI alumnus, Drew Endy, now at MIT, like to talk about "open-source" biology, drawing an analogy to the open-source software movement that has produced, most famously, the Linux operating system. In most commercial software, the source-code at the heart of programs is kept confidential. Making it available would allow people to copy and modify it gratis. In open-source software, appropriation that improves things is welcomed. Anyone is free to point out flaws in Linux source code, suggest fixes, or make improvements. The technology becomes the shared responsibility of the community that forms around it, manages it, and updates it.

Brent and his colleagues imagine a similar future for an ever more engineering-like biology. Predictive biology—using computer models to look at biological systems—will make it possible to design organisms that process biological information in new ways. That novelty could be managed on an open-source basis, with the tools for describing and building biological systems freely available to all, rather than hidden away in labs or locked down as intellectual property in companies too inept or undercapitalized to use them. Open-source biology would open up possibilities for small players currently not able to take advantage of biotech tools.

The idea of spreading biological know-how so widely may sound crazy—what would the bad guys do with it? But Brent and his colleagues don't see open source as a problem for biodefense so much as part of a long-term solution. The open-source approach, they argue, would make biological systems more robust and defensible. In the computer world, they note, proprietary software is typically as buggy as the market will allow, while open-source software is as low on bugs as the pride and ability of the community can make it. Ironically, proprietary software ends up more vulnerable to attack; the chinks in its armor may be hidden, at least a little, but they are there. Its complexity is brittle, not resilient. In open-source software, on the other hand, the potential failings are open to the world. This means they can be quickly noticed and fixed, and so systems tend to become robust and resilient in short order. As society becomes ever more reliant on engineered biological systems—first in agriculture, later in our bodies themselves—robust and resilient is the way to go. Brent and Carlson summed up their position in a letter to DARPA in 2001 that outlined their ideas about open-source biology: "We think it would be a shame if, in 2009, most of the wheat in this country was dependent on an operating system of the quality and stability of Windows 95."

Even beyond software and biology, the sweep of this idea—that the greatest number can achieve the best results—is relevant in many of the places where terrorism is a threat. Terrorism is often described in the military jargon as an "asymmetric threat," a threat that doesn't depend on having the same might as the country you're threatening. For generals who've come up through the ranks thinking the answer to enemy tanks is more or better tanks of your own—preferably both— this sort of asymmetry is a fearful thing. But looked at from the right standpoint, the asymmetry is a powerful resource for the defense. The extraordinary, ever-accelerating biological progress that makes new weapons possible also makes new therapies conceivable—and though the therapies, constrained to do no harm, are much harder to develop than crude weapons might be, there are vastly more people working on them, with far greater resources.

Terrorists may be powerfully if unconventionally armed, hard to spot, innovative, and animated by unpitying fanaticism. But they are also few in number and poorly resourced. The civilization terrorism would seek to attack has far greater resources, and the number of people willing to use those resources for good is huge. The challenge is to find ways for those decent intentions and distributed resources to be brought to bear on the problems—to make the complexity and scope of an open, technologically advanced civilization a source of strength, not weakness.

An inspiring example of this sort of asymmetry is flight 93, the hijacked aircraft that demonstrated that the most valuable defensive technology on September 11 was the humble cell phone. Like tools honed in an open-source environment, those phones were dependable and accessible to the populace; they connected the passengers to the world, and allowed them to decide to act and to use the advantage of their greater numbers decisively. In the context of terror, arming a larger number, whether with biological insight, security information, easily usable resources, or the power of authority, makes all sorts of sense.

Biological engineering, as Brent sees it, should be a resource available to humanity at large, something with which we can protect ourselves against enemies both human and natural. Abuses, he believes, should be thwarted not by sequestering the technology but by sharing it openly with a "self-confident, self-aware community." That community, larger and better equipped than any terrorist group could be, will find efficient ways to protect society that today's experts can only dream of. What's more, Brent believes that if biological engineering is taught and practiced in a spirit of openness and responsibility, it could be largely self-policing, and less open to abuse by sociopathic drop outs, or more organized forces, than a system where knowledge is hidden away.

Whether or not they're developed on an open-source basis, new biomedical technologies are undoubtedly going to offer magnificent possibilities for protection and more general progress. What is lacking is a system for bringing the possibilities together, for directing basic

research in such a way that more possibilities are constantly generated, and for getting ideas out of the lab and into the clinic. A system
such as that would require significant basic research spending. At
NIH, $1.5 billion is a lot by some standards. But in a total defense
R&D budget of some $60 billion it's fairly meager. As well as needing
more money, though, an ambitious biodefense research strategy would
require two things in shorter supply—a commitment to making practical use of the wonders of well-funded research and inspired leadership that attracts first-rate thinkers.

The germs, the techniques for enhancing them, and the terrorists
with a bent for mass destruction all exist today, though not necessarily in the same place. So does the capacity for launching research projects that could lead to defenses against them. But the research system
able to leverage a vast store of biological talent into a "self-confident,
self-aware community" with the capacity to take biodefense seriously
doesn't. Creating that system and that capacity, in some form or other,
is what O'Toole, Brent, and others are trying to achieve.

Can they do it in time? Brent, when asked this question, looks
grim and then quotes the techie creed of Gene Krantz, Apollo 13's mission controller: "Failure is not an option." Of course, Brent knows it is
an option—just as he knows that MSI may not, in fact, be ready to
model all the horrendous complications of yeast's signaling pathways.
But he will not allow himself to say so, or believe it.

Tara O'Toole, perhaps, is more realistic. She pauses for an unchar-
acteristically long time, then says softly: "We're trying. It's really
hard. I don't want to place bets."

SEVEN

Living Clues

O N MAY 3, 1992, a young man riding a dirt bike in open country northwest of Phoenix, Arizona, discovered a naked female
body face down in the dirt near a cluster of trees. The dead woman,
later identified as Denise Johnson, had been tied up, strangled, and
dumped. A witness told police that he had seen a white pickup truck
heading down Jackrabbit Trail Drive heading away from the crime
scene at around one-thirty in the morning.

The most useful piece of evidence found at the scene was a pager
that was registered to a man called Earl Bogan but mostly used by his
son, Mark—who owned a white pickup. When questioned Mark Bogan
said that he'd given Johnson a lift in his pickup that evening, that
they'd had sex, and that she had then tried to steal his wallet. He said he'd
retrieved the wallet from her with a struggle, in which she'd scratched
him, but that she must have gotten away with the pager. Bogan said he
hadn't been near the site where Johnson's body was found.

Charlie Norton, the detective interviewing Bogan, didn't believe
him; but he didn't have clear evidence that contradicted him either.
What's more, Bogan's story meant that physical evidence that he'd had
sex with Johnson or been scratched by her—the sort of evidence that
then-new-fangled DNA fingerprinting might have established—could

be explained away. So Norton went back to the site where the body was found to look for signs that would show Bogan had been there. The takings were meager; all the detective found was a branch that had broken off one of the trees. The next day, though, he saw that the inventory taken of Bogan's truck included seed pods from a paloverde tree—the sort of tree that had been damaged at the crime scene.

The pods themselves weren't all that incriminating. There must be thousands of paloverdes—Arizona's green-barked wispy-branched state tree—around Phoenix. But if the seed pods could be shown to come from the site where the body was dumped, Bogan's claim never to have gone near the place would fall apart. So while a "genetic fingerprint"—the common term for a pattern of DNA that differs from individual to individual—drawn from Bogan's hair or skin might not help the detective's case, the genetic fingerprint of the paloverde tree might.

With that in mind, Norton contacted a number of biologists and geneticists to ask whether there was any way of tying a particular seed pod to a particular tree. Tim Helentjaris, a professor of molecular genetics from the University of Arizona, Tucson, said that such matching was indeed possible. Helentjaris agreed to compare the seed pods in the pickup truck to seed pods from three dozen paloverdes, carefully gathered by the sheriff's office from around Maricopa County, and to the seed pods from the tree at the site where the victim was dumped.

What the professor found was that paloverde trees—which no one had studied in this way before—were quite genetically diverse: all those in the sample had distinctly different fingerprints. All those, that is, save for the seed pods belonging to the damaged tree at the crime scene and the ones found in Bogan's pickup, which Helentjaris found to be an identical match.

Bogan was arrested and charged. The court, realizing that the genetic fingerprinting of the state tree was going to be a somewhat unusual form of evidence, held pretrial hearings into Helentjaris's techniques and conclusions. One of the expert witnesses who testified was Paul Keim, a friend of Helentjaris's who studies molecular genetics at Northern Arizona University in Flagstaff. Keim pointed out that the technique used to compare the sequences of the various seed pods was not as accurate as other techniques then available. He felt Helentjaris's estimate of a one-in-a-million chance that the match was a coincidence was rather too high; Keim put it at 1 in 136,000. But that was good enough. Bogan was convicted and sentenced to life in prison.

For Helentjaris, the Johnson case was an odd opportunity to use his skills as a plant geneticist for something far out of the ordinary; today he works at Pioneer Hi-Bred, DuPont's Iowa-based genetically modified seeds company, where he is immersed in the corn genome. But Keim, the slightly skeptical expert witness, has found himself drawn into more and bigger cases. The expertise in genetic fingerprinting he offered to the Arizona court was honed on plants but can be applied more widely—and applying it to bacteria has made Keim a leader in what's known as "microbial forensics." In this new field, the DNA to be studied doesn't come from recalcitrant suspects and tragic victims. It comes from the murder weapons themselves. If you want to fingerprint a sample of anthrax and find out where it came from, Keim is the man to see.

To a molecular biologist, a genetic engineer, or for that matter a bioweapons maker, a DNA sequence normally serves as a beginning. The information it carries determines the capabilities of the organism, its ability to do things or have things done to it. From another point of view, though, a DNA sequence is a conclusion, a summing up. The information it carries is a record of what has gone before. Each gene, on each strand of DNA, is a copy of an earlier gene, which is in turn a copy of a yet earlier gene, in a chain that leads all the way back to the dawn of life.

The most recent links in any one of these chains will be all but unique, which is why a genome can be used to tell the difference between one person—or one paloverde—and another. The chains now separate, though, were once tangled together. So while comparisons between genomes allow you to detect how organisms differ, they also allow you to see how organisms are the same. Comparing similar genomes opens a window back onto the past: roughly speaking, the more similar two genomes, the more recently they shared a common ancestor. This approach can be turned to the study of relationships within a species or between species; it can reveal the history of a dangerous mutation in a human family or the history of life on Earth. The stories of the past recorded in DNA are of great practical importance to the study of emerging disease, whether the disease in question is emerging from the environment or from some fiendish laboratory. If you want to know what's causing an infectious illness and where it came from, the first place to look is in the genes.

Take SARS. When the 2003 outbreak of the disease, which had originated in China's Guangdong Province in late 2002, made the headlines, the hunt was immediately on for the microbe that caused it. Chinese scientists believed the bug was a bacterium, while researchers elsewhere suspected the culprit was a virus—specifically, a paramyxovirus—that some SARS patients were found to be carrying. A lab in Hong Kong, however, isolated microbes whose genetic sequences were typical of a family of viruses—coronaviruses—that get their name from the crown-like halo that seems to surround them when seen with an electron microscope. Other labs agreed that the offending viruses carried gene sequences suggesting a coronavirus, though one unlike any other previously reported. The coronavirus's culpability was confirmed when monkeys injected with it developed a SARS-like disease.

Once researchers had the virus's genome, they were in a position to find its relatives. That several early infections had hit workers in Guangdong restaurants specializing in the meat of wild animals was a strong clue, and soon a number of animals in the area's meat markets were being tested for the virus. It turned out that some of them, especially civets—smallish, predatory mammals whose flesh is said to bring stamina and health—carried a range of coronaviruses, including some nearly identical to the one causing SARS. It's not yet certain that civets are the primary reservoir of the infection—the virus might be something that civets catch from rodents they prey on, for example—but it does look as though meat markets were the places that ushered the virus into the human world.

There's no evidence that the SARS virus came into the human world by design rather than by accident. But in some ways that is unimportant. David Heymann, the WHO official who coordinated the global response to the disease, is adamant that it makes no difference at all if such an outbreak is a natural emergence or a deliberate attack. The public health response will follow the same principles in both cases: find and analyze the pathogen; treat and, if necessary, isolate

the infected; define sensible travel restrictions and hygiene precau-
tions that will stop the spread; and so on. Indeed, Heymann points
out that although there's no evidence of malice, it's still impossible to
say with absolute certainty that the SARS outbreak was natural.

However, there is one crucial difference between a natural emergence and a deliberate attack. In a natural emergence there is no need
to bring a perpetrator to account. True, in 2003, a large number of
civets paid with their lives for the role their species played in the
SARS outbreak, but the cull was a matter of pragmatism, not justice.
In a deliberate attack, justice demands that the assailant is found.
And on a more pragmatic level, deterrence demands the same thing.
Without a way to attribute blame, you can't frighten would-be attackers into restraint. It is sometimes argued that terrorists are less
vulnerable to deterrence than nation states, and this may be true. But
even if it's not a perfect tool, deterrence is still a useful one; deterring
some enemies is better than deterring none.

Bioweapons are the opposite of the cold war's nuclear threat. If an
ICBM were launched against the United States the location from
which it had been launched would be known within seconds, thanks
to surveillance satellites permanently scanning the face of the earth
for the infrared signatures of rocket exhausts. Retaliation would most
likely be swift and awesome. With biological weapons there is no such
clarity. Finding out where those sorts of weapons may have come from
requires serious scientific detective work. And before you lay claim to
deterrence, you have to convince people that you can do that science.

Microbial forensics is a catchall title for this sleuthing: think Sher-
lock Holmes with a magnifying glass powerful enough to search for
bacterial clues. Randy Murch, a retired FBI investigator who took the
lead in setting up the bureau's Hazardous Materials Response Unit,
which deals with the scenes of biological and chemical attacks, sees
this work as a crucial missing piece of biodefense. He's one of a
number of researchers involved in a grassroots campaign to develop
the expertise and the processes necessary to find out where a dangerous microbe originated. The techniques Murch envisions range from
the mundane (what packaging did the agent come in?), to the technical (are there any telltale traces of the medium in which the
bioweapon was grown?), to the speculatively high-tech (can the ratios

of certain chemical isotopes found in the agent be matched to those in a particular part of the world?). But the techniques he talks of most are those that use the unique characteristic of biological weaponry itself—the fact that the weapons are shot through with information written in their genes, right there for the investigator to study.

At the moment, the ability to bring the perpetrators of bioattacks to justice doesn't look particularly robust. Consider the two most significant biocrimes in recent American history: the Rajneeshee salmonella attack in Oregon of 1984 and the anthrax attacks of October 2001. In 1981 the Bhagwan Shree Rajneesh, a mad, drunken guru with a truly impressive Rolls-Royce collection, moved the headquarters of his cult from India to the Big Muddy ranch in The Dalles, Oregon, taking thousands of followers with him. Relations with the locals were far from cordial, and a couple of years later some of the Rajneeshees tried to rig a local election for judgeships by means of a food-poisoning epidemic. Sick people, they reckoned, wouldn't vote, and a low turnout would help the candidates whom the Rajneeshees supported. By contaminating salad bars and the like at 10 restaurants in town with a "salmonella salsa," they made hundreds of people very sick. Seven hundred fifty-one cases were linked to the outbreak, and 45 people were hospitalized.

The Rajneeshees didn't win the judgeships they were after, but they weren't prosecuted for the attack either. Indeed, though there was some immediate suspicion, for more than a year it wasn't clear that there had been an attack at all. Food-poisoning outbreaks are common natural occurrences and normally not the work of malicious religious loonies. Only after the cult fell apart and some of its members testified against its elders did prosecutions begin. If no one had dropped the dime, the attack might never have been recognized for what it was, and it's unlikely there would have been any arrests.

In the case of the anthrax attacks, there was no need for an informer—the crime was announced as it was committed. The same is likely to hold true for many other forms of bioterrorism: most terrorists are trying to send a message, one way or another. But though the warnings that came with the anthrax letters made it clear that attacks were occurring—and did a lot to diminish the attacks' lethality—they didn't expose the villain. At the time of this writing the precise source of the anthrax and the identity of the perpetrator or perpetrators are unknown, although a massive investigation continues.

A great deal, however, is known about the anthrax itself, thanks in part to work done by Paul Keim of the paloverde tree case. In 1995, Keim took a sabbatical and worked at Los Alamos National Laboratory with a friend from graduate school named Paul Jackson. Jackson was involved in researching anthrax and enlisted Keim's help in finding ways to fingerprint the bacteria. The aim was to show how different anthrax strains around the world were related and to build a system that could quickly and easily compare new samples—called "isolates"—to existing strains. Since then, anthrax has occupied more and more of Keim's time. (He does still take an interest in other subjects—such as the biodiversity of magnolia trees in Costa Rica, where he spent part of 2003 and caught dengue fever for his pains.)

One of the fascinating aspects of anthrax is its peculiar history. Until fairly recent millennia, Keim explains, the anthrax bacillus was an innocent soil microbe that did little harm to anything. Then, sometime in the past few thousand years—or possibly the past few ten thousand—it mutated into a pathogen. And since making this crucial lifestyle choice, anthrax has evolved very little, mainly because it spends most of its time as dormant spores in the soil, patiently waiting for a vulnerable vertebrate to come by. This highly intermittent approach to making a living—or a killing—is why it's hard to say when anthrax first became a pathogen: the normal methods for dating evolutionary change by looking at mutations won't work on a microbe that can spend decades or centuries doing nothing at all. The strains of anthrax that kill elephants in South Africa's Kruger National Park and elk in Canada's northern territories are, Keim says, "remarkably similar" in their DNA. If you look at all the known strains you'll find that the most dissimilar will differ in the sequence of a typical gene by just 1 percent, if that; the bacteria's genome is more homogenous, says Keim, than that of any other human pathogen. Consequently, in many cases it is almost impossible for researchers to tell anthrax strains apart. Were we humans to suffer the equivalent uniformity, we'd exhibit no racial differences the world over, and would be hard put to differentiate between a spouse and a stranger.

The fingerprinting tools Keim has developed have changed all that. The long strings of DNA that make up the anthrax genome are punctuated by easily identified DNA sequences called "markers." Through careful lab work, Keim was able to find places in the anthrax genome where the presence or absence of a certain sort of mutation changed the length of string between two fixed markers. The best use of this technique depended on mutations called variable number tandem repeats. A VNTR is a place in the genome where a short and biologically pretty insignificant sequence of DNA reappears a few times in a row (hence "tandem repeats"). When the enzyme responsible for copying DNA into the next microbial generation reaches these sequences, its chances of getting bored and confused increase, like a copy typist asked to work on repetitive material. This confusion will sometimes result in a copy of the DNA that contains one more, or one less, of the repeats than were in the original (hence "variable number"). Even in genomes such as anthrax that otherwise show very little variation, there can be quite a lot of these little VNTRs, and once you know what you're looking for, it's fairly simple to find them. When he was at Los Alamos, Keim was able to track down the VNTRs he needed just by pasting sequences from the anthrax genome into Word documents on his PC and using the search function. The anthrax genome, Keim thinks, probably contains about a thousand of the VNTRs; he and his colleagues have characterized a set of 20 that vary enough from bacterium to bacterium that they can be used both for assigning a new anthrax sample to an already known strain and for exploring the relationships between different strains.

Keim's VNTR fingerprinting technique has revealed fascinating historical details of the way anthrax has spread around the world. Martin Hugh-Jones, an acerbic British-born epidemiologist now at Louisiana State University in Baton Rouge who, among many other things, tracks down anthrax outbreaks and sends new samples to Keim, points out that the strain of anthrax that is widely used in laboratories, called the Ames strain, has a particularly intriguing past. Most anthrax in the Americas can be traced back to forms of the disease found in Europe, which can in turn be traced back down the silk road to central Asia. But the Ames strain, originally isolated from a dead cow in Texas before being whisked off to a lab in Ames, Iowa, in 1981, is more closely related to those from eastern China, which are in turn related to strains from southern Africa. Hugh-Jones suspects this lineage was brought to China from Natal in the fifteenth century, when vast Chinese fleets roamed far and wide across the Indian Ocean, before moving from China to the Americas courtesy of the wool trade.

Another set of samples that Hugh-Jones provided to the Los Alamos team has a more disturbing recent history: it consists of fragments of DNA from the preserved tissue of people who died in a 1979 outbreak of anthrax in Sverdlovsk, a Soviet city in the eastern foothills of the Urals now known as Yekaterinburg. The Soviet authorities claimed that the outbreak came from diseased meat. In fact it came from a 30-year-old biological weapons plant that had been among the first in the Soviet Union. The plant was based on designs seized from the Japanese when Soviet forces invaded Manchuria in 1945 and found the remains of Unit 731, the empire's biowarfare operation. Unit 731's grisly experience and its industrial approach to bioweapons were among the foundations for the postwar program that Stalin created.

The Sverdlovsk anthrax outbreak was a biological Chernobyl, a very Soviet mixture of dangerous technology, poor working practices, cover-up, and callousness. As Ken Alibek tells the story, a technician on the day shift at the Sverdlovsk bioweapons production facility removed a clogged air filter from an exhaust pipe in the building where liquid anthrax cultures were dried to form a powder. When the next shift came on, the filter wasn't replaced, and anthrax spores spread downwind across the city in a thin plume that stretched about 60 kilometers. Using bus and tram timetables to reconstruct the movements of people who got infected, Hugh-Jones was able to ascertain that the leak continued for an hour or so, pumping out maybe a pound of spores.

As people started to sicken and die (the night-shift workers in the factory across the road from the facility went first), Western intelligence sources and Russian émigrés realized what was up, denouncing the outbreak's source as a bioweapons plant banned by international treaties. The Soviet authorities came up with, and stuck to, the contaminated-meat story despite clear evidence that most of the cases were due to spores that had been inhaled, rather than eaten. In the

early 1990s, various Russian sources—including, tangentially, Boris
Yeltsin, who had been governor of the Sverdlovsk region at the time—
admitted the contaminated-meat story was spurious. What was more,
although the Soviet authorities had tried to remove all evidence per-
taining to the accident from the city's hospitals, some pathology specimens taken from the dead had been saved by local doctors and passed
on to experts in the United States. At Los Alamos, the samples offered
an opportunity to find out just what kind of anthrax the Soviet
weaponeers had been working on. According to Alibek, the anthrax on
the plume was a particularly vicious strain called anthrax 836, isolated from a sewer rat that had been infected after a previous leak at another plant in Kirov. Anthrax 836 would have been the strain used in
missiles directed against Europe had the decision to field such biological weapons been made.

Getting microbial DNA out of the preserved tissues from the victims of Sverdlovsk was not easy, and the resulting fragments were
short. But they were long enough for Keim and his colleagues to
find, to their surprise, that there were a number of different genomes
involved—at some of the VNTR sites they were getting repeats of four
different lengths. Either anthrax 836 was not the only strain released
or it had been modified in various ways. Did this mean that the anthrax being produced at the plant had been modified to resist various
antibiotics or to avoid immune responses? (Such modifications would
change the genome, and thus the fingerprints.) At the moment, no one
knows—or at least no one is saying.

Another sample that came Keim's way for forensic identification
was from Aum Shinrikyo, the Japanese cult that released sarin gas in
the Tokyo subways in 1995. Two years before the chemical attack, cult
members sprayed an anthrax-containing liquid into the air from the
roof of their eight-story headquarters in Kameido, near Tokyo. At the
time, no one in the neighborhood noticed anything other than some
nasty smells, and no one got sick; the incident only came to be known
for what it was when the cult was investigated after the sarin attacks.
But the nasty smells at the time had led to a visit from public health officials, who collected a fluid oozing down the side of the building. The
samples were stored away, and in time they got to Keim's lab in
Flagstaff. VNTR analysis quickly produced a fingerprint that could be

checked against the database of strains already studied, and the sample was identified as belonging to the so-called Sterne strain, which does not produce disease and is safe for use as a vaccine for animals.

Aum's "minister of health and welfare," Seichi Endo, had previously worked at the Virus Research Center at Kyoto University and was a graduate-level molecular biologist. He would presumably have known that the Sterne strain—which seems likely to have been the one used in nine attempted attacks carried out by the cult in the early 1990s—was basically harmless, and why he went on using it is something of a mystery. Keim lists three possibilities: "One is that they screwed up, that they didn't know any better; but I don't think that's true. The second possibility is that they were doing a trial run; maybe they were just testing their equipment out and were going to use a real strain, a hot strain, later on. The most likely possibility is that they had the vaccine strain, and Asahara, the cult leader, said, 'Carry out the attack now,' and everyone was so afraid of him they were unwilling to say no, despite the fact they knew it didn't work."

Keim and his colleagues put the finishing touches to their Aum work in the summer of 2001 and sent it off to the *Journal of Clinical Microbiology*. By strange coincidence, it was accepted for publication on October 5—the day the media first reported a case of anthrax in a newspaper office in Boca Raton, Florida. Days later Keim was asked by the FBI to start work on a much more pressing piece of microbial forensics. Keim's VNTR fingerprinting quickly showed that the spores discovered in Florida and during the next weeks in Washington were examples of the much-traveled Ames strain, which, in the 1980s, had been used in American biodefense research and had been shared with other laboratories around the world. This was helpful, but only to a point; knowing the strain of a sample, especially one so common, is sort of like knowing that a person's name is Smith. What you really want is better resolution, more detail, such as the ability to differentiate between Archibald Smith and Archibald's brother and each of their distant cousins.

It just so happened, however, that such detail was on the way. Just as the deadly envelopes were reaching Senate mail rooms, researchers at the Institute for Genomic Research (TIGR) were two years into sequencing an entire sample of the Ames strain. In time, and with a boost of funding inspired by the scares in Washington, researchers would have the first complete anthrax genome sequence to refer to, and this data would allow Keim and his colleagues to take their forensic analysis to an unparalleled new level of detail.

The room at the heart of the Joint Technology Center, a one-story building in the Maryland suburbs of Washington, D.C., is a strange sort of holy of holies—larger than a laboratory, smaller than a factory floor, but somehow reminiscent of both. At the same time, it's hardly a workplace at all; it's dimly lit, sparsely staffed, and quiet enough to make you feel you ought to whisper. There are about a hundred benches in it, arranged in 10 aisles, and on each bench there sits a beige box about the size of a large dog kennel. On the front of the kennels there are little glass windows opening onto the workings inside, which look like tiny labs for pixies, with sinks and arrays of little wells for liquid samples. Like the space outside, the miniature labs are still and silent.

These are top-of-the-line genome-sequencing machines, worth about a third of a million dollars each, and to look at them you wouldn't even know they were working. Indeed, on the day before Thanksgiving, it seems quite obvious that they aren't. The other rooms in the building, the rooms where the DNA is prepared for sequencing, emit a low-level bustle, but in the main room everything is so calm the place seems clearly in a pre-holiday hiatus. Except it isn't. These machines work all the time, methodically churning their way through an internal store of a couple of days' worth of DNA samples. When you've invested in some $30 million of high-tech equipment, you don't let it stand idle because people are going on holiday. You fill it up and leave it to get on with its job, silently reading out the messages that evolution has passed down through the ages.

The Joint Technology Center (JTC) is a facility shared by TIGR and the Institute for Biological Energy Alternatives (IBEA), all three of them parts of the empire of nonprofits founded by geneticist Craig Venter—most famous, ironically, for having led the commercial effort to sequence the genome. The JTC is one of the largest sequencing shops in the world. And among many other things, it's the face of microbial forensics to come.

In the fall of 2001 a previous generation of sequencing machines at TIGR had just about finished producing a complete anthrax genome, using a sample of the Ames strain the company had received from the British Ministry of Defense's biological warfare lab at Porton Down, in rural Wiltshire.

When the U.S. anthrax attacks began, Rita Colwell, the head of the National Science Foundation and a board member at TIGR, suggested to Claire Fraser, TIGR's president, that on top of finishing its mapping of the Porton Down sample, the lab also sequence the anthrax used in the attack. Keim's work showed that the weapon was an example of the Ames strain, but a full genome sequence could turn up more clues—and would reveal if, as some people feared, the attack strain had been subtly genetically modified. "No one had done anything like it before," Fraser recalls. A decade earlier there had been no complete sequences of anything bigger than a virus; five years earlier there'd been just a couple of bacterial genomes; and until the last few years, each new bacterial genome had represented a million-dollar investment. The idea of going to all that effort to produce a second sequence from a strain that had already been analyzed was unheard of. But so was a lethal bioterrorist attack coming hard on the heels of an atrocity like 9/11.

Sequencing the anthrax from the Florida attack took a few weeks, according to TIGR's head of bioinformatics, Steve Salzberg. Finishing the sequence of the Ames strain sample that had come from Porton Down actually took a little longer, in part because getting fresh samples to work with was suddenly subject to a great deal more red tape than previously. Then Salzberg and his colleagues had to apply a stringent new level of quality control to the finished sequences. TIGR's sequencing method normally yields one error in every 100,000 DNA letters, which is pretty good. But it still means that a 5 million letter sequence such as that of anthrax will typically contain 50 errors. If only a handful of letters were different in two isolates from the same strain of this extremely uniform species, those real differences would be lost in the 50 errors. So every time there was an apparent disparity between the sequences, technicians went back to the raw data to assure themselves that it wasn't an error. The result was probably the most accurate sequencing ever: in all likelihood, those two sequences are perfect.

The sequences showed that the Florida and Washington anthrax had not, in fact, been genetically modified. And the anthrax used in the attacks was remarkably similar to the Porton Down genome; the two differed in just four places. Those differences, though, are real, and if samples of DNA linked to a suspect in the anthrax attacks are ever found, identifying the same differences in it will be a crucial piece of evidence. It's possible that further work comparing the Florida sample to other versions of Ames from other labs may also have been done; Salzberg will only say it hasn't been done with NIH grant money. According to Fraser, "We're finding things that are intriguing enough that everyone's still enthusiastic."

Although the details of the investigation into the anthrax attacks of 2001 are obviously being kept quiet, other work at TIGR that will have a profound effect on microbial forensics is being done out in the open. Since 2001, TIGR's capacity for sequencing has increased phenomenally. A task the size of the anthrax genome is just 12 hours work for the silent machines in the JTC's sequencing sanctum; add in the time taken to prepare an organism's DNA for the machines, do the computation needed to stitch thousands of short overlapping sequences the machines originally generate into the complete genome, and add some finishing touches to the work by hand and you can imagine getting a genome mapped from start to finish in just a week. And indeed that's pretty much what TIGR is doing; a new biodefense contract from the NIH requires the facility to provide 44 genomes every year—40 from bacteria, 3 from other microbes, and 1 from a more complex organism, such as a disease-carrying insect—for the next five years. Almost all this work will involve sequencing new strains of creatures for which a sequence already exists, such as the plague bacteria; the new sequences can be compared with the existing one as a reference, to get a sense of the range of variation found in nature.

The most advanced such project, funded separately by NIH, is the sequencing of 15 different strains of anthrax from around the world. When this is finished TIGR will have unprecedented insight into the few places in the genome where variations occur; armed with that knowledge, it will be possible to design tests, focused on those key regions, that will instantly distinguish between thousands of different isolates of anthrax and establish within seconds which known isolate a new sample, perhaps someday taken from a crime scene, most closely resembles. Similar projects for most of the other potential bioweapons, such as plague and tularemia—and for most major infectious diseases—will soon follow. Fraser talks about using the information about what strains come from where "to build databases that we can tap into not unlike DNA criminal databases that are being discussed, so when a disease outbreak occurs you can quickly ask where it came from and when did we most recently see it."

The ideas are ambitious but they are not outrageous. As Steve Salzberg puts it, "To think about looking at 10 or 20 strains of a species—that used to be ridiculously expensive. Now you can do it. It's a million bucks, which is a big grant, but we get grants for a million bucks all the time. It used to be a million bucks for one genome. Now for the same money we can do 10 or 20." In other words, while a genome-sequencing machine costs as much as a nice house in the suburbs, a single microbial sequence only costs as much as a Lexus to put in the driveway. And the price is still dropping. The machines in the JTC are far more capable than the machines with which the first anthrax sequences were created, and they are not the end of the line. Machines 4 to 10 times faster are expected on the market in a year or so.

And at some point, sequencing costs will plummet so low that it will become cost effective to look for the subtlest of effects. Fraser imagines running experiments to see what sort of laboratory handling produces what sorts of genomic change; for example, can you tell what sort of mixture of nutrients a bacteria has been grown in, or at what temperature it's been incubated? That would not only be scientifically fascinating, it would also have implications for forensics. It would mean that when you found a new sample you would not only be able to say what strain it was descended from—you'd be able to say how it had been treated along the way.

When the technology gets really cheap, you can imagine starting to sequence not just human pathogens but representative microbes from all the ecosystems of the earth—from the soil and water, from the air, from mulch and dung gathered in the forests and savannas. That may sound absurd, but compare that to what's happened in computers, where a couple of decades has taken us from Space Invaders to game consoles that offer real-time cinematic photorealism. Map that sort of progress onto genome sequencing, and then imagine a world where SARS researchers—or any other set of genome detectives—no longer have to spend months searching through the meat markets to find what they're looking for. Instead they'll have a huge family tree of coronaviruses from around the world sitting in their databases. They'll have a sense of how the genomes of the world are interconnected—and thus of how any emerging novelty relates to what has come before. Every emerging pathogen could be found in the process of emerging, and in the context of that from which they have emerged.

The sequencing capacity to have the world's rich diversity of life pretty thoroughly sampled may not be that far off. But to put that technology to use will mean automating a lot of other processes, too. Putting samples into the right form for DNA sequencing, which involves, for instance, growing the bacteria and extracting their DNA, takes longer than the sequencing itself at a place such as TIGR. And many other sorts of tests that would be needed to make sense of a large-scale biological attack—tests that look at the effect of a pathogen on a tissue sample, say, or some sorts of straightforward genetic fingerprinting—are barely automated at all. The anthrax attacks of 2001, which led to fewer than 30 cases of disease, resulted in more than 100,000 laboratory tests of samples from the places that were attacked, places close to attacks, places the contaminated mail might have passed through, and so on; the work almost swamped the capacity of the CDC and its network of helper labs, running the technical staff ragged.

Scott Layne, a professor of public health at the University of California in Los Angeles (UCLA), worries about this bottleneck among the bottlewashers. And his worry is of the peculiarly frustrating kind that accompanies a belief that you know the solution to the problem, but you can't quite make it happen. Layne wants to get a new generation of public health laboratories built. He imagines using the highly automated techniques pioneered at genome factories such as TIGR and the sumptuously equipped research divisions of big pharmaceutical companies to establish a network of largely automated laboratories; this would hugely increase the available capacity. Such "high-throughput" laboratories, in which hundreds of tests are run in parallel, would be vital, he thinks, for dealing with clinical tests required during a large outbreak, be it an attack or a natural epidemic.

Layne's involvement in the field, like Paul Keim's, owes a lot to time spent at Los Alamos. But in Layne's case it wasn't the specific work that he did in the secluded lab on the mesa outside Santa Fe that changed his way of thinking. It was the style of the work getting done there. In the 1980s, the Department of Energy's weapons laboratories—Los Alamos and Sandia in New Mexico and Lawrence Livermore in California—were weirdly feudal and very rich. A fixed percentage of their billion-dollar budgets could be used to fund research at the director's pleasure, which meant the barons at the top had tens of millions of dollars of discretionary spending with which to attract the great minds in weaponry and the bright people in other disciplines to keep them company. When Layne arrived at Los Alamos in the early 1980s as a newly qualified M.D. interested in the science of how anesthetics work, he realized he was in a world most medical researchers and academics never get close to. "I saw science being done on a different scale," he remembers. "I saw science being done with leading-edge technology. I saw people getting five or ten million dollars of funding with one-page ideas."

The sight of all that state-of-the-art equipment at the disposal of researchers with "wild-man money," Layne says, made a lasting impression that would inspire him in later years to think big. Another experience that would influence his career was his work on AIDS; in the mid-1980s, Los Alamos started modeling the AIDS epidemic, and Layne became part of the effort. As a result, he became interested in public health, and in his late thirties he decided to put himself through the 100-hour weeks of a hospital residency to get board-certified in internal medicine before getting a position at UCLA. But he kept up his association with Los Alamos. With a colleague there, computer scientist Tony Beugelsdijk, he began to think about what Los Alamos–scale science could do for his new profession.

In the mid-1990s, Layne and Beugelsdijk started to argue that public health workers simply didn't appreciate the potential of automation for increasing the amount of work done in laboratories. With properly designed automatic systems in which robot arms move racks of samples or trays of Petri dishes from station to station, with every sample bar-coded and tracked by computers, the number of tests a lab could do might be increased 100-fold. High-throughput technology was already in use in the pharmaceutical industry, where there were labs devoted to screening thousands and even millions of possible drugs against cell cultures of various kinds, looking for the next big breakthrough.

To appreciate the sort of technology he's talking about, Layne says, you must go to deeply drab-sounding events such as the annual exhibition of the Association for Laboratory Automation. There you will see technology both mundane and wonderful. Take fridges. Every biology lab needs a fridge for storing samples. But a top-end fridge, these days, is a masterpiece of automation. These are fridges that can reach out an arm, pick up a sample, or a tray of samples, read the relevant bar codes, and deposit the samples in an appropriate slot of one of their carefully cooled internal carousels, there to be stored at exactly the right temperature alongside as many as 100,000 other samples, all accessible in seconds should they be needed.

What's more, this sort of top-of-the-range lab equipment could be used at a distance, with the help of computer tools that instruct the equipment to conduct a particular test. (The programming know-how that makes this possible is Beugelsdijk's area of excellence—he is an acknowledged expert in getting laboratory machines to coordinate their actions and talk to each other constructively.) Most researchers would never need to visit such labs; they would write a script, send off properly labeled samples, and wait for the result. Layne sees a system of such labs as doing for physical samples what the Internet does for data—creating a system for moving them around the world and manipulating them at a distance.

The problem, at the moment, is expense. High-throughput labs are not "big science" in the way that a space telescope or a particle accelerator is. But developing demonstration systems—Layne imagines starting off with two facilities, one customized for growing samples and the other for fingerprinting—would cost well over $10 million, much of it spent on happy vendors at laboratory automation exhibitions. And getting such a sum of money is hard. While small in the overall scheme of things—the National Institutes of Health had a

$1.5 billion biodefense budget in 2003—a $10 million or $20 million commitment is big compared with most grants given to individual scientists, grants aimed at people with already existing labs, rather than people trying to build one. There are only a handful of grants in the $10 million range given out every year across all of the U.S. government's R&D agencies. The wild-man money that was available at the National Labs in the late cold war has long since dried up.

Layne points to a request for proposals in the biodefense area put out by the National Institute of Allergy and Infectious Disease in 2003. Researchers were asked to submit grant proposals that might "lead to the development of new vaccines, adjuvants, therapeutics, immunotherapeutics or diagnostics . . . focused on . . . priority pathogens." He says the opportunity was the nearest he'd seen to a chance for funding research in mass fingerprinting processes and mass diagnostics. But one of the terms of the request was that grantees not ask for more than $300,000 to spend on equipment. That's less than the cost of a single sequencer at TIGR.

Hence Layne's frustration. He can see what his automated labs could accomplish. Sampling the food chain thoroughly for pathogens would be one application. Another would be flu surveillance. In a normal year, influenza kills half a million or more people worldwide, 30,000 or so in the United States. That's bad enough, but there is also the ever-present risk that a new flu strain might emerge that is far worse, killing millions or tens of millions. This happened three times in the twentieth century, and there's no evidence whatsoever to suggest it won't happen again. Details of the way the flu virus's genes are arranged, along with the fact that flu viruses from one species (such as chickens) can infect other species (such as people), make the threat of a new pandemic from flu much greater than from other diseases. All it takes is for someone somewhere to be infected simultaneously by a human strain, which will be good at passing between people but not very dangerous, and a more lethal animal strain, such as the avian flu found in chickens throughout Asia in the winter of 2003. If the two strains combine their strengths, the result is a deadly disease easily transmissible between people—the sort of thing that makes SARS look like a minor inconvenience.

(Research on this possibility is one of those areas where public health can look worryingly like weapons development. Many scientists argue that the best way of invesitigating the possibility that bird flu could combine some of its traits with an existing human flu and thus unleash a pandemic is to allow such hybrids to develop in laboratories, under the strictest of controls. Others worry that the risks of such a creation escaping, or being released, could outweigh any benefits that studying it might bring—an argument strengthened by the fact that SARS has twice struck within labs that are studying it. By the middle of 2004, regulators had not approved any such experiments.)

To stay on top of the flu risk, the WHO has a worldwide system for gathering flu samples so that it can see which strains are active where. This information is used for, among other things, choosing which strains to vaccinate against. At the moment, though, a lot of flu slips through the net and therefore is never incorporated into vaccine programs. Over 100,000 samples are gathered up every year, but only about 6,000 get thoroughly analyzed. In 2001, Layne, Beugelsdijk, and a number of flu experts got together in the pages of the journal *Science* to suggest that massive, automated laboratories could make the process of flu surveillance both more thorough and much swifter. Surveillance could be applied to birds and pigs as well as people, and the information needed for vaccination could be made available to pharmaceutical companies far faster. Such a system could save quite a few lives even in a normal year. If it provided just a few weeks extra warning at the start of the next pandemic, it could save hundreds of thousands of lives.

Labs built to deal with flu samples from around the globe would have been able to handle the 2001 anthrax attacks without breaking into the robotic equivalent of a sweat. And for biodefense purposes, their capacity for keeping watch on more pathogens could be invaluable. While tracking nuclear proliferation has proved hard enough, the surveillance of deadly biological weapons is still harder. Making the metals needed for nuclear bombs—plutonium and enriched uranium— uses a lot of energy and generates a lot of heat, and that heat can be spotted from a distance (indeed from orbit) with infrared cameras. In addition, the equipment needed to make nuclear weapons is distinctive, even if some of it is legitimately dual-use, and if you monitor people's imports with a well-trained eye you stand a chance of spotting a bomb program in the making. What's more, some equipment might give off signals that can be monitored from outside a facility; high-speed centrifuges, for instance, may give off either vibrations that can be felt by sensitive instruments or faint but distinctive radio emissions, or both. In contrast, there are very few ways of knowing what's going on in someone else's biology lab.

Labs give little indication on the outside of what's going on within: the ingredients needed to design and test a weapon will be exactly the same as those for developing a vaccine, and once a strain has been chosen, lots of different industrial processes could be used as cover for manufacturing the weapon itself. There was much media mirth in Britain when workers at the Bruichladdich distillery on the Isle of Islay found that the webcams they were using to show the making of their scotch whisky to clients around the world were being monitored by the Pentagon's Defense Threat Reduction Agency. But why should this be surprising? A weapons plant might look very much like a whiskey distillery; learning how the place that makes Bruichladdich works would be a sensible piece of intelligence-gathering.

The best way to know what biological processes are going on in a lab, or a brewery, or a distillery, is to look at the genes of the things doing the processing. This is a little beyond the capability of a webcam, but it's something a monitoring and surveillance system that used Scott Layne's high-throughput labs might excel at. Such knowledge factories could be used as part of a monitoring system, perhaps one such as that envisaged by the Chemical and Biological Weapons Treaty. In the nuclear world it's accepted that for all the cleverness of satellites and high-tech centrifuge spotting, the best way to know what's going on is to have access to the nuclear facilities and take samples. At the moment no such monitoring system has been put in place for bioweapons, in part because pharmaceutical and biotech companies fear that such inspections could be used for industrial espionage. The right sort of monitoring system could assuage those fears, in part: high-throughput labs running an array of quite specific tests—looking for the genomes of infectious pathogens, for antibiotic resistance genes, maybe for some other signatures—would not be in the business of finding out what a given sample is—just what it isn't. Such sampling might become a simple fact of doing business—the price

you pay, in Layne's words, "for the privilege of working with dangerous microorganisms." Strains of infectious pathogens and the like
would be routinely fingerprinted, so that any outbreak could be
quickly checked to see if it came from a known lab. One worry about
the vast ramp-up in biodefense research currently underway in the
United States—eight new government-funded centers at universities,
each with the sort of "hot zone" where researchers in one-hole-and-
you're-dead hazmat suits study really nasty things such as Ebola,
were funded in 2003—is that they provide more places for things to go
wrong. Twice in late 2003, new cases of SARS broke out in laboratories
trying to study the disease and make vaccines against it.

No monitoring system can hope to pick up everything, just as no
deterrent will deter everyone. And microbial forensics is unlikely ever
to have the judicial precision and certitude that human DNA fingerprinting now enjoys. The legal system is quite good at guiding the development of new forensic techniques—they get critiqued all the time
until they become reliably convincing, and then they get accepted.
But the sort of trial-by-trial evolution that turned human DNA fingerprinting into a reliable, defensible, and eventually routine form of evidence is unlikely to be available to the developers of microbial
forensics. Unless biocrime becomes nightmarishly prevalent, microbial forensics will always find itself dealing with the rare, the unusual,
the unforeseen. Human DNA turns up in all sorts of cases. Microbial
DNA doesn't. It turns up more often than the seed pods of the
paloverde tree do, but that's not saying much. However skilled the
practitioners of microbial forensics, and however careful they are
about procedures and chains of evidence, the first conclusions reached
through microbial forensics are likely to be less than legally water-
tight.

But if the capacity to monitor the world's DNA ever more closely
can't guarantee cases that stand up in court, it can no doubt help in
identifying the perpetrators in the case of deliberate attack. More important, the knowledge of the world's DNA that hugely automated sequencing and testing operations can provide as a matter of course
would be the basis of an early-warning system for emerging natural
diseases and give researchers a head start in looking for vaccines and
cures. Helping the great system of planet Earth know itself, helping

make connections between places and times, makes that system more understandable and more robust. And such work also has a value beyond the utilitarian. New ways of seeing the world are wonderful in and of themselves. These technologies provide a way for science to see the world of life anew, to see it all at once, as if from a satellite or the moon. It might allow them to get a handle on the diversity and activities of microbes around the world, and perhaps even see it changing. In a way, there is already a precedent for this global vision. By far the most detailed information on the Earth's interior is currently being gathered by a network of seismic sensors called the International Monitoring System, which is part of the Comprehensive Test Ban Treaty. The monitoring system is designed to stop proliferation by picking up nuclear tests—even small ones equivalent to just a few hundred tons of TNT—anywhere on the planet. The effort to spot new nuclear threats is giving scientists an unparalleled window into the solid Earth; the search for new biological danger could do the same for the living Earth.

EIGHT

Common Sensors

I N   A   H O T E L   R O O M seven blocks from the White House, Jay Boris sits at a laptop and wreaks havoc on downtown Washington. At his command, a deadly gas is released in a corner of Lafayette Park. An invisible cloud spreads quickly among and above the trees of the one-block park and sets off down Pennsylvania Avenue. Surprisingly, its tendrils avoid the White House itself, only 100 yards or so away, but quickly start to fill up the spaces between the termite mounds of bureaucracy in the Federal Triangle area. As it moves forward, the cloud of death unfurls like an ink drop in a slow stream of water, swirling and perverse: it turns back on itself; it rushes forward; it laps up against the facades of some buildings with a powerful illusion of purpose; it leaves others almost unscathed, protected by a simple whim of the wind.

Outside the hotel room's windows, of course, there's no such poison cloud. Just a normal Washington morning. The attack exists only within Boris's laptop and is visible only on the computer's screen. The computer is running a program that produces models—unprecedented in their detail—of how the wind can spread toxic gases or radioactive particles or pathogens such as anthrax spores. At this sort of fidelity, such modeling normally requires hours, even days, on a supercomputer. Boris, a sharp, intriguing, and slightly arch physicist in his sixties, is a chief scientist at the Naval Research Laboratory in Washington who has spent decades on supercomputer simulations, and the program on his PC is an ingenious new way of making the results of such modeling available through a simple system that works on a laptop. Boris hopes it's going to be picked up as a new tool for dealing with one of the most intractable and unsettling aspects of the threat of terrorism—invisibility.

The attack on Boris's computer screen is only a scenario. The problem is that even if this horrendous event were playing out for real and a plume of anthrax was spreading past the White House, the scene outside the hotel room's windows could well look precisely the same. A normal Washington morning. Nothing to be seen until people start dying.

Among the ominous qualities that make the "war on terror" different from other wars is the invisibility of the terrorist enemy and his weapons. In this conflict, there are hidden leaders, disguised combatants, and undetectable intentions; and there are weapons you cannot see: radiation, gases, and, most terrible of all, germs. A threat that can be seen may also be unstoppable—ICBMs remain so to this day, despite tens of billions of dollars spent on Star Wars and its successors—but at least it is defined. If you can muster the fortitude, you can face it, seek out its weaknesses, look it in the eye and formulate a strategy to minimize its effects. A threat that cannot be seen is something else entirely. It's not just that invisibility gives a tactical advantage by denying a defender information. It can also amplify the fear the threat engenders until that fear dominates everything else, disrupting and distorting the norms of daily life. Terrorism thrives on these effects at least as much as on damage or death.

So among the many challenges for experts such as Jay Boris is the urgent need to invent ways of piercing that invisibility, of capturing the imperceptible and making it known. Put another way, technologies that augment the human senses, of interest in numerous fields, have huge additional relevance in a world threatened by terrorism. The challenge for scientists and engineers in this area is to expose the weapons of terror—radiation, environmental chemicals, and germs—to a new kind of seeing.

•   •   •

In the cold war, the invisibility to be surmounted was geographical. The problems were not things that couldn't be seen in principle, just things that couldn't be seen from the United States. The solution was to use space flight and extraordinarily powerful cameras, capable of perceiving detail from a great distance, to do an end run around geography.

We take the view from space so much for granted now that it's easy to forget it had to be invented. At the beginning of the cold war, Russia's interior was a complete blank—a situation that Richard Leghorn, an MIT physicist who had been in Air Force reconnaissance during World War II, found unacceptable. Leghorn had scouted out German defenses in Normandy prior to D-Day and had witnessed the first postwar atomic tests, the Crossroads explosions at Bikini Atoll. The two experiences convinced him that the incipient cold war required a new form of reconnaissance, a system that could see into the enemy's country all the time, keeping tabs on the military, on the industries—on the targets.

A decade later, Leghorn was a key player in the technological revolution that created that system. In 1957, under the shadow of Sputnik, he founded a company called ITEK, a scrunched-together form of the new term he had coined to describe the company's business: information technology. Backed by Laurance Rockefeller, ITEK became a hot stock, landing Leghorn on the cover of *Business Week*. In a beguiling foreshadowing of the Internet boom, the press lapped up his ideas about "information technology," about devices that could store and analyze "books, pictures, maps, letters and memoranda," even though no such technology yet existed. There *was* a revolutionary technology being developed at ITEK, however, just not one the company could tell its investors about: Leghorn's ITEK provided the cameras for the first U.S. spy satellites.

This system—which collected photographic film in a canister that was then ejected from the satellite and retrieved (with any luck) in midair by a recovery aircraft—made the invisible heart of the Soviet Union visible in a way no enemy had ever been before. The development program was incredibly ambitious, and the first dozen flights all failed. But the first successful satellite sent back more pictures of the

Soviet Union than all the U-2 reconnaissance flights ever undertaken. Sometimes covering more than 1 million square miles with a single film canister, the CORONA program showed that the "missile gap" was a mirage (or, rather, a gap with the Soviets on the lagging side) and gave the Kennedy administration a powerful edge in superpower politics. These remote sensing satellites were soon followed by others that provided not just photographic imagery but also readings of energy emitted from nuclear tests and the capability to detect infrared radiation that would signal the launch of an ICBM. The early warning that this system provided was vital to the doctrine of deterrence on which the cold war's strategic balance was based.

Faced with the new challenge of catastrophic terrorism, it's natural for decision makers to want some equivalent to the all-seeing gaze of the sensors in the sky—some way to make the threat visible and thus tractable. One way to satisfy this wish is to start looking, not down at the world from outer space but into the world that only exists behind the computer screen—the world of models. Let a model, based on the fundamentals of physics and chemistry, work out where those invisible substances are going. If such models are good enough, and you trust them enough, they offer a powerful window onto the invisible world. That's the solution Jay Boris and a large number of other "transport modelers" and "plume modelers" are working on. To see a cloud of death wander across a city on a computer screen is eerie—but also a bit reassuring. At least it is *there,* constrained by limits and contours. The problem with this solution is that Jay Boris's laptop can only simulate, not see. Unlike Leghorn's cameras, models can't gather information; they can only process it.

To feed those models with the data they need in order to make the invisible visible—and predictable—we have to stop looking and start feeling: to use sensors that interact directly with threats in the places where they're looming, rather than inspecting them from afar— sensors that respond to their own immediate environment. At the moment, only a few places are alerted to invisible threats by such sensors, but developments in both the technology of sensors and the ease with which they can be linked to each other promise to change this. Sensors of many different kinds are set to become nearly ubiquitous. Half a century after satellites revolutionized our perceptions of the world, sensors arrayed and interconnected in networks of all scales—from networks that cover a building to networks that cover the globe—may do the same thing all over again. Satellites changed the way we see the world, offering us an outside to look in from; the generations of sensor networks to come could change the way that we feel the world from the inside out.

The technologies for feeling will vary according to what they're feeling for. Sensor technologies span a range of capabilities, sensitivities, locations, shapes, and sizes. Typical sensors detect heat, electromagnetic energy, pressure, motion, light, the presence of a chemical, the weight of a microbe. The thermometer on the wall is a sensor. So is the device that keeps the elevator door from squashing you and the instrument beneath the seat of your car that triggers a beeping until you buckle up. In industry, sensors all around the factory floor check the positioning of parts on conveyor belts. They monitor the makeup of liquids in vats. They watch for pollutants escaping from pipelines and tanks. Sensors can be as big as SUVs and as petite as jellybeans. They can look like nuts and bolts or like boilers; they can be as obtrusive as the tools of a mad scientist or almost as invisible as the things they are meant to detect.

Radiation, while in some ways the most feared of the invisible threats people would like to be warned of, is also the easiest to detect, thanks to the fact that physicists have put a lot of thought into radiation sensors both as ways of studying the stuff and as ways of staying safe while they do so. Unlike a nuclear bomb, which creates a great deal of new radioactive material at the same time that it blasts a city off the face of the earth, a radiological weapon or "dirty bomb" disperses recycled radioactive materials that have been packed around an ordinary explosive. One upshot of this is that while hugely disruptive, and extremely expensive to clean up after, a dirty bomb would not necessarily lead to mass casualties. Another, and a very helpful one in terms of detection, is that dirty bombs are radioactive before they go off, so radiation sensors stand a chance of spotting them using well-established technologies for detecting radiation—neutrons, electrons, and gamma rays. For many applications, basic Geiger counters will do the job. Discreet radiation detectors at transportation hubs and choke points where traffic comes into cities could go a long way to thwarting such attacks, and some such systems appear already to be in place. In 2002, the *New York Times* reported that a patient receiving treatment for an autoimmune disease involving the thyroid gland was stopped by police twice at Penn Station: the iodine-131 in his thyroid had triggered some sensors. A woman trying to drive out of Manhattan after a heart test that used a radioactive marker was stopped at a tunnel entrance.

Chemical threats are a different sort of problem. They can't be detected until the cat is out of the bag—and once that has happened, they can spread quickly and unpredictably. "Nerve agents" such as sarin and VX (they're called nerve agents because they work by blocking the body's ability to break down acetylcholine, a neurotransmitter, and thus leave nerves permanently switched on) are highly lethal. Whereas in the open air they might dissipate quite quickly, in an enclosed space they do a great deal more damage. What's more, as we've witnessed at least once, using them is not especially hard.

On March 20, 1995, the Aum Shinrikyo religious group released sarin into three different subway lines in Tokyo. The Aum members were not the first people to see the possibilities of underground railways as killing grounds for invisible weapons. In the 1930s, a former editor of the *London Times,* Henry Wickham Steed, published an article claiming that Germany was targeting underground railways for biological warfare, which stirred up concerns in both the United Kingdom and France. (There is a certain irony in the fact that when the Germans captured the French laboratory responsible for germ warfare in 1940, they found that the Wickham Steed scare had propelled French capabilities in the area far beyond those of the Germans.) Whether or not the Germans studied the transmission of airborne microbes through the Paris Métro, as Wickham Steed claimed, the British certainly ran such experiments in the London Underground in the 1960s, measuring how easily anthrax-like spores could be moved around the system. U.S. Army researchers carried out similar experiments in the New York subway, concluding that it would be feasible to get enough anthrax spores into the tunnels, where speeding trains would act as pistons forcing bacteria-laden air through the system, to infect up to 90 percent of commuters.

The madmen of Aum would have been quite willing to have tried anthrax in the Tokyo subway; given how poor their attempts to use anthrax were (their nine releases of the Sterne strain of the bacteria, which, it so happens, doesn't actually cause disease, were hardly rave successes), it would have been less deadly if they had. But in March 1995, they felt a need to do something that would kill immediately, rather than after incubation. The Japanese police were planning a large raid on Aum's headquarters for March 22; 500 police officers were undergoing training in defense against chemical weapons with the country's Self-Defense Force. Cult members learned of this training and quickly put together the subway attack to forestall the forthcoming raid by incapacitating as many police officers as possible. The idea was to hit trains going in both directions on the three subway lines that meet at Kasumigaseki station, which serves Tokyo's police headquarters. The attacks would be on trains timed to pass through the station at around eight o'clock on Monday morning, just as the officers on the day shift were reporting for duty.

Each of Aum's five assault teams consisted of a man who would release the poison and a getaway driver. The release men got on trains headed toward Kasumigaseki from the outskirts of Tokyo, carrying bags with containers of liquid sarin inside. At a prearranged station along the way, each of them stabbed at the sarin containers with a sharpened umbrella before leaving the train to join his getaway driver. The sarin leaked out and began to evaporate. Of 11 containers taken onto the trains, 8 were punctured, releasing 159 ounces of sarin. Three of the trains reached Kasumigaseki; two stopped short when passengers complained of the smell and started to get sick. Eventually, 5,500 passengers ended up in the hospital. The vast majority were unhurt, "worried well," but almost a thousand had genuine if minor symptoms of poisoning. Fifty-four were critically or severely injured, and twelve died.

The Aum attack was not as terrible as it could have been. The sarin, which was made in a rush, was very impure—pure sarin would not have given off the tell-tale odor—and also rather dilute. The dispersal method—puncturing little plastic containers on the floor—didn't put very high levels of the deadly vapor into the air. Most people walked off the trains, through the stations, and out into the fresh air without

harm; the deaths all seem to have involved people who came into contact with the concentrated liquid form of the agent. Straightforward
bombs or incendiary devices might have killed many more, as the coordinated attack on Madrid commuter trains in the spring of 2004
showed. Indeed, such mayhem does not need the resources of terrorist
cells or religious cults—it is within the grasp of a single individual. In
2003, a mentally ill former taxi driver killed almost 200 people on two
crowded trains in Taegu, South Korea, with a carton of gasoline and a
cigarette lighter.

Less lethal than the cult expected, the Aum attack also failed to
achieve its goal. Two days later 2,500 soldiers and police officers
raided the sect's headquarters and more than 20 other sites. The
attack's most enduring effect, in fact, was to raise the possibility of
catastrophic terrorism from a marginal issue to a mainstream one. In
the United States, Senators Sam Nunn (later to play the president in
the Dark Winter smallpox simulation) and Richard Lugar launched in-
quiries into the incident and introduced legislation to instruct emergency teams in the nation's biggest cities on the danger of catastrophic
terrorist attacks. Another response was to stir interest in setting up
sensor systems that could pick up an attack as quickly as possible.
The Washington, D.C., Metro system, spectacular and vulnerable, was
an obvious place to start.

The details of the system of sensors that now monitors the Metro in
Washington, D.C., are kept deliberately obscure, but Duane Lindner,
the deputy director of chemistry and biology programs at Sandia National Labs, and one of the system's architects, is happy to talk about
the general principles. The Metro sensors are based on those used by
the military on battlefields. They had to be toughened up a lot for city
life, though. Military sensors get a lot of cosseting from the troops
looking out for them. In the Metro the sensors had to be able to get by
on their own, with many of them working nonstop around the clock,
and to deal with all sorts of things that you don't find on battlefields—
such as perfume, cleaning fluids, and "rail dust," a miasma of dirt and
grit and grime and tiny particles of metal constantly milled from the
rails and wheels. This meant using better, tougher components.

Perhaps the most pressing modification the Metro project entailed
was that the sensors' failure rates had to be reduced—and so did false

alarms. A false-positive rate of 1 percent means that once in every 100 times that your sensor looks at a clean air sample, it will think it spots something nasty. One percent sounds low, but if such a sensor samples the air every 15 minutes or so, it will give you a false alarm every day. If 100 such sensors are sampling the air every 15 minutes, you'll get a false alarm every 15 minutes.

On battlefields, a certain level of false positives can be tolerated. The devices are used only some of the time, and the people they're protecting know what to do when the alarm goes off. Soldiers will grumble like hell if the sensors go off by mistake too often, but they know it comes with the territory. Commuters are less phlegmatic. "In battlefield situations," says Lindner, "you'd rather be safe than sorry. But you can't be evacuating subway systems on false alarms."

Reengineering the sensors allowed Lindner and his team to drop the false-positive rate. Their most important work, however, was not the tinkering with individual sensor devices. It was designing the system as a whole: seeing the network of sensors, the tangle of tunnels in which it resided, the human beings maintaining the equipment, the false positives, the false negatives, and the commuters who would re-spond to all this as a single functioning entity. The network could not be simply built into the walls of the tunnels and the ceilings of sta-tions; it had to be embedded in the way the Metro system actually worked. For instance, when thinking about the purposes of their sen-sors, Lindner and his colleagues distinguished between "detect to warn" systems, which set off an alarm that allows people to avoid a threat, and "detect to treat" systems, which tell them about a threat so that some sort of remedial action can be undertaken. The classic everyday examples are the smoke alarm, detect to warn, and the heat-sensitive sprinkler, detect to treat. On the battlefield, the design of the system will often be weighted toward detect to warn, because soldiers are trained and equipped to respond when threatened. In the subway, a strong component of detect to treat was needed; the system had to be able to tell you where paramedics were needed, for instance, as well as which stations train operators should use and which to avoid.

At the same time, work continues on improving the sensors them-selves. Lindner is interested, for instance, in developing sensors that can address a wider range of threats than a few specific chemicals terrorists might deploy. On the battlefields for which today's detector technology was originally designed you might expect to come up against a small number of sophisticated nerve gases (and if your intelligence is good, you might even know which one was most likely). But a terrorist operation in a city is not so predictable. It might well use something less lethal but available in much larger quantities. Methyl isocyanate, an industrial chemical, is not as lethal as a nerve gas—but not much more than a tanker-trailer's worth of it killed 2,000 people when 40 tons of the gas were accidentally released from a Union carbide plant in Bhopal, India. That night, 170,000 people were injured. If you're worried about bags of sarin but ignoring tankers of commercial chemicals, you may be in for a shock. At the moment, chemical sensors tend to look for specific agents, or specific families of agents. The long-term aim, says Lindner, is "to move beyond asking 'Is X there?' to asking 'What is there?'" His hope is to make chemical sensing, with its touchlike qualities of reacting to one local thing at a time, just a little more like seeing, which reveals everything at once.

Lindner stresses the importance of keeping humans in the sensor system. Though they can't sniff out sarin or spot a gamma ray, humans offer flexibility and powers of discernment that machines can't match. That's in part because we're sensing machines ourselves, evolved to deal with the inputs from vast arrays of sensors—a skin stuffed with nerve cells responsive to heat and pressure and pain, retinas sucking endless megabytes of information through the eyes, a nose ever alert for a hint of the sweet or a whiff of the rancid, ears that vibrate ceaselessly to the rhythms of the world. However many sensors there are in the 50 miles of the Metro's tunnels, there are a million times more wired into every commuter.

Could an artificial system ever begin to match this sort of complexity? In the near term, no. But systems in which the number of sensors climbs from hundreds to thousands to millions are now being developed. This future is the vision of two scientists, chemist Michael Sailor and electrical engineer Kris Pister, for whom "small and plentiful" has become something of a mantra. Their aim is not necessarily better sensors, but more of them. Each device will gather meager bits of data, but the devices will be so numerous that nothing will escape them. The loose net of old sensor systems will close into a tight mesh. Space will become sensor-saturated.

Kris Pister, a self-confident and slightly mischievous professor at the University of California in Berkeley, began in the mid-1990s to call this technology "smart dust." "I coined the phrase as a joke," he explains. "At that time everybody was talking about smart houses, smart highways, smart buildings, smart bombs. Everything was smart." Smart dust wasn't a description of anything Pister could actually build—it was more an aspiration. Today, he and Sailor joke that they are sort of halfway there: Sailor makes smart dust that's not really smart, Pister makes smart dust that's not really dust.

Sailor's technology is sufficiently dustlike to be hard to pick out with the naked eye. A little vial in his lab, easily held between finger and thumb, seems to be filled with just enough clear liquid for perhaps a couple of days of contact-lens rinsing. Shake it, though, and you'll see something inside—tiny specks of greenish gold suspended in the liquid at the edge of visibility, motes of dust that seem to be catching the sun indoors on a cloudy day. And here's the astonishing part: each of these glittering specks is a precisely engineered and carefully tuned sensor, capable of reacting to subtle changes in its environment.

The glittery property of Sailor's dust is key to the invention's powers. The smart-dust particles, made in Sailor's lab in the chemistry department of UC San Diego, are specks of silicon shot through with a latticework of tiny pores. The lattice interferes with light waves, reflecting back some wavelengths and not others; it's the same principle that gives a butterfly's wings their iridescent sheen or the layers of abalone shell their glisten. Except here, the lattice responds to its environment. The pores in the silicon will suck up chemicals from the atmosphere the way a wick sucks up wax from a melting candle. As the pores fill up, the chemicals cupped within them change the wavelength of the light the lattice reflects, and the dust specks change their hue.

To understand how Sailor could possibly have created sensors almost too small to see, it helps to know that he did it by accident. He and his fellow researchers were working on building pore-based sensors using silicon wafers an inch or so across in which they drilled

holes using carefully controlled electric currents and scaldingly un-
pleasant acids. By changing the size of the pores, they could change
the things that got sucked into them and the wavelengths of light that
they would interfere with. All the while, though, they found themselves irritated by bits of silicon flaking off and floating around the
lab. It took some time for it to dawn on them that perhaps these annoying specks of glitter could be useful. Today, they create smart dust
by smashing their carefully designed, micromachined wafers to
smithereens—which is gentler than it sounds, given that it's done
with finely tuned vibrations. Each sliver will change color on its own,
just as the chemically treated wafer as a whole did: a puff of ethanol
will change it from yellow to red. Sailor can shine a laser beam on the
dust motes and analyze the wavelengths they're reflecting—and thus
the contents of their pores—with some accuracy. And if he needs to,
he can do it from a distance. In woods a few miles from their lab,
Sailor and his students have shown that they can track the spread of
organic chemicals from a distance of 20 or 30 meters by using lasers to
illuminate barely visible motes of smart dust they've scattered around
and see them change color when they come into contact with the
chemicals.

As Sailor says, his iridescent flakes are dustlike, but they're not
very smart. They can react and change color, but they have no way of
understanding that they've done so, and unless someone's watching
with a laser they're not going to pass the information on. Kris Pister's
smart-dust sensors, on the other hand, are remarkably clever. Some
recent models are capable of sensing objects, temperature, or accelera-
tion; turning those perceptions into quantifiable data; and sending the
data via radio waves to a receiver while sipping so modestly from the
power stored in their batteries that they can go on doing so for years.
At the size of a small matchbox, though, they're just too ungainly to be
called dustlike yet.

Pister sees lots of room for shrinkage. Although it's "not ready for
prime time," as he puts it, his team built a device in early 2004 containing a light sensor, a converter that could encode the light reading
as a digital message, and a transmitter that could send that message a
distance of 50 meters—all in a package so small that, placed atop a
penny, it was no larger than Abraham Lincoln's cheek.

Pister's mini-instruments are not only capable of performing all these functions solo; they're also built to team up. To offset their weaknesses—their fairly short range of communication and their paltry power supply—the individual "motes" can organize themselves into networks through which both instructions and data flow freely by means of short hops from mote to neighboring mote. Cheap, simple, and plentiful, the devices can expand their range of communication by joining forces and radically reduce their power needs by sharing tasks. Their abundance is also at least a partial remedy for the problem of false positives. A system can be instructed—and eventually, perhaps, it could learn without instruction—to ignore a rogue mote that pipes up and take note only when sensors respond in a pattern that makes sense, revealing some change or threat spreading realistically through the space that they're monitoring. Pister is clear that some of the early applications for his networks will be military. His work has, in part, been funded by DARPA. And like Sailor, Pister is a member of the Defense Science Studies Group, set up in the 1970s to give academic scientists a feeling for what the military does and what new technologies it might need. The group's members get to visit military installations and play with some very neat toys, and when they see something in their own line of research with a military or security angle, they take note and know whom to call.

In 2001, Pister demonstrated some of the things that a sensor network such as his could do in a trial at a military proving ground in California. Prototype smart-dust motes about a centimeter on each side, carrying sensors that could detect magnetic fields, were dropped along a roadway by a small, unpiloted aircraft. Once on the ground, they automatically established a radio network among themselves and developed an idea of where they were in relation to one another. When trucks rolled along the road the network picked up their movement. By sending very compact messages back and forth to compare notes, the motes were able to work out the trucks' speed and direction.

To Pister, a handful of motes such as this is just the beginning of the military possibilities; he imagines his motes monitoring battlefields from below as thoroughly as satellites watch from above, the sensors' perceptions and the satellites' overview complementing each other to provide a multifaceted view of what the enemy's up to. And current rates of technological improvement make such visions plausible. The radios that allow sensors to communicate with one another are getting smaller and cheaper all the time, and the power of computer chips is increasing; by 2010, a smart-dust mote's radio should cost only 10 cents, and its miniature processor could be as powerful as that of an early PC. Electrical power could come from batteries or solar cells or the microwaves given off by nearby cell phones, or in some cases from the vibrations in the environment; Pister estimates that, using foreseeable technology, a cubic-millimeter battery could provide a mote with enough energy to take a measurement and send a message every second for a decade. And there are many different sensor technologies already available to combine with the motes' other powers—accelerometers that tell you if the sensor is moving, magnetometers that tell you if something else is, thermometers that tell you temperatures, and chemical sensors such as Sailor's.

The manifesto on Pister's website extolling the possibilities of smart dust evokes the strange and wonderful ways in which he sees these networks changing the world far beyond the battlefield. This is not a dual-use technology—it's an all-use technology. Hence Pister's high hopes for his start-up company, Dust Inc. In a smart-dust world, all possessions of any value get labeled with smart dust and yell out, electronically, if anyone tries to mess with them. People are labeled with smart dust, too, so buildings—filled, of course, with networks of smart-dust sensors—can keep track of their occupants in order to accommodate them by opening doors, turning lights on and off, and so on. Smart dust in fingernails ties hand movements into the network so that gestures can be used to control all of the devices not already opening and closing and turning themselves on and off automatically. Typists no longer require keyboards; air guitarists make real noise. The individual nodes of the network may be quite stupid, but tens of thousands of them added together have the same power as a large computer—a supercomputer spread out through, and responsive to, the environment. In smart dust–rich environments, the barriers between the real world and the electronic one start to break down.

This picture of a world transformed is not without its worries. What about privacy? What about keeping the network unhacked? What about just breathing the stuff in? On the last of those concerns,

Pister's manifesto tries to reassure: "Consider the scale—if I make a million dust motes, they have a total volume of one liter. Throwing a liter's worth of batteries into the environment is certainly not going to help it, but in the big picture it probably doesn't make it very high on the list of bad things to do to the planet."

Sailor, a slim man with a runner's physique whom you could easily mistake for an Air Force officer (not least because of a plaque on the wall of his office thanking him for having refueled a B-2 stealth bomber as one of his Defense Science Studies Group activities) is not as gung-ho about the smart-dust future as Pister. Whatever he thinks about Pister's grander dreams, though, he argues that fairly simple smart-dust environmental sensor networks used for homeland security could be the basis of a new detection paradigm: detect to protect. If a smoke alarm is "detect to warn" and a fire sprinkler is "detect to treat," then "detect to protect" is a thermostat—a system that adjusts to changes in the environment looking for balance.

In such a detect-to-protect system, as Sailor sees it, a network of environmental smart-dust sensors would be linked to a building's air-conditioning and ventilation systems. Sensors the size of a pinhead would be nowhere near as specific or sensitive as the best that Lindner and his colleagues can do at Sandia, but they would be cheap and plentiful. If they picked up an increase in, say, organic chemicals, or chlorine, or anthrax-sized particles, they would not broadcast an alert, because the chances are that it would be a false positive ("You can't evacuate the building when someone shakes out a rug," Sailor says with a smile), but instead would trigger the ventilation system to increase the rate at which air is pulled from the affected area and pass it through special filters. If you wanted to guarantee an absolute minimum of casualties, you'd pass the air through such filters constantly, and in some government buildings—the White House is probably at the head of the list—that's already being done. But the cost—in terms of the energy needed to move all the air around and the endless replacement of filter materials—means that such a system is hardly applicable to every public or private building. "It's probably a reasonable approach for the FBI building," says Sailor, "but it's not a reasonable approach for the subway or the shopping mall."

Even cheap technologies, when distributed broadly and networked tightly, can provide the flexibility to respond with some specificity, with intelligent awareness of context. Eventually, it's possible to imagine a system that does at the network level what Lindner's sophisticated chemical sensors might someday do on their own: produce ever more complicated patterns of perception in reaction to the environment—patterns that go beyond flagging the presence of X or Y substance to describing the situation in its entirety. Networks of different sensors spread around the place could begin to match the capabilities of the best single sensors—but would apply them much more widely. As Pister has it, "the network becomes the sensor."

When invisible terror actually struck Washington, D.C., it was not on the air swirling around Federal Triangle or through the deep conduits of the Metro tunnels—and it was not picked up by any sensors. Nor was it made of radiation or nerve agents or other types of chemicals. It came in an envelope opened in the office of Senator Tom Daschle, and it came in the form of living organisms. The envelope contained a few grams of extremely dangerous anthrax particles that could drift in the air with ease and which, once settled, would loft themselves back up at the gentlest disturbance. There may have been enough anthrax in the envelope to deliver 100,000 lethal doses; spread through the building it could have killed hundreds of senators, staff members, and building staff. But it didn't, for one simple reason. Along with the fine white powder in the envelope was a letter that said:

09-11-01

YOU CAN NOT STOP US.
WE HAVE THIS ANTHRAX.
YOU DIE NOW.
ARE YOU AFRAID?
DEATH TO AMERICA.
DEATH TO ISRAEL.
ALLAH IS GREAT.

Anthrax mailed to the *Boca Raton Sun* had already claimed its first victim on October 5 and had been recognized as a deliberate attack; a mailing to Tom Brokaw at NBC News in New York City had also been identified as anthrax. So the claim that the letter to Daschle contained anthrax was highly credible, and a quick test of the powder showed that it was in all likelihood true. As a result, everyone who'd been anywhere near the letter was given a course of Cipro. It was a highly effective defense, and no one in the building died.

The warning in the anthrax letter more or less ensured the recipients' survival. The problem is what to do when the attacker is not so forthcoming. Biological attacks are much more difficult to detect using environmental sensors than chemical attacks, and no amount of miniaturizing and multiplying of the sensors is likely to change that for now. The stand-alone biological-agent detectors available today are able to distinguish possibly suspicious biological particles in the air with the use of antibody tests, but these tests are far from reliable enough for routine use. The best they can do is send a signal for someone to come and take a sample to a lab for full analysis if there is something in it that looks suspicious. But lab results can take hours. There may be some value in having these current sensors around a military base during a war, or around vital government buildings as a routine matter (and indeed such machines are already found in some sensitive parts of Washington). But they're a very long way from a system that can spot a biological attack whenever and wherever it takes place.

Calvin Chu, who used to work on biological sensors for the military and was at the Civilian Biodefense Institute at Johns Hopkins while he pursued doctoral research into the Venezuelan equine encephalitis virus (one of the more obscure of the usual suspects on bioweapon lists), is quite straightforward about the dearth of technologies: "At the moment, I can't see any automated device working in a routine kind of way." George Poste, chair of the Defense Science Board's bioterrorism task force, agrees: "Sensors are absolutely vital; the problem is that there isn't the technology to make them feasible. Ten years from now there might be—I'm not saying we should give up. But right now if you rely on sensors, you might as well not bother."

This is something of a surprise. To the imagination, a little device that somehow senses bugs is a much more plausible idea than, say, a car-sized camera in orbit. On *Star Trek* episodes in the 1960s, the crew of the *Enterprise* already had handheld detectors called tricorders that identified any sign of life. The tricorders went along with the communicators that let Kirk and his crew talk to each other at the flick of a wrist and the space-age doors that opened with a pleasing swish when someone approached (presumably activated by smart dust). So why has real-life science had no difficulty bringing us swishy doors and communicators but failed utterly in providing tricorders?

There are two essential barriers to developing a tricorder. One is that there is a bewildering range of living things in the environment to sense. A detector that just picks out anthrax or smallpox is not enough. You need one that can pick out anything threatening, which means you need a device that can run a wide range of tests in parallel. That is a fiendishly difficult proposition. The other problem is sorting out bad bugs from normal ones. The air is full of small biological particles—innocuous bacteria, bits of dead skin, pollen, tiny wax particles from leaves. And pathogenic bacteria look almost identical to benign ones. The *E. coli* bacteria that live harmlessly in our guts, for example, and the vicious 0157 strain of the same bug that can kill in a matter of days are vastly different at the level of their genes, but almost identical from the outside. The air is also full of things that interfere with delicate biological tests—a little bit of soot, for example, is a great way to mess up certain lab tests. An unruly industrial civilization interacting with a biosphere that's been growing more complex for most of the past 4 billion years produces too much mess for any tricorder we're capable of building today.

It's not an impossible problem; there's even some consensus on how to tackle it: use the systems nature uses. For millennia, biological systems have been used to detect poison—there's a reasonably direct link from the slaves who tasted Caesar's food to the 10 mice that, according to the UK press, performed the same duty for President Bush on his 2003 trip to Thailand. And biological systems have been used elsewhere, to detect a variety of threats: the Japanese police who finally raided Aum's headquarters went in with caged canaries to detect danger; U.S. troops, during the 2003 invasion of Iraq, strapped caged chickens ("Kuwaiti Fighting Chickens," or KFCs) to their Bradley Fighting Vehicles as a low-tech way of knowing if there was something nasty in the air; and biologists at Sandia are working on using bees to detect landmines. In the future, though, whole animals may be less and less necessary. Cells and cell networks have extraordinarily fine powers of distinction and are great at doing lots of things at once. Chu is sure that, in the long run, incorporating more biology into sensors is the way to make them work. DARPA is already funding a range of projects based on that hope, such as research into how to make cells fluoresce in response to environmental changes.

The lack of good technologies for detecting biological attacks can lead to the promotion of poor ones. Chu points to a project called Biowatch being implemented by the Department of Homeland Security. Biowatch takes environmental air samples in cities around the United States by piggybacking on systems used to provide pollen counts. The samples are then gathered up and studied extensively, undergoing as many tests as the labs can think of. Yet the chances of a deadly plume happening past one of those pollen detectors—there's typically only one collection point in a city—are minimal. As Chu points out, labs have other jobs to do without having their resources used up frivolously. The Biowatch system, its critics say, is designed to feel like an early-warning system and sound like an early-warning system, but it's very unlikely to provide an early warning.

At the moment, there's really only one feasible way to put together a sensor network for detecting biological attacks: assign the job not to technologies but to people. Today, as throughout the history of public health, most disease outbreaks are spotted when a clinician recognizes something unusual or out of place; the anthrax attacks were recognized because a doctor in Florida had recently been trained in recognizing the symptoms. The trouble is that clinicians in any given hospital or clinic won't see the big picture, just the patients they're treating. That's why public health systems gather information from clinics and hospitals—typically a tedious and bureaucratic process— and try to find patterns that escape local notice. But in the age of the Internet, if people outside the formal system keep their eyes and ears open, they can do this same information gathering directly, picking up on emerging diseases and alerting the world faster than has ever been possible before. That's how the system called ProMED works.

ProMED is the most old-fashioned of Internet innovations: a mailing list. Every day, its 30,000 subscribers receive a handful of e-mails addressing specific disease outbreaks. This information, which comes from the news media, from public health systems, from doctors who have spotted something odd or heard something curious or have an answer to someone else's question, has first been read, assessed, and perhaps commented upon by one of ProMED's 10 moderators. After that, it's been passed up to the top moderator, who sends the items out. ProMED's editor and assistant editors share the hot seat on a one-week-on, four-weeks-off rotation, filtering and assigning incoming information to the moderators. It's a hard week's work, says ProMED editor Larry Madoff, an unassuming infectious-disease doctor at Brigham and Women's Hospital in Boston (and also an associate professor at the Harvard Medical School). "My family hates it. We don't pretend to do it 24 hours a day, but we do it all the time we're not sleeping. When I'm on, I get up in the morning and immediately start reading my e-mails and assigning reports to different people. Throughout the day I receive and send reports and post things onto the mailing list. I come home at night, have dinner, and then disappear into the study for another session. I don't stop until I go to bed at night. It's fun when it ends."

The result of all this work is a deeply informative day-by-day digest of unwonted interruptions to the planet's health—the health of its plants and animals, as well as its people. "We don't really separate animal and human diseases," explains Madoff, "because so many emerging diseases [come from animals]"—11 of the last 12, according to the CDC—"and also because bioterrorism is, in our view, at least as likely to occur in crops and animals as it is to people." Take a typical day in January 2004. Of 10 mailings, 6 were devoted to different aspects of the growing avian flu crisis in Asia, mostly comprised of various media reports on the possibilities and pitfalls of vaccination. But there was also a report from Singapore's *Straits Times* on a foot-and-mouth disease outbreak in Pahang, Malaysia, with helpful commentary by a moderator pointing out that in the run-up to the Muslim festival of the sacrifice, Eid al-Adha, such outbreaks need careful monitoring, because livestock is moved in larger numbers over greater distances than normal. There was a report on a fungal disease that attacks plants, called brown rust, being found in sugar cane crops in Kimberley, Australia, and an e-mail from a veterinarian in the Canadian Food Inspection Agency giving new details on an outbreak of anthrax on a cattle farm in Saskatchewan, along with a note from anthrax expert Martin Hugh-Jones comparing this rare winter occurrence of the disease with outbreaks in Oklahoma in the 1980s and France in the 1920s. And there was a newswire report on a mysterious disease that had infected 60 and killed somewhere between 14 and 20 in the Bangladeshi district of Rajbari; it seemed to be an encephalitis, and the moderator ended her notes asking for more information.

The Bangladeshi outbreak was eventually identified as a virus apparently passed by bats—something already known to science. Most such reports are. However, some aren't. On February 10, 2003, Madoff himself posted a moderator's note asking for more information: A ProMED subscriber and bioterrorism expert had sent in an e-mail about a possible disease outbreak in southern China, having learned through a former neighbor—who in turn had picked it up in a chat room for teachers—of a message that began, "Have you heard of the terrible sickness in my city?" That posting was the first public recognition anywhere of the disease in Guangdong Province that a month later would become known as SARS. The first WHO investigations started a day later, prompted in part by the ProMED mailing. Throughout the crisis, ProMED offered subscribers breaking news and informed commentary, providing an open forum for researchers to exchange news.

ProMED was created in 1994 under the aegis of the Federation of American Scientists, an organization started in the late 1940s by Manhattan Project scientists who were having moral and political misgivings. The mailing list was intended from the first to be a bioterror and biowarfare early-warning system as well as a way of keeping up with natural disease outbreaks. It now operates under the aegis of the International Society for Infectious Diseases. It is not a large organization—there are 20 people on staff, none of them full-time. "Everybody does something else for a living," says Madoff, "and that's partly intentional—to keep us grounded." It's also partly because ProMED, which is funded philanthropically, is not exactly rich. The volunteers are paid a modest stipend, not a wage. Nor do they necessarily reap great professional rewards: "It is not entirely clear to me whether this is good or bad for my career," Madoff says. "I don't publish as many papers or get as many grants when I'm doing this. But on the other hand, I feel like I'm doing something worthwhile."

There's something very inspiring about ProMED, a small private initiative that binds expertise together with fellow feeling in an attempt to make the world both better and better understood. Its success— from 40 subscribers in the beginning to the current 30,000, and with a lot more people reading the posts from time to time on the organization's website—has given it a certain amount of clout; its moderators are able to cajole recalcitrant health officials into divulging what needs to be known with a mixture of moral suasion and the implicit threat that they'll look bad if they don't. ("A lot of public health is done in private," Madoff explains.) And ProMED has received the accolades of emulation. The WHO itself has developed something similar in its Global Outbreak Alert and Response Network, though this is able to go much further than ProMED, since the WHO has the power to send people to any of the 50 or 60 outbreaks a year that it deems of interest. Another WHO initiative, also supported by the government of Canada, has been a fully automated system for trawling quickly through the Web for news of outbreaks. The output from this project—the Global Public Health Information Network—is not in itself openly available; it goes only to a restricted list of subscribers. But since some of them are members of ProMED, anything of importance is quickly shared.

As far as Madoff is concerned, the more such networks and tools the better; he himself now uses the Google News Alerts feature to track outbreaks of dengue around the world. The added complexity of more and more sources and connections makes the system more complete, and at the same time it stays well-regulated. Each network is a check on the others, pointing out data that have been missed. "One of the things that comes out of the infectious disease community is that no single system is sufficient," Madoff says. "There's room for official report systems; there's room for automated systems and regional networks and topical networks that just focus on, say, tuberculosis. And there's value to all of these overlapping and redundant systems. That way, if we miss something we can see it through another system." Asked what ProMED has missed recently, he pauses: "I can't think of an example. We're pretty good." He gives a shy laugh. "But I'm sure it's happened."

• • •

Networks of experts are one possibility for tracking disease; networks of the infected are another. The need to get the earliest possible warning of biological attacks has led to the development of experimental systems that try to spot the patterns of disease even faster than doctors can. Their hope is that modern data-gathering systems may be able to find unusual patterns in the incidence of disease just a bit more quickly than traditional methods, and thus save lives. Even a little bit less time between a large-scale attack—the release of a big cloud of anthrax, say, or of some other aerosolized killer—and its discovery could, if used properly, save lives. Infected people fall sick at different speeds depending on the dose of pathogen they receive, their immune systems, and other quirks of fate. If the first to sicken after an attack are quickly identified as being part of a pattern, then it may be possible to get early warning of what's afoot. At that point, further work could show where and when the pathogen was released, which would give health authorities a chance to trace other victims, treat them if possible, and if appropriate vaccinate those who have been in contact with them.

There are a vast number of attempts under way to make such "syndromic surveillance" systems work, especially in the United States. A system that actively gathered information from the city's emergency rooms was set up in New York in the aftermath of 9/11; another, called Essence, which started in the Washington, D.C., area, works with data from military clinics. Researchers at the University of Philadelphia are developing a system to monitor sales of over-the-counter pharmaceuticals on the principle that the earliest symptoms are more likely to be treated at home than in hospitals. Other systems are looking at the data gathered by HMOs. The urgent desire for an early-warning system—*any* early-warning system—against biological attack has prompted dozens of pilot projects. The challenge, according to Dan Sosin, who keeps track of these developments for the CDC, is trying to figure out which, if any, are really effective.

The technique of syndromic surveillance also has inherent problems. In the case of a large outbreak, for instance, overwhelmed emergency rooms might make what was going on obvious with no need for any syndromic surveillance system. On the other hand, small outbreaks would be difficult to pick up at all. Martin Hugh-Jones—anthrax specialist, ProMed stalwart, and gadfly—enjoys provoking colleagues working on syndromic surveillance by posing the problem of a small-scale anthrax release in Metro Center, the busiest interchange in the Washington, D.C., subway system. If 20 people became sick and all ended up treated in different places, would any system notice? He doubts it.

You might argue that, regrettable as they may be, small attacks that no one notices are not a huge problem. They're not catastrophic, and if they go unnoticed they won't provoke terror. But to take that attitude would mean not just ducking a duty of care. It would also be strategically shortsighted. Some terrorists may do trial runs—what the military calls "reconnaissance by fire"—to find out how well their systems work. It has been suggested that the anthrax letters might fall into this category. Others may just make mistakes, as Aum did with its anthrax. The best way to forestall a large attack may be to pick up small ones.

Another problem inherent in this sort of surveillance is specificity. A syndrome is, in medical-speak, a cluster of symptoms to which a specific cause has not been ascribed. And with most bioweapons, the early symptoms look very like those of much more common diseases. Like picking out shipping containers for inspection from among the thousands that pass through a port every day, discerning which early symptoms require more investigation is a huge hurdle. A vast range of diseases start off with "flu-like symptoms"—they're the body's default response to lots of infections. All but one of the CDC category A bioweapon threats first make their presence felt with "flu-like symptoms." So does SARS. So does initial infection with HIV. So, of course, does the flu—which would make an attack during flu season, let alone an attack using a modified flu virus, particularly hard to spot.

A study carried out at Johns Hopkins University's Center for Civilian Biodefense Strategies made this point vividly. Researchers there considered a scenario in which an anthrax attack in Baltimore took place during a professional football game. On a November evening, an unmarked truck releases a plume of anthrax for 30 seconds while driving on an elevated highway upwind of the stadium, infecting 16,000 fans without being detected. At the university hospital's emergency room, John Bartlett, head of infectious diseases at the medical school, asked the staff how quickly an influx of affected patients might be noticed at that time of year. The physician in charge, one of only five in Maryland to have completed a recent course on bioterrorism, asserted that patients turning up with the symptoms of anthrax would be diagnosed as having flu and sent home. The radiologist, shown an x-ray of someone with inhalation anthrax, said ER staff would interpret the picture as a "widened mediastinum," a condition that might be a symptom of aortic aneurysm, tuberculosis, or sarcoidosis; anthrax didn't come up. The ER's lead lab technician said he had not seen *Bacillus anthracis* isolated from a patient in all his 25 years on the job. If it turned up in a blood test, staff would consider it meaningless contamination caused by one of anthrax's benign relatives. Only if it turned up in three different tests would the staff conduct further investigations, and those tests would take 48 hours. Forty-eight hours after the first blood tests, a large number of the infected would be dead or dying. This scenario's final death toll was about 4,000.

None of this should be seen as a critique of Johns Hopkins. Most people with flu-like symptoms in November really do have the flu or something similar; the vast majority of widened mediastinums aren't a result of anthrax; and almost all *Bacillus* species in blood cultures are contaminants. George Poste explains this in terms of an old medical school adage designed to warn students away from recondite but unlikely diagnoses: "When you hear hoofbeats it could be a zebra—but it's probably horses."

The problem is that in a bioweapon attack the hoofbeats will indeed be zebras—quite possibly vast trampling herds of them. This is why Poste is pressing for new diagnostic technologies. Although we do not yet have the miniaturized versions of canaries, food-tasting mice, and Kuwaiti Fighting Chickens that will one day harness pure biology to detect and identify pathogens, we can combine elements of biology with electronic information processing. A combination of biotechnology and semiconductor manufacture techniques makes it possible to put hundreds of thousands of probes capable of recognizing DNA sequences and protein structures onto a single silicon chip. It's a technology such as that with which the biotech company Arrow monitors jumping genes while looking for new proteins against which to develop antibiotics; some companies are now offering chips with probes that can see the whole human genome. Poste suggests that chips be made that would recognize the protein and gene signatures of 50 or so of the most plausible bioweapons. Such chips could then be used to test blood from patients on a routine basis. All those people with their flu-like symptoms, all the hypothetical fans feeling flu-y after an attack at the football game, would be tested with these chips upon showing up at the hospital. Most of the time the tests would show nothing unusual, because the steed that had brought the patient to the clinic would be a horse. Occasionally, though, these novel blood tests would turn up a zebra—which is why Poste calls the technology the Z-chip.

The idea would be to make Z-chip systems easy to use and, by buying in bulk, very cheap. "They have to have a low-tech front end," explains Poste, meaning that feeding in a sample has to be simple, "and they can't interfere with the 7 to 15 minutes per patient [doctors and nurses have] or the chaotic procedures of the Emergency Room." The system would use already developed techniques for handling small samples of fluids to get the blood to the relevant portions of the chip automatically. Like a network of smart dust, or a trawl through databases crammed with details of human relationships, the chip would produce patterns of data. The patterns formed on the chip would be highly reliable, because the chip would be testing for various different subcomponents of various different organisms. Those patterns would be quickly interpreted—and at the same time as they were delivered back to the clinic, they'd also be put into a national system looking for greater patterns, a system such as those imagined for syndromic surveillance, but working directly off test data and designed for speed from the ground up. Cobbling things together ad hoc, Poste says, is not the way to solve the problem. "You need a total end-to-end solution. You need to treat this like an industrial campaign." The Pentagon takes him seriously and is allocating millions of dollars to Z-chip research.

In some ways, the Z-chip looks like the sort of biodetector you might want plastered all over towns, cities, subway stations, and, indeed, the Baltimore Ravens' M&T Bank Stadium—something that can pick up the patterns associated with all sorts of pathogens. But it would never be able to work in such environments. The Z-chip needs the pathogens it detects to be processed by the human body before it can get to work. The human body puts a lot of effort into making sure that its bloodstream is very stable and clean, its acidity and temperature tightly controlled. It is a nicely noise-free environment in which to look for a signal. So the Z-chip only serves as an all-purpose biodetector if you include an infected person as its front end.

Once Poste's "industrial campaign" gets underway, advanced Z-chips will look for more and more things at once: "This chip," Poste boasts, "can contribute enormously to improved public health and rational therapeutic decisions by giving us a much better profile of the horses that are with us all the time." Later versions of the technology, he imagines, could leverage their intimacy with the body's biology further still by looking not just at the physical manifestations of a weapon in the blood, but also at the information the body is generating about it. The things that cells in the immune system do and the genes they express while doing them are very precisely tailored to the threat the system is facing at any given time. Z-chips that looked at the genes expressed in white blood cells could listen in on the body's own surveillance system, a far more subtle and perceptive judge of what's going on than human technology can yet manage. The technology might end up as something that simply makes the information that you already have inside you accessible to your eye and mind. Making the way things work more visible, letting the whole system know itself so that it can talk to itself, is the high road to resilience.

Like Kris Pister's tiny, ubiquitous sensors and Michael Sailor's flakes of glitter, Poste's Z-chip would be a new way of sensing the world—not from afar, but from within; not from a massive camera high up in orbit, but with intimate traces of technology up close.

One of the most striking images of the cold war was a fictional one: the Big Board in Stanley Kubrick's classic film *Dr. Strangelove*. This magnificent black comedy about a mad general setting off irreversible nuclear apocalypse is played out in scenes at Burpleson Air Force Base, on board a B-52 bomber—and in a cavernous, shadowy Pentagon War Room, a theater of the absurd where the sacrilegious becomes sacred.

Suspended above the circular conference table and polished black floors is the Big Board. A gigantic screen, worthy of a state-of-the-art Cineplex, leaning in toward the center of the room, it presents the assembled politicians and generals with a map of the hidden half of the hemisphere, a real-time image of the Soviet Union penetrated by the tracks of U.S. bombers on their way to start the destruction. To Kubrick's cold warriors gazing up at it, the Big Board was a holy icon, the thing they most wanted to defend from prying eyes; it was a representation of the world so valuable as to be worth risking the world for.

Tomorrow's sensor networks have none of that iconic force. Made up, purposely, of cheap, minuscule components, they may be insidious but they won't be imposing. In other words, they would never have satisfied *Dr. Strangelove*'s worshipers of the mythic Big Board. But that's no bad thing. The new sensor systems understand the world by finding ways to be part of it, not by fostering the illusion of standing apart from it. To swap the God's-eye view for the view from within— within bodies and spaces and atmospheres—is to swap the sensation of power for the knowledge of connection.

NINE

A Storm in Any Port

IN LATE APRIL OF 1986, Lieutenant Stephen Flynn was serving as the captain of the Coast Guard patrol boat *Point Arena,* stationed in Norfolk, Virginia. It was late Sunday night, shortly after the Reagan administration had ordered the bombing of Libyan leader Muammar al-Qadhafi's military headquarters. The air strike, carried out in response to evidence that Libya had sponsored the bombing of a West Berlin disco, was the first ever by the United States in response to terrorism. Flynn, assigned to the *Point Arena* two years before at age 23, was the youngest Coast Guard commander in the United States. He was both ambitious and green. "When I look back on it," he says now, "I wonder how the hell they ever gave me the keys."

Born in Salem, Massachusetts, Flynn grew up in nearby Newburyport, the birthplace of the Coast Guard. As a child he relished the sea stories told by his father, a former enlistee in the service, but as the third of six children, his decision to attend the academy was more financial than romantic. "There weren't deep pockets to send us all to school," he says. "The price was right." A year and a half out of the academy, Flynn was a capable commander and respected by his sailors. His cutter, with a crew of 10, conducted routine search-and-rescue, law enforcement, and drug interdiction operations along the

Virginia and North Carolina coasts. That April night, however, Flynn received an unusual classified message from the Navy, requesting that he report to fleet headquarters the next day.

Arriving the following morning, Flynn found himself surrounded by an imposing array of military brass: half a dozen Navy captains, a Navy intelligence officer, and a senior Coast Guard commander he had never met. When the meeting began, the intel officer announced that the Navy had information that terrorists were planning to attack the USS *Yorktown,* a missile cruiser, as it approached Virginia on its return from the Libyan operation. (Large naval ships such as the *Yorktown* are great at lobbing missiles at faraway lands and other ships, but not so great at protecting themselves against a sneaky civilian boat, as the attack on the USS *Cole* would prove more than a decade later.) The officer then turned to the senior Coast Guard commander and asked what steps he planned to take to protect the *Yorktown.* "Well, that's why we have Lieutenant Flynn here," Flynn remembers his senior officer saying. "He's going to run that operation for us. Flynn, at what point would you allow a vessel to approach the *Yorktown* before you fire upon it?"

All eyes turned on the unsuspecting lieutenant. "I'm listening to this, and it is sort of surreal," he recalls, hunched over a plate of scallops at the trendy Washington, D.C. restaurant Zola. Flynn is now a 41-year-old homeland security expert at the Council on Foreign Relations, a nonpartisan think tank in New York City, and the author of 2004's *America the Vulnerable: How Our Government Is Failing to Protect Us from Terrorism.* He's a stocky man, with deep-set, piercing eyes, and a wry sense of humor that often seems to accentuate his Massachusetts accent. Back in his Navy meeting, he tried to use humor to lighten his response to the commander. "I said, 'Well, sir'— this is before I went to my diplomacy school—'I would fire on that vessel after it has assaulted the *Yorktown,* turned, and decided to pursue me as I'm moving at flank speed in the opposite direction. I don't know what your war plan says, but I'm not a counterterrorism asset.'" The commander, not amused, ordered Flynn to devise a plan to protect the cruiser.

Assessing his options back at his station, Flynn realized first that there was little point to escorting the boat in, since he had neither the speed to keep up with it nor the firepower to protect it. At the time, there was a bullet shortage for his patrol boat's 50-millimeter guns, and the *Point Arena* hadn't fired its weapons even in practice for months. He also knew that randomly plowing the waters of the Chesapeake Bay in search of threatening boats was beyond futile. On an average day the bay would be teeming with vessels of all kinds. Without any more specific intelligence, he had no idea how to distinguish which one was the terrorists'.

So Flynn turned the problem around. Instead of focusing on the vessels that might belong to terrorists, he decided to concentrate on making sure he knew which ones didn't. On the day of the *Yorktown*'s arrival, he stationed Coast Guard cutters at each of the five entrances to the lower Chesapeake and—with no explicit authority—had them instruct all legitimate boat traffic to stay out of the bay. "We had a relationship with these boaters, so it wasn't a big deal," he says now. Because they regularly worked the bay, Flynn and his crews could easily identify the good guys and direct their behavior. In network terms, they turned the lower Chesapeake into an "island," separating it from the rest of the maritime system: without the cover of a crowded bay, potential terrorist vessels would become easy to spot. The result, Flynn says, was that "I had nothing on the water. I called the captain of the *Yorktown* and I said, 'Look, here's the deal: I basically emptied the bay. Anybody that comes at you, I don't know who it is and it's your job to take care of it.' "

Whether deterred by the show of force or, more likely, nonexistent to begin with, the Libyan terrorists never materialized, and the *Yorktown* steamed home without incident. In *America the Vulnerable*, Flynn recounts the anecdote to illustrate the reluctance of the military to deal with homeland threats, while agencies like the Coast Guard are underfunded and undertrained. But the episode also helped Flynn formulate one of his most valuable principles in confronting terrorism. In a world where threats can come from anywhere—where every boat, airline passenger, or shipping container seems to present a potential danger—the solution is not about scrutinizing every person and object, looking for the one that may be a risk. It's about knowing which ones aren't dangerous at all.

•   •   •

Counterterrorism is an endeavor plagued by grim probabilities. The perpetrators of asymmetric warfare, the small groups seeking to inflict damage on governments and nations, succeed by taking advantage of the near-impossible odds of finding a few determined terrorists hidden among a population of ordinary people. Any effort to prevent terrorist acts inevitably requires applying limited resources to a problem that is limited only by imagination—thus the oft-stated adage in intelligence circles that "the terrorists only have to get lucky once, but we have to get lucky every time." Like Steve Flynn in his defense of the *Yorktown,* we are faced with an intractable task of identifying a shadowy enemy in the context of countless friendly distractions. The result is that in the realm of terrorism no foolproof protections exist, whether technological or otherwise. Rather, there are only combinations of defenses that serve to improve the odds by providing a greater chance of distinguishing a few dangerous people (or weapons, or vehicles) among a population of millions.

The problem is similar to that faced by radiologists evaluating a series of routine mammograms. Simply knowing what a cancer "looks like" is not enough. Clinicians must be able to confine their field of possibilities with an equal understanding of what a normal, healthy mammogram *should* look like, or else every spot on the x-ray is flagged as a potential cancer. Counterterrorism requires the same ability to distinguish the benign from the malignant. In airport security— where Rafi Ron and others are attempting to meld technological searches with the human ability to detect suspicious intentions— proposed trusted-traveler and computer-profiling systems are technological means to try and separate harmless passengers from potential threats. Finding the proverbial needle in the haystack is a challenge that involves two distinct but inseparable notions: winnowing down the amount of hay to search and learning to spot the needle.

Nowhere is that fact more evident than in the vexing challenge of securing the global cargo network, the thousands of ports, tens of thousands of ships, hundreds of thousands of companies, and millions of people that combine to transport the world's goods. More than 90 percent of all trade moves by sea—a great deal of it in 20- and 40-foot steel containers that pass by truck and rail to ports, where they are loaded onto giant container ships and sent across the oceans.

In the United States they are offloaded at one of some 360 ports and then hauled over land, the contents eventually finding their way to our store shelves. That system, because it involves moving containers through multiple modes of transportation, is known in the industry as the "intermodal supply chain." Two hundred million of those metal boxes travel the world each year, and more than 7 million of them enter the United States by sea and over land, carrying the goods and materials we use every day—from clothing, to toys, to car parts. For Steve Flynn—and for many experts in and out of government—the most haunting terrorism nightmare is the prospect of a nuclear weapon delivered to its target in a shipping container full of cocktail drink umbrellas or plastic bobble-head dolls.

Depending on who you ask, U.S. Customs physically searches somewhere between 2 and 5 percent of the millions of shipping containers imported into the country. One can argue about how easy or difficult it is for terrorists to gain access to the materials and know-how to create a nuke. But if terrorists do obtain such a weapon and decide to use a container to deliver it, they have a simple, almost overwhelming, mathematical advantage. The detonation of a nuclear weapon in a U.S. port—many of which are located in highly populated areas—could dwarf the September 11 attacks in its consequences. And for a terrorist group looking to import weapons of mass destruction to targets within the United States, container shipping offers the best opportunity to do so. In both 2002 and 2003, ABC News crews succeeded in using shipping containers to smuggle a suitcase full of harmless unenriched uranium—from Turkey and then from Indonesia—into the United States without detection. In each case, the container was never opened by any authority.

Cargo security, then, becomes a matter of finding a single dangerous object hidden among a vast amount of benign ones. The hijacker lurking among the millions of innocent airline passengers offers the same quandary, as does the suicide bomber among everyday café patrons, the car bomb among the innocuous vehicles on the street, and the bag with sarin gas among thousands of subway passengers. The problem is compounded by the fact that the cargo system itself, like the Internet, the power grid, and other infrastructure networks that we all depend on, was never designed with security in mind. Instead it evolved through the efforts of the companies that make and send goods (known in the industry as "shippers") those that transport them ("carriers"), and a host of middlemen to find the most efficient way to move cargo from one place to another. In today's era of globalization, container shipping is a form of controlled chaos, where razor-thin profits drive fierce competition, encouraging ever greater efficiency and speed. The system has evolved organically toward the goal of standardizing trade and thereby moving as many goods as possible, as fast as possible. It is a network where "just-in-time" manufacturing strategies—in which products are assembled only when they are needed, and companies try to keep as little inventory of their goods as possible—squeeze margins out of every hour and every container move. Just as in other market-driven systems, latency has been banished in the name of efficiency.

Products in this system can originate almost anywhere in the world, and every shipment involves a dizzying number of people and companies. End to end, a typical cargo transaction can involve 20 to 25 parties, including manufacturers, truck drivers, freight forwarders, customs brokers, middlemen, government agents, retailers, and carriers. The result is a system that doesn't care what it is transporting, only where it is going. And like the Internet or the public health system, no one company or government owns this network; it is perpetuated by the collective interest of its participants. By the same token, no one party has the means or incentive to secure the whole system. From the point of view of getting things from all the point A's in the world to all the point B's—which the system does very well—security is a drag. Finding out who has touched every box, keeping track of which containers have waited where, checking and ensuring that nothing has been tampered with—all this takes time. So people secure their own containers as well as they think makes sense—which may be with as little as a 50 cent cargo seal—and accept some level of theft and pilferage. The rather more consequential risk that someone might slip in a nuke is left unaddressed.

Securing the entire system is, like assuring global public health, one of those things that economists call a public good: if it benefits anyone it benefits everyone. But one shipping firm tightening its security doesn't solve the problem; it just drives up that firm's costs. So although cargo companies collectively have an incentive to put up barriers to terrorism, individually they have little incentive to act unless everyone else does. (As it happens, the classic example of a public good, lighthouses, also involves the safety of shipping: the provision of lighthouses benefits every ship if it benefits any one, but you can't charge individual users according to the benefits they receive.)

The resulting gaps in the cargo security network are starkly illustrated by an incident that occurred on the evening of October 18, 2001. In the Italian port city of Gioia Tauro, a police officer patrolling the waterfront heard noises coming from a shipping container aboard a docked ship from Egypt. When the police opened the box, they made a startling discovery: a well-dressed and clean-shaven Egyptian-born Canadian citizen named Amid Farid Rizk. He had apparently been at sea in the container—which he had outfitted with a laptop, a satellite phone, and air holes—for five days. Italian authorities arrested Rizk on suspicion of terrorism, and media reports announced ominously that he had been carrying airport maps, security badges, and an aircraft mechanic certificate. After a hearing in which his lawyers claimed that he was a Christian fleeing persecution in Egypt—a dubious claim given that he held a Canadian passport—an Italian judge ordered him released on bail. He subsequently disappeared, and has yet to resurface.

The discovery of Amid Rizk was no surprise to those close to the cargo industry, in which smuggling, organized crime, and stowaways are a permanent fixture. And there are, of course, easier ways for terrorists to get around than shipping themselves in a metal box with a bucket for a toilet. But Rizk did drive home a frightening series of questions: If a person could hide inside a container, why not a weapon? What would happen to the free flow of commerce in the aftermath of a nuclear incident involving the cargo system? Steve Flynn, for one, believes that failing to fill the security holes now will produce a paralyzing and draconian response if an attack does occur. "Right now you can blow some smoke," he says, "but when we really have an event involving a container and you tell the American people, 'Well, yeah, as a matter of routine anyone with 30 tons of material can load it into a box, throw on a 50 cent seal, put it off to the races, and nobody checks it,' they probably will say: Stop that. You will get the post-9/11 aviation response: Assume every box has terrorism in it. Check every single one." The ripple effects of that response—an obsessive focus on confiscating tweezers and checking shoes—are the same ones Rafi Ron is still trying to overcome in airline security.

That, after all, is our natural reaction to any extreme threat. Whether the issue is food safety or airline security, we want guarantees. Stop the terrorists from slipping through, stop the mad-cow beef from getting to our table. One way to try and provide that certainty is to test every sample—or in the case of terrorism, to hand-search every passenger, open up every container, and keep a file on every citizen. But absent unlimited time and resources—and a privacy-free surveillance state—it is impossible to track down every terrorist, keep tabs on every immigrant, or search every corner of society. Worse, such a blanket search avoids concentrating efforts and resources where they might be more effective. Flynn likes to cite what he calls "an age-old axiom of the security field: If you look at everything, you will see nothing." Treating every case equally is, as Israeli airport security experts would argue, yielding the statistical advantage to terrorists.

So how, instead, do we improve the odds? Flynn's answer is the same one he employed in protecting the USS *Yorktown:* turn the cargo security problem on its head. Stopping the "bomb in a box"—as the scenario is called in security circles—is not just about searching for the dangerous container. It's about trying to identify and then ignore the vast majority of containers that are completely safe. "Your real problem," he says, "is how do you validate low risk as low risk. There's no such thing as a low-risk container in the current environment." The goal of Flynn's scheme is not to figure out how to search more containers. It's to figure out which containers *not* to search.

The solution—of which Flynn is not the only architect, but perhaps the most articulate spokesman—involves creating "smart containers," boxes that contain tracking devices and sensors that he says will bring order, consistency, and transparency to the cargo system. The data from those containers, supported by a rigorous inspection system, could feed software systems to keep tabs on the location and status of nearly everything moving through the network. In such a system, one could quickly identify containers that deviate from their prescribed path or are somehow tampered with—just as network security people use traceback to identify rogue packets.

Such technological tools, whether in cargo security or any other area of terrorism, are an attempt to take the advantage away from terrorists, pushing them into smaller and smaller piles. The goal is not to permanently secure the network—an impossibility, given its global scope—but to make life more difficult for an adversary that is both relentless and highly imaginative, constantly searching for new vulnerabilities. "If they see that the containers are all sensored-up, they are always looking for the easier path," says Charles Massey, a port and nuclear material security expert at Sandia National Labs in New Mexico who is helping test technology for the Ports of Los Angeles and Long Beach. "It's like Jell-O: squeeze it in one place and it comes out the other. I think unfortunately what 9/11 showed is that [terrorists] are pretty good systems analysts. We've just got to be better."

A smart container network, however, could potentially serve to turn the odds against the attacker. A group that has spent millions of dollars and years of effort obtaining a single nuclear weapon is less likely to stick it in a container and send it off if they know that there is a high chance that it could be detected. Even an imperfect defense can provide deterrence, or at least ensure the viability of the system if an incident does occur. "It won't be perfect," says Flynn, "but the chances are that when it fails, its credibility will survive."

Reducing the searchable pile becomes a constantly evolving process of revalidating low-risk containers—a process that is repeated throughout our efforts to defend against terrorism. "As I succeed in determining where the risks are, focusing on them, and putting aside the ones with no risk," says Michael Wolfe, a logistics consultant who has studied the cargo security problem extensively, "what I have done is increased proportionally the attractiveness of the supposedly secure ones to the bad guys." Just like airline trusted-traveler systems, the technological solution can itself provide new targets for those looking to do mischief, by declaring some types of containers or people benign. "Clearly that's where I want to be if I'm on the terrorist side," says Wolfe. "If I can figure out a way to crack that, boy, am I home free. There are games within games within games, and it's a matter of continually making it harder and diverting and pushing the opportunities into tougher places."

The story of that idea is the tale of our most vexing problems in the "war on terrorism." To find what is suspicious, we have to first filter out some large portion of the world that we judge to be outside the realm of suspicion. In that process, technology becomes an assistant, helping to verify the everyday from the hazardous and feed us information about both. Like card counters in a Las Vegas casino, our hope is to use that information to turn the odds in our favor.

You need only drive across the Vincent Thomas Bridge, arched high above the Los Angeles Harbor, to get a sense of how bad those odds are. Below lies the vast expanse of the Ports of Los Angeles and Long Beach, the entry point for 45 percent of all the goods coming into the United States. At over 10,000 acres combined, the two ports are a beehive of activity stretching in all directions. Ships stacked high with containers sporting labels such as "K" Line, Hanjin, and Maersk Sealand glide through the port, while trucks and tractors zip among the concrete yards, between row after multicolored row of stacked boxes. The ports handle more than 8 million of those containers every year, in addition to the tons of oil, salt, fruit, cars, and machinery that aren't shipped in containers but also pass through en route in and out of the country.

At one end of the bridge lies Terminal Island, home to the ports' three largest container terminals and one of the busiest cargo through points on the planet. That's where you'll find Pier 400, the world's largest shipping container terminal. John Ochs, Pier 400's chief of security, is sheriff of this domain, responsible for securing an area the size of 366 football fields. A former Coast Guard shipmate of Steve Flynn, Ochs is tall and imposing, with a thick moustache and a dry wit. "When people first come here," he says, as he extends a hand adorned with a large Coast Guard academy ring, "they usually have a 'holy shit' reaction."

Indeed, Pier 400, operated by Maersk Sealand, a division of the Denmark-based cargo behemoth A.P. Møeller-Maersk, is home to a staggering array of activity. Maersk's distinctive sky-blue cranes grab containers off docked cargo ships—the largest of which can carry over 8,000 boxes 20 feet long, 8 feet wide, and 8 feet tall, stacked 9 deep into its hull and 8 high on deck—at a rate of one every two minutes. Crane operators deposit them onto trailers that drivers whisk to rows of stacks nearby or directly to rail and truck depots. Twenty-five hundred trucks come in and out of Pier 400 each day, through 56 lanes of gates, and the terminal can load two mile-long trains simultaneously. On any given afternoon around 15,000 containers are either moving through the terminal or stacked in wait. If you could see the inside of an Internet router as data packets zip in and out, this is what it might look like.

And all that before Maersk is even finished building Pier 400. When completed, it will move nearly 2.5 million containers a year, 60 percent of which come from China. "At any one time," says Ochs, "there are probably 20 that we don't know where they are." Those missing containers, he says, are usually tracked down in a matter of hours by so-called optical character readers—pickup trucks outfitted with cameras that automatically capture the serial number of each box. The trucks drive through the stacks, beaming the numbers back to an operations center that maps the box's location. It's an impressive system, but the ultrasophisticated Pier 400 is far from the norm in an untidy cargo network where containers are often lost, stolen, or misdirected.

Pier 400, after all, is merely one hub in a global cargo network that has employed container shipping since the 1950s. Before that time, all ocean-bound commerce traveled on ships in loose crates and stacks of materials, called "break bulk," which were loaded on and off using the manual labor of longshoremen. In 1955, a North Carolina trucker named Malcolm McLean, looking for an easier way to move goods from ships to trucks, bought a steamship company and began using a standardized container to send goods across the Atlantic. The containers could be loaded and then transferred directly from one mode of transport to another using cranes, vastly increasing efficiency at ports. As McLean's company, Sea-Land—which would later merge with Maersk—grew into a global trade behemoth, other shipping lines began adopting the container system, and by 1972 over 6 million containers were in service around the world. By 2000, the number of boxes (containers come in standard 20- and 40-foot lengths, and ship-

ping statistics are measured in 20-foot equivalent units, or TEUs) in existence had reached 200 million, growing at a rate of 5 percent a year.

The network that evolved to accommodate McLean's containers
allows for an infinite variety of routes by which components are
brought together to assemble final products for store shelves or show
floors. Home Depot, the builder and home-supply retailer, obtains
products from over 40 countries alone. Mammoth international ports
in Hong Kong, Rotterdam, and Singapore serve as hubs for the globalization of production, taking in goods manufactured cheaply in
remote areas of countries such as Slovenia, Pakistan, and China and
pushing them through toward the world's vast consumer markets.

To see how difficult it is to secure such a complex system, imagine
a batch of cheap plastic toys, ordered by an American retailer. The
goods might be made in a factory in a remote area of China, boxed in
wooden crates, and driven to a nearby town, where they are combined with products such as hand-carved chairs and fans. The load is
then trucked to a larger city, where a foreign exporter—one of the
freight consolidators that handle 30 percent of all container shipping—
loads all the shipments into a container, attaches a thin 50 cent metal
seal with a serial number, and sends it on its way. The container travels by truck or train to a nearby port, such as Shanghai, where it is
loaded on a ship. That ship sails to Hong Kong, where the box is transferred by crane to the transoceanic container vessel that arrives at the
docks of Pier 400.

A typical shipment through that system encounters little security
along its route. While each carrier is required to electronically trans-
mit manifests describing the contents of its containers to U.S. Customs
24 hours before the U.S. bound ship is loaded, there is little effort or
time to verify that what is in the box matches the manifest. Containers,
in fact, are already used as a vehicle to fund terrorism. According to
Interpol, in October of 2003 Beirut authorities confiscated containers
full of fake auto parts intended to help fund the group Hezbollah.
Consolidated cargo shipments are often labeled vaguely. Theft and organized crime are rampant. Although the container industry loses an
astonishing $50 billion a year to pilferage—most of it through inside
jobs—the loss for each individual company is considered largely a
standard operating cost.

The majority of shipments into the United States are carried by large companies, on container ships that originate from major ports such as Hong Kong, Singapore, and Rotterdam. But the sea is also a lawless frontier of smaller merchant vessels, many of them registered by mysterious owners, under "flags of convenience" in such countries as Liberia and Cambodia. The seemingly simple issue of knowing where all the ships are located is remarkably intractable. Although most ocean-going vessels over 65 feet are required to carry GPS-equipped transponders that let authorities identify them, many of them are out of compliance, since standards are only enforced by the country in which they are registered. And as of February 2004 only five U.S. ports had the equipment to read signals from international ships.

The threats to ports from ships, trucks, and cargo are as diverse as the materials that move through them. "The port is just a target-rich environment," says John Holmes, the former captain of the Ports of Los Angeles and Long Beach. "We have gas. We have oil. We have chemicals. We have bridges, roads, and containers." Complicating matters, ports are a jumble of law enforcement jurisdictions, between the Coast Guard, Customs, the local authorities, the port authority (which operates essentially as a landlord), and the port operators (which lease their terminals from the port authority). Before 9/11, Holmes says, "I'm not overstating it when I say that the airports were Fort Knox compared to the ports."

When it comes to containers, there are two basic ways that terrorists could exploit the chaos of the cargo network. As the nonprofit Markle Foundation's Task Force on National Security in the Information Age pointed out in a 2002 paper titled "Turning the International Shipping Industry into a WMD Delivery System," the first is to pose as a legitimate shipper, placing the weapon within what is purportedly a valid shipment. The second is to compromise a legitimate shipment along the way, either by enlisting inside help or simply breaking in at some point when the cargo is stationary. Millions of empty containers also pass through the network alongside full ones, each of them offering a tantalizing target. "As commercial airliners can be turned into weapons by terrorists," the report says, "so can this international delivery network be turned against the United States, with great efficiency."

An attack using the cargo network would only need to be partly successful to have massive effects. Even a minor incident could disrupt the system, setting off a network cascade effect not unlike the failure of an electrical grid. A terrorist event in one U.S. port would lead authorities to shut it down, along with perhaps all of the ports in the country. Those shutdowns would instantly trigger a massive cargo backlog that would overload other ports and halt the flow of cargo all over the world. There are precedents for such a failure: when the West Coast ports in the United States were shut down for 10 days in October 2002 because of a labor dispute (an outcome that had been anticipated and planned for by the shipping industry months in advance), the repercussions cost the U.S. economy an estimated $10 billion.

The impacts of a terrorist attack via the ports could be many times more devastating. A simulation exercise conducted by the consulting firm Booz Allen Hamilton imagined the effect of "dirty bombs"— nasty radioactive material wrapped around a conventional explosive— discovered in the cargo system. In the exercise, one dirty bomb was discovered in the Port of Los Angeles and another in Minneapolis, while a third exploded in Chicago. The tab for the resulting 12-day shutdown of U.S. ports ran to $58 billion. More damaging than the immediate disruption of the system itself would be the complications of restarting the gears. "People really just don't understand the consequences if something were to happen through the ports and they were shut down the way the airports and the airlines were shut down on September 11," says Jack Riley, a terrorism expert at the Rand Corporation, a nonprofit policy think tank. "Restarting the global shipping system is going to be logistically hugely more complicated, by multiple orders of magnitude."

The obvious solution to the threat is to increase the number of inspections of containers, but searching cargo is no simple task. Inspecting all the boxes on a fully loaded large container ship by hand would take more than 800 consecutive days.

Weaving through the traffic in a beat-up pickup truck, Ochs points out the row of containers from the day's unloading that Customs has selected for inspection—about 20 out of several thousand. Later that day, Customs agents will drive machines called Vehicle and Cargo Inspection Systems (VACIS) alongside the containers. A VACIS, mounted on the back of a truck with an arm that extends out over the container, sends gamma rays through the box to generate a picture of the interior—like an x-ray but capable of punching through thick metal. The whole process takes roughly 30 seconds. But VACIS does not, as the ABC News operation demonstrated, detect radiation or nuclear materials. If a decision is made to search a container by hand, the process can take a team of five up to three hours.

Every Customs agent is also outfitted with a radiation detection pager, ostensibly to help detect potential nuclear materials. But the pagers—notorious for their limited range—are largely feel-good measures. They might detect a large amount of radiation material being used in a dirty bomb, but they'd be unlikely to pick up the subtler signals of a genuine nuke. More sensitive radiation sensors called portal detectors, now in place in some U.S. ports, are useful. But they are plagued by false positives, set off by the low-grade radiation from shipments of bananas (all that potassium) and electronics. Some have suggested that instead, advanced sensors could be placed on the cranes that remove containers from ships, but the technology is not there yet. Ochs, as someone who deals with security on the ground, is wary of trying to apply high-tech tools in the chaotic setting of a port. "Whatever this gadget is, it's going to be a fairly sophisticated piece of technology," he says. "Now we're going to put it in the most hazardous environment, from a vibration standpoint, that you can imagine."

Many counterterrorism technologies risk the same fate: what works in the lab may not necessarily work in the field. In the case of containers, more powerful and robust sensors are undoubtedly on the way. The Department of Homeland Security wanted and got its own DARPA, the Homeland Security Advanced Research Projects Agency, which is funding research into a variety of radiation detectors—everything from backpack-sized passive sensors for the Coast Guard to radioactive material detectors that can scan a container in fewer than 20 seconds. Faster VACIS-like machines are already being deployed, and new variations using everything from neutron beams to cosmic muon rays are under development. Such technologies will no doubt improve our ability to detect nukes at our ports and borders. But even the most advanced detector won't let inspectors stop and check out every container, at least not without slowing down the global cargo system to the point of collapse. Even if it were possible, relying on sensors to catch threats at the port, when they have already arrived in the country, means gambling to get lucky every time. "A lot of money is being spent on sensors," says Flynn. "But none of that solves the problem of the needle in the haystack. It's better than nothing, but only as a last line of defense."

The 300-plus ports in the United States, after all, are only nodes in a worldwide cargo network that offers limitless opportunities for a terrorist attack. Confining our defenses to places such as Pier 400, says Flynn, "is like a network security expert protecting the server next to his desk."

It's a mild summer afternoon in Washington, D.C., and Steve Flynn is in the midst of one of his typically dizzying visits to the capital. Much of his day is spent rushing between government buildings, from one briefing to the next, rolling his overnight bag behind him. With his short but comb-resistant hair and lumbering stride, even in a suit he gives the impression that he may have just stepped off a Coast Guard cutter. Flynn spends much of the day giving a kind of stump speech on port and cargo issues to congressional caucuses and policy groups, proselytizing the dangers of ignoring the terrorist threat. As he walks the halls of the Capitol, congressional staffers greet him as "the busiest man in Washington." One congressman introduces him as "the world's foremost expert in port security."

It's not a designation that Flynn would have imagined in his Coast Guard days. After his two-year stint as the captain of the *Point Arena,* he went on to teach at the Coast Guard Academy and pursue a Ph.D. in international politics at the Fletcher School at Tufts University, and then worked for the National Security Council under the first Bush administration. It was during the mid-1990s, between stints at the Brookings Institute, back at the helm of a Coast Guard buoy tender, and back at the National Security Council during the Clinton administration working with the future counterterrorism czar Richard Clarke, that Flynn became interested in studying ports and borders. In particular, he was curious about the effects of organized crime and the international drug trade on the cargo world following the collapse of the Soviet Union. What he found was an increasingly fluid system of commerce that failed to take security into account. "The drug problem was really like the diet that cardiologists prescribe to see how your arteries are doing," he says. "It was telling us something about globalization, that globalization is value-neutral. It facilitates bad as well as it facilitates good." Globalization, in other words, is dual-use.

Flynn began to see the confluence of asymmetric warfare and the lack of cargo security as a disaster in the making. "Like a well-designed computer software package, the transportation revolution is striving to create a largely transparent underpinning to global economic life that belies a complex and interdependent infrastructure that facilitates it," he wrote in a paper for the Department of Transportation in 1997. "Just like software, a single 'bug' could conceivably throw the entire system into a tailspin."

In 1999, Flynn landed at the Council on Foreign Relations, where he combined his study of ports with an examination of the increasing threat of terrorist attacks. There, he began to envision what would later evolve into the design for a technology-based, multilayered approach to port security, one that employed the principle of reducing the amount of hay to search. The first element would be a policy to "push the borders out," confronting the container threat closer to the source by sending agents overseas to inspect containers as they are loaded. The second would be to apply technological hardware, in the form of a "smart container" outfitted with electronic seals and built-in sensors. Containers that kept track of their location and status, he reasoned, would create what he calls "in-transit visibility"—containers that are easily verifiable as low risk. The twin thrusts of Flynn's plan would counter the two threats outlined later by the Markle Foundation report: terrorists posing as shippers and compromising containers en route. "The first part is, can we feel confident that what is loaded into a box is legitimate and authorized," he says. "The second piece is, once it's on the move, has it been intercepted and compromised? If we have a system in place that gives us the confidence on those two things, that most cargo is legitimate, we could in fact say, 'That stuff we don't need to worry about.'"

Before 9/11, Flynn was virtually unknown outside of the obscure realms of port and border security policy. Even in those worlds, he says, he was viewed as "a little wacky." He wrote a portion of the now-famous Hart-Rudman report on the vulnerability of the United States to terrorist attacks, but still his ideas got little traction. "It was met with this big laissez-faire reaction, which was that's somebody else's problem to handle. 'My job is to make a buck, the government's job is security.' The government wasn't interested in doing security because they ran into the marketplace. So it was this constant exercise—and a lonely business—of saying there is a real problem here, but there is a solution."

Then came the attacks on the World Trade Center and the Pentagon. "When the second plane hit," he says, "my reaction was 'I have failed.' That wasn't a rational reaction. I mean who am I to say that I can save the world?" A few months before 9/11, he had given a presentation in which he juxtaposed a picture of Osama bin Laden with a container ship, describing the possibility of a weaponized container shipped from Karachi, Pakistan, to the Port of Long Beach. "After September 11," he says ruefully, "I was the only guy who didn't need to change his slide show."

The tragedy, however, kicked Flynn's ideas into high gear. In the summer of 2001, a U.S. federal marshal in New Hampshire named Raymond Gagnon had seen one of Flynn's presentations on port security. Captivated by Flynn's description of the problem and technological solutions, Gagnon approached him about using the Northeast as a test bed for his cargo security and information technology ideas. Gagnon rounded up local contacts in law enforcement and industry, and the Massachusetts–based light bulb maker Osram Sylvania volunteered one of its shipments for tracking. The group, calling itself Operation Safe Commerce, hired the Volpe Center, a research institute under the U.S. Department of Transportation, to develop the technology for the project. The Volpe engineers quickly discovered that they didn't need to invent the tools to track containers. They already existed.

After scouting out the shipping route, Volpe technicians picked up an off-the-shelf GPS receiver, which uses satellite signals to determine its position within 20 meters. The engineers wired the GPS and an intrusion-detection sensor to a car battery, and placed the setup inside an Osram shipping container full of headlight bulbs. The idea was that the two devices could keep track of two pieces of information—where the container was located and whether or not it had been opened—and relay them to an electronic seal on the outside of the container. Volpe team members placed laptops at pre-assigned depots along the way, and as the container passed through, the seal would send status updates via radio frequency signal. The result was a bulky, Rube Goldberg setup on a $200,000 budget, but in May 2002 the team was able to track a container at various points along its route from Slovakia, through Czech Republic, and into the port of Hamburg. From there it sailed into Montreal and was shipped across the border by truck and into New Hampshire, sensors intact.

It was a limited demonstration—with no control group, and no adversaries—but to Flynn and the team members it hinted at the possibilities for sorting benign containers from potentially threatening ones. With a test now in hand, Flynn visited the major seaports to sell them on the idea of using a more sophisticated version of Volpe's setup to track containers. Officials at both Los Angeles/Long Beach and Seattle—which, as Captain John Holmes points out, were struggling just to convince terminal owners to put up fences and guards—were enthusiastic about the prospects for a larger security program.

The federal government, while reorganizing agencies such as Customs and the Coast Guard into the new Department of Homeland Security, had just announced a plan to place U.S. Customs inspectors in foreign ports. This Container Security Initiative involves inspecting containers as they are loaded, an idea akin to Flynn's concept for pushing the borders out. The government had armed Customs agents with radiation detection pagers and instituted a program that gave a special expedited status to shippers that provided information and complied with security regulations, called the Custom Trade Partnership Against Terrorism (C-TPAT). C-TPAT is essentially a cargo version of an airline trusted traveler program. But to Flynn's mind, the government did not take the smart-container idea he offered seriously enough.

Meanwhile, a group of more than 50 private companies in the shipping industry created the nonprofit Strategic Council on Security Technology and began to test and promote its own version of a smart container network, taking Flynn on as an adviser. A more sophisticated version of his original Operation Safe Commerce test, their system employs radio frequency–based electronic container tags created by Savi, a Silicon Valley company, in combination with software from technology giants Sun Microsystems and Qualcomm, to keep track of containers' whereabouts and contents. By the beginning of 2004, the technology was up and running on multiple trade lanes connecting the United States to Asia, Europe, and Latin America.

The proliferation of independent efforts spurred the federal government to take action of its own, and in 2003 the Department of Homeland Security began funding a national Operation Safe Commerce initiative, doling out money to ports to finance the testing of new technology. Flynn's ideas—later encapsulated in *America the Vulnerable*—had now formed the basis for both the public and private cargo security efforts, making him one of the single most significant individual influencers of homeland security policy. "I have the vision of it as being like the Genome Project," he says. "Let the private sector pursue it alongside a public program."

The overlapping efforts of government and the private sector, however, also epitomize the confusing jumble of projects that underlie attempts to protect the cargo system. As a discipline, cargo security essentially didn't exist prior to 9/11. There are no experts trained specifically in its practice, no academic field to back it up, and no journal in which to publish carefully controlled studies. It is uncharted country, and thus tends to attract experts and companies of all stripes, whose skills and resources are shoehorned into a line of work no one had ever imagined.

At Los Alamos National Laboratory in New Mexico, Roger Johnston operates in one of the most bizarre niches of the new field: he makes his living demonstrating the failings of cargo container seals. Standing in a dusty park in White Rock, New Mexico, in a Hawaiian shirt, he has a digital pager clipped to his belt, a calculator watch strapped to his arm, and large square glasses that dominate his face. Johnston is the director of a group at the lab called the Vulnerability Assessment Team (VAT).

Johnston and the VAT fall into a long tradition of attempts to develop better security systems by simulating attacks on them. In the military, those pretend armies are called the "red team"—the physical security world's version of "white hat" computer hackers. (Confusingly, the bad guys are sometimes called "black hatters," but that moniker can also mean those who, like Johnston, are *pretending* to be bad guys—as in "putting on the bad guy's black hat.") Johnston's black hatters started off testing the security for shipments of nuclear materials across the country. But in the post–cold war world, the national laboratories have scrambled to find new relevance—and new funding—for their various capabilities. The skills Johnston honed while exploring ways of stealing nukes are now used to outwit the much simpler security on commercial containers.

The main goal of Johnston's current brand of black hatting is to raise awareness about just how insecure the seals on containers are. According to a paper published by Johnston and the VAT, "Were Ancient Seals Secure?" the idea of outfitting trade containers with devices to indicate their trustworthiness dates back at least 7,000 years. In ancient times, the sender or maker of a piece of merchandise would carve an impression into a block of wood, bone, or stone, and then press the pattern into wet clay that could be used to seal everything from jars, to baskets, to documents. If thieves breached the security of the package, the seal would be broken, telling the recipient not to trust the contents.

Most modern containers are similarly secured—with a 50-cent numbered lead or plastic strip that fastens through the door handle, a version of a basic railcar seal developed in the 1880s. The intent is not to prevent break-ins but to reveal that they have occurred. They are tamper-indicating rather than tamper-proof. "In the cargo world," says Johnston, "people put seals on because insurance companies tell them to." Johnston's team has hacked over 200 commercial cargo seals, all successfully—meaning that they were able to open the container and return the seal undetected. For most, they used tools no more sophisticated than what's found at a local hardware store. Their median attack time is 60 seconds. Johnston says he has ceased to ask himself, Will this attack work? "The answer is yes. That's not an interesting question. The interesting question is, Does this attack make the most sense—is this what the bad guy is likely to do?"

Johnston isn't sure that a more sophisticated electronic seal is the

answer. In his view, everything is hackable—technology particularly
so. But container seals have nonetheless become the lynchpin in plans
to secure the cargo system. Steve Flynn and others see it differently, arguing that a more transparent network is a more secure network. One
way to create that transparency, they say, is to replace the maligned
metal or plastic seal with a high-tech electronic one. These small devices come in many varieties, but in their most basic form consist of a
small wire loop fastened to a radio-frequency identification tag
(RFID). Some e-seals would be disposable attachments clipped off
when the container arrived at its destination. Others might be built
into the container door itself. In either case, the tags can be used to
store information about the container and its contents, including the
cargo manifest and a list of which parties have handled it along the
way. When the container departs or arrives at a port, the tag would
pass the information to an electronic reader in the container yard—
through radio, cellular, infrared, or satellite signals. Like the rudimentary smart container used in Flynn's Operation Safe Commerce
test, electronic seals could also be linked to global positioning systems that would keep track of their location at all times. Flynn has
also proposed that the seals include digital photographs of the container, taken during the loading process.

The end goal is to allow cargo carriers, and potentially Customs,
to track each container's movements, using software to keep up with
the data generated by electronic seals. Safe containers, Flynn points
out, typically follow regular patterns. Just as in the airport security
equivalent—behavior pattern recognition—the goal is to detect those
people or things that are out of the ordinary. To do so, however, one
must have some sense of what constitutes ordinary behavior. "How
you transport something often says something about whether you are
a legitimate commercial actor or not," he says. "Generally, legitimate
commercial actors take the most direct routes." Smart containers, by
keeping track of those movements, create "this sort of constant picture
of, Okay, this traffic leaves on Thursday, it arrives at Friday morning at
this time." If a cargo carrier moves that same commodity in an entirely
different sequence—say, takes a detour through a smaller port—the
anomaly stands out against the usual pattern.

At the same time, being able to track elements of the network, he says, means that "if you have intelligence, you can act on it without causing disruption to the system or putting critical transportation infrastructure at risk. If you know that this thing has a weapon of mass destruction in it, you want to be able to figure out the best place to intercept it." If a terrorist attack using the container network does succeed, authorities could easily trace the attack to its source. Just as with the computer virus traceback hunters and bioforensics sleuths, the ability to find the perpetrator helps not just with law enforcement but also with deterrence.

Besides allowing the container to be tracked, the electronic seal would be attached to a variety of interior sensors. Initially they would consist of intrusion detectors, based on light or motion. Eventually they could also include inexpensive radiation, chemical, or biological sensors. The sensors would report their status to the seal, which in turn could inform the network that the container wasn't a danger.

In a parking lot in Arlington, Virginia, Jay Brosius, the thin, mid-fifties chief technical officer for SkyBitz, a small technology company, shows off one possible method to do just that. In November 2002, Brosius and his team outfitted each of four shipping containers parked in the lot with a white plastic box about the size of a slim textbook. The terminals are part of a proprietary global location system (GLS), a tracking method that the company spent six years developing under a grant from DARPA. To save on battery life—the critical limitation for any smart container electronics—GLS uses only a snapshot of data from GPS satellites, sending it via satellite back to a central server to calculate the precise position instead of doing the calculation in the device itself. SkyBitz sells the system to trucking companies, who use it as a way to keep track of their trucks and minimize theft. But since the institution of Operation Safe Commerce, the engineers at SkyBitz have been working to adapt the technology for containers. "Our units weren't really built for the container environment," Brosius says, "so it's very challenging in terms of just, Where do you stick it so it won't get smashed?"

The company attached its terminals to the roofs of the four containers, connecting them to light and door sensors inside. The containers were then shipped south from Norfolk, Virginia, through the

Panama Canal, and eventually into the Port of Oakland. The system—which was able to follow the containers' progress nearly the entire way—was designed so that if the sensors were alerted at any point, the terminal would send a satellite message to the central system at Sky-Bitz's headquarters announcing the intrusion. That message could then be routed to e-mail, cell phones, or pagers. When Brosius met the containers in Oakland, cut the seal, and opened the door, his phone beeped 20 seconds later with a message: "Terminal 5607, sensor 2 container security breach."

Brosius and company ran another test in the fall of 2003, this time with eight containers—outfitted with both light and radiation sensors—that they tracked for a month as they traveled from the United States to Japan and back. In several cases, the containers were lost or misdirected for days at a time, problems that could be identified and corrected because Brosius could pinpoint their exact location and alert the local rail company, trucker, or port operator.

Like Steve Flynn's Operation Safe Commerce test, SkyBitz's demonstrations were just that. They showed that the technology could work, but not necessarily that the system could thwart the efforts of real-life black hatters. The makers of such devices have a variety of hurdles to overcome, including building devices that can handle the harsh environment—saltwater, extreme weather, and constant jolting—and settling on standards that will allow boxes that are sealed in China to be read in Los Angeles. They also must avoid the problem of false positives. In SkyBitz's second test, a forklift punctured one of the containers as it was removed from a train in Chicago; for the rest of its journey, during daylight hours, its light sensor transmitted constant false intrusion alarms. Such false positives, when multiplied over the millions of containers shipped every year, could quickly overwhelm the system. Once again Othello's error—Paul Ekman's bugaboo—is a factor to be reckoned with. Just as you can't question every person who is stressed out in an airport, you can't automatically halt the shipment of every container that has a hole in it.

Electronic seals, tracking systems, and interior sensors, however, are a way to help human inspectors, rather than supplant them. It's another example of the symbiosis between humans and our machines. The dilemma is no different than the one we face in combining automated surveillance with human observation, or metal detectors with behavior profiling. Electronic seals and sensors—and even the concept of the smart container itself—are not solutions unless they are part of a system that connects to people, everyone from longshoremen to Customs officers. Both Charles Massey, the systems expert at Sandia, and Roger Johnston, at Los Alamos, emphasize that training humans is as important as adding technology to the container network. "In the end security really has to be about paying attention and asking questions," says Johnston. "It just can't be massively automated if you want good security. So there has to be an intelligent human being in there at some point saying, 'Wait a minute, I never saw this guy on the loading dock before. What the hell is he doing?'"

A smart container, in essence, is simply an attempt to make the network aware of the objects moving through it. All that awareness is useless, however, without the ability to analyze the massive amounts of data captured by the network. Already, each container shipment generates at least 30 to 40 documents and over 200 types of data per shipment, according to Rob Quartel, the former U.S. Maritime Commissioner and now CEO of FreightDesk Technologies, a company that develops software to track cargo information. Sophisticated software like that being designed by a company called Natural Selection might help find the needles in all that hay.

If you travel to the company's headquarters, in a nondescript brick office park just off Highway 5 in La Jolla, California, you are almost certain to run into a Fogel. And probably not just one Fogel, but several. There is, first, Lawrence Fogel, the company's 76-year-old Brooklyn-born founding genius, who happens to closely resemble Mel Brooks. There is Gary Fogel, the younger of his two sons, with a doctorate in biology, and David, the older son and CEO, with a doctorate in computer science. Then there is Eva Fogel, Gary and David's mother, who is the bookkeeper and technically the owner of the company. Together, the Fogels constitute one of the world's leading outfits in evolutionary computing, a method of applying the rules of the natural world to complex problems. The elder Fogel is one of the godfathers of the field, and Natural Selection has employed it in everything from a breast cancer detection program for mammograms, to autonomous tank battle simulation for the military, to software that identifies molecular targets for drug development, to a gaming program that taught itself to become an expert in checkers.

Now they are working on a piece of the container security puzzle, with a program that will automatically profile shipping containers based on commercial data. In 2002, Natural Selection was part of a team of three companies—including FreightDesk Technologies—that won a $1.9 million grant from the U.S. government to develop a global container profiling project. The goal of the project, the first phase of which the companies completed in the fall of 2003, was to develop a more sophisticated version of the software that U.S. Customs currently uses to target containers for additional scrutiny.

The idea behind that software, called the automated targeting system, is to make sure that when Customs does choose containers to screen using VACIS machines or hand searches, they are targeting the most suspicious ones. "Even if it were 2 percent, if it's the right 2 percent, I think we are all cool with that," says David Fogel. Fogel, an avid surfer and professional jazz piano player in his free time, is trim and fit in his early forties, with slicked-back hair and tortoise-shell glasses. "It's really a resource management problem. How do we take the resources that we have and apply them best to the threats as they are perceived?"

The current container targeting software—like the original CAPPS system for airline passenger screening—consists of a rules-based system, meaning that it sorts the data based on predetermined criteria. Customs won't reveal the rules it uses, but reportedly there are less than two dozen that help decide which containers deserve special attention—likely conditions such as where the shipment originated, its size, and its listed contents. The problem with such a rules-based approach is that, as with the CAPPS system prior to 9/11, it's easy for adversaries to find a way around it. "Even if they had 10,000 rules, covering all sorts of exceptions and so forth, any operation that is only using rules is going to be limited," says Fogel. "People have to maintain that rule base. Are the rules still appropriate? How often do you want to change them?" Then there is the problem of an intelligent enemy. As with CAPPS, terrorists can employ trial runs to find out which methods—in this case, which containers on what routes—

regularly elude detection. "If the rules don't change, and you can successfully probe the system and get things through, as contraband or simulated contraband, two, three, four times in a row, it would give you some confidence that on the fifth time you are going to be successful."

That, says Fogel, is where adaptive machine learning comes in. The goal of Natural Selection's system is to group data into clusters of normal occurrences, so that anomalies are more easily detected. Those anomalies, he says, include "things that you don't have to associate with a particular threat, just things that are out of the ordinary." The software does with data what Logan Airport security personnel do with behavior pattern recognition and automated surveillance cameras do with blobology: it sorts the threatening from the unthreatening.

The way evolutionary computing works, in its simplest form, is this. A computer software program is assigned to attack a highly complex problem, often one with massive amounts of data that no human could possibly analyze. The program first randomly generates a set of solutions to that problem, and then ranks them according to how well they solve it. In the case of Natural Selection's checkers program (which David Fogel named Blondie24, so as to attract more opponents in online games), the software looked at the position of the checkers on the board, then generated a random set of strategies for evaluating their positions, trying out those strategies in a series of games.

In the next step, the software culls the worst of the solutions, keeping some percentage of the best ones. It then takes those solutions and creates random variations on only the best answers. In the checkers game, Blondie24 would take its initial set of strategies and randomly alter some of the parameters, playing more games to see if it had generated a better player. When it did, it would take those improved strategies and "seed" the next set of parent strategies. The process is parallel to biological evolution, in which random mutations allow some creatures to survive—thus spawning a new generation with those traits—and others die off. In evolutionary computing, the software creates generation after generation of solutions, each one offering an increasingly accurate evaluation of the problem. Except in this case, it does it in seconds instead of millennia. In checkers, Blondie24 started with no knowledge of tactics beyond that of a novice player, and evolved into an expert that beat real online players all over the board.

With containers, as it did with Blondie24, the Fogels' algorithms start with little knowledge of what exactly they are looking for. No one knows what a nuclear weapon in a box would "look" like in the data. Does it originate in Pakistan or China? Is it hidden among regular goods? Shipped in an empty container? There is no precedent, and thus no data to use to train the program. So instead, the software examines information—data extracted by FreightDesk from commercial shipping documents such as the manifest, the company, the origin of shipment, and the schedule—for patterns of normalcy, generating subcategories of data and evaluating them by how anomalous they are from the norm. In each generation of solution, the patterns of normal commerce become increasingly well modeled and anomalies become increasingly easy to detect. Car parts shipped directly from a Chinese company to the United States over multiple weeks on the same schedule, for example, might become clustered together as the algorithm evolves. If one week that company instead shipped a container through some indierect route, for example, it could raise a red flag.

Using that information, the software is doing exactly what Flynn did in his defense of the *Yorktown:* it is determining first who the good guys are. "Eliminating things that are not a problem," Fogel says, "is just as important as trying to detect the things that are really a problem, if not more so."

The initial prototype of the system, by the companies' accounts, was highly successful—although the Department of Homeland Security hasn't decided whether to implement it on a large scale. "One of the biggest things that came out of it is just how much data there is," says Quartel, "and how much you could collect if you had the right systems and devices to do it. Using David's software, you can absolutely show the anomalies." Quartel and Fogel are currently working on a similar system to profile food shipments across the border for the Food and Drug Administration.

Eventually, Fogel says, the evolutionary algorithms could be used to simulate both sides of the terrorism equation—the defenses of the screening system and the attacks of terrorists. Just as it evolves the best way to protect the cargo system, the program could evolve the best way to defeat it. Under a grant from the National Science Foundation, Natural Selection is adapting its software to be used in video games. Soon, Fogel says, they will be changing the subject from imaginary bad guys to real ones. "What you can do—which is usually not done—is you can turn it into a game, where you have one side simulate how to defeat the system at the same time you try to evolve what the system should be." If, as arms control experts say, it's a fallacy to think there's a last move, it's nice to at least know what the next one is.

Talk to many people in the cargo industry or government, and you are likely to find grumbling complaints of technological solutions being oversold. While both groups are eager to show they are doing something about the problem, some are equally eager to disparage any proposed technology as unwieldy and—more to the point—too expensive. Even determining what constitutes a "smart" container—what kind of sensors will it have?; what information will it keep track of?—is a challenge. Compounding the problem, the container shipping threat has spawned a raft of new companies pitching their high-tech wares as cure-alls for terrorism. "We've got vendors out the ass," says John Ochs, "trying to sell us this stuff." As always, Ochs injects a note of practicality into the smart container debate. Flynn, he recalls from his Coast Guard days, "has always been kind of a big-picture guy." And his smart container ideas, Ochs says, "make sense from a national security, unlimited budget, standpoint. But I think even he appreciates that when you come to a terminal and see what the process is, the cost associated with slowing down commerce, he understands that whatever the gadget is that is going to sniff, or scan, or whatever has to be transparent to the operation both mechanically and functionally."

Flynn, however, believes the same smart containers that help winnow down the searchable pile will actually help both cargo carriers and their customers become more efficient. After an initial outlay of cost, he argues, the system will pay for itself over time through the increased savings that would come with knowing where everything is at all times—and being able to plan business accordingly. Cargo security could be a rare case where efficiency and security coincide. For Flynn, the calculus is simple: "We can't afford to operate in a world of mystery boxes anymore," he says. "We have to solve the mystery in advance of something bad. And there are the tools out there that can get us to that."

At the end of Flynn's day in Washington, he huddles at a table with Representative Jerrold Nadler, a New York City Democrat, in the wood-paneled Rayburn Room across the hall from the floor of the U.S. House of Representatives. Over the din of a roomful of Congress people, lobbyists, and constituents, Nadler asks Flynn about a technology he's heard about—something that would "blast neutrons" into containers from a ship cruising around a harbor. One of his contributors, Nadler says, claims that the device "could detect plutonium, uranium, and maybe even explosives" at a distance. Does Flynn, the congressman wants to know, think it might have potential for port security? Flynn pauses to compose a diplomatically skeptical response. "The physics of that would be really tricky," he finally says. "I'm not saying it's impossible. I'd love to see it."

Nadler then rolls out a large map of New York Harbor and gets down to his real business. He wants Flynn's support for a plan to use federal money to build a rail tunnel from Port Elizabeth, New Jersey, under the Hudson River to Brooklyn. Because almost all of the containers that arrive at Port Elizabeth depart by truck over New York–area bridges, Nadler says, the potential of terrorist disruption is dangerously high. It's an old-fashioned merging of national security interests and pork-barrel politics, but Flynn is happy to lend his name to the project. "This is an area that I'm willing to support," he tells Nadler. "The zapping of a ship is something I'll have to drill down on."

Later, as he ambles to the Metro en route to his last appointment of the day, Flynn reflects on Nadler's request. "What's quirky about that is the reverse lobbying," he says. Before 9/11, "it was like I was out in the wilderness. But I'm equal opportunity. If someone wants to tackle this problem, I'll listen to them." He only hopes that cure-all ideas like the neutron zapper don't distract from more comprehensive answers. "If we spend all this time looking for the holy grail," he says, rolling his luggage onto a subway train, "we'll miss out on some really smart things we could do."

TEN

The Network
of Networks

IN LATE 2001, DEAN JONES was working on railroads. Mark Ehlen was working on chlorine. Federal security officials had asked researchers at the Sandia National Laboratories some questions— what-if questions that could only be resolved with the right kind of math. For Jones, the question was what would happen if an important railroad bridge were bombed: How would such an event affect the millions of tons of commodities—coal, corn, consumer goods— transported by train tracks around the country each day? Ehlen was focused on what would happen if the government suddenly halted all shipments of chlorine. On a routine basis, freight trains pulling 90-ton tank cars of chlorine pass just four blocks from the Capitol Building; a single such tank exploding could spread a plume of poison over a 14-mile radius, enveloping the Department of Justice and Supreme Court buildings, FBI headquarters, Congress, and the White House. What if authorities were to suddenly stop the movement of chlorine?

Jones is a preppy guy, though much more cerebral and intense than his khakis suggest. His talent is hard-core mathematics, which lends itself perfectly to the problem of calculating shortest paths for freight cars routing around a disruption. The chlorine question, meanwhile, was a natural for Ehlen. A Cornell University Ph.D. well-versed in the economics of manufacturing, he can trace the financial ripple effects of a missing raw material (without chlorine, for instance, you can't make plastics or process paper) through dozens of suppliers in dozens of relevant industries.

Within 90 days, both Jones and Ehlen, aided by their small band of infrastructure researchers at Sandia, the Department of Energy research facility in Albuquerque, New Mexico, had produced computer models that could help their higher-ups make security decisions. Ehlen, who bears a striking resemblance to Dick Sargent's Darren Stevens of the 1960s TV show *Bewitched,* had gotten a sense of the economic consequences of life without chlorine, and Jones understood, among other things, that union contracts could prevent train conductors from taking certain alternative routes. In both scenarios, the team saw an obvious impact on the economy. Yet they also discovered a more troubling problem. It was an outcome no one had anticipated, and it turned on the fact that Ehlen and Jones worked side by side. In the case of railroads, they discovered, the secondary consequences of a terrorist attack on the nation's railway system looked bad—trains stranded in rail yards, raw materials arriving late to manufacturing plants—but the tertiary consequences looked worse. What the engineers hadn't anticipated was a rail attack's impact on the nation's drinking water. Most water systems use chlorine for treatment purposes, and it turns out that most of the chlorine transported in the United States travels by rail. If the chemical were withheld, whether by fiat or because it was stranded somewhere along sabotaged tracks, whole communities would be deprived of potable water.

To Lillian Snyder, one of the managers of this particular cadre of engineers, economists, and mathematicians at Sandia, the connection between freight trains and drinking water was a perfect example of her team's ability to penetrate a tangle of interconnections, in particular the tangles of infrastructure that are key to everyday life, and find life-and-death information. "We're really good at this stuff," she says, leaning in over the group's conference table, her face brightening with a broad, muppet-like smile. "We consider the whole picture, in all its complexity." She lifts up her fingers, as if holding a globe or an imaginary Rubik's cube, and twists. In the space between her hands is a world of many dimensions. "It's our forté," she explains, "to look at things in toto."

These Sandians—as some at the labs like to be called—have a special mandate. While infrastructure experts such as earthquake engineer Tom O'Rourke or security guru John Ochs at the Ports of Los Angeles and Long Beach grapple with understanding and protecting previously defined systems, the Sandians are charged with studying the ways in which all these systems are inextricably bound together, in the process creating new systems that have yet to be defined or understood. Data networks control coal plants; cables and pipes run across bridges; without corn, you can't make sparkplugs (it's true); without telecommunications, you can't get cash. And then *everything*—from air travel to radio broadcasts to cattle farms—relies on electricity.

This interdependency of networks has permeated the world of engineering. Hardhats and health officials, transport chiefs and 911 dispatchers—all are beginning to recognize that achieving optimal performance of their own machinery is just part of the challenge. At numerous points, their networks touch someone else's, and at each one of those junctures, their perfectly humming operations can be compromised.

A simple example was the historic blackout of August 2003, when a power loss that had already spread across eight U.S. states prevented electric pumps from delivering drinking water to Ohio residents. Had that shortage persisted, it would eventually have shut down hospitals, which can't function without thousands of gallons of clean water, and all sorts of manufacturers that use water-cooling systems to absorb the heat produced in their plants. One truly tragic example of such interdependencies occurred when the data network of a gas company in Bellingham, Washington, malfunctioned, failing to alert managers about a gas main leak alongside a creek. The spilled gas caught fire and killed two 10-year-olds. In the case of 9/11, according to a study conducted by Al Wallace and others at Rensselaer Polytechnic Institute in Troy, New York, of the 244 disruptions to New York infrastructure in the ensuing nine months, 51 were part of chain reactions—or "cascades"—from one system to another.

"We don't *understand* these interdependencies," warns Paula Scalingi, who was the founder of the DOE's Office of Critical Infrastructure Protection and whose résumé includes a stint at the CIA. Scalingi, who comes across on paper as an inscrutable, dispassionate bureaucrat but turns out to have a richly expressive voice and maternal demeanor, is particularly lively on the bone-dry subject of mutually dependent infrastructures: "Major electrical power companies say, 'We deal with outages all the time!' The telecom people say, 'We have excellent redundancies!' But interdependencies can exacerbate the problems of recovery. We don't have a handle on these scenarios."

After many years employed by the government, Scalingi has recently gone out on the road, holding workshops in different regions of the country for utilities, emergency workers, and governments. At these events, which have names such as Purple Crescent, Blue Cascade, and Golden Matrix, the participants come up with a scenario based on what they fear most from a terrorist attack. Repeatedly, the exercises have exposed the players' ignorance about relationships among infrastructures. In San Diego, for example, the scenario that concerned the participants most was a physical attack on electric power facilities, complicated by a cyber attack that shut down computers at three key utilities—just the sort of coordinated assaults that even those skeptical of Internet terrorism take seriously. The 200 attendees played out the events that would follow, interjecting the hypothetical responses of their respective organizations. When they got to the realities of shortages of emergency generators and labor power, limited hospital capacity, and gridlocked traffic, they witnessed the impact of infrastructure interdependencies firsthand. Ultimately, the scenario led to prolonged blackouts that no one had planned for. And beyond illustrating the physical realities and vulnerabilities caused by so many interconnections, it drove home the point that if infrastructures are interdependent, then so are the people who run them: without disaster-proof communications plans among utilities and emergency officials, without agreements for sharing information and coordinating responses, the Blue Cascade players were at a loss.

While the power grid alone is gigantic and the phone system has tentacles everywhere, what Paula Scalingi is waving her arms about is more vast than both of them and dozens more in the bargain. It's the sum of all networks together, the network of networks, a phenomenon far beyond our cerebral reach.

Complex systems have qualities that are especially confounding for humans. For one thing, their numerous interactions are difficult to track and can set off multiple, successive interactions. As a result, massive systems can hide behaviors behind layers and layers of cause and effect. Just as attackers on the Internet, where data crisscrosses among thousands of computers, have such extraordinary cover that even the most destructive hackers are rarely apprehended, the catalysts of events in all complex systems can be extremely difficult to locate. Not only that, the events can be truly deceptive. Witnessing two failures side by side, for instance, might reasonably suggest that one caused the other, but it might also be that a chain of mishaps followed a great loop and came back toward its origins. Very complex systems can produce illusions of connection and disconnection, the way Groucho Marx moving "his" image in a "mirror" is in fact looking at Zeppo, who is moving in synch, wearing the same cap and nightgown. And intuition, with its powers of persuasion, can trap even the best scientists.

The beauty of computers is that they can do things we can't. You provide them with great quantities of data, and they provide the memory. You give them a task, and they provide the processing. A less lauded characteristic of these machines, though, is that unlike human beings, computers are *not* intuitive. They don't have hunches, and they don't make assumptions. So ask a computer to make sense of a complex system, a network of networks, say, and it'll be very well-suited to the task.

The process of using computers to better understand systems is what geologists, meteorologists, physicists, economists, and all their friends call modeling. In its broadest sense, modeling is simply the creation of a framework, a set of symbols, that help us comprehend the real world. With computers, biologists such as Roger Brent can model the behavior of proteins inside cells. Construction engineers such as Eve Hinman can model the loads that shift as high-rises collapse (or stay standing). Civil engineers such as Tom O'Rourke can model the ground shifting during an earthquake and the effects on underground water pipes. And someone such as the Naval Research Laboratory modeler Jay Boris can spend a career applying related techniques to modeling everything from nuclear explosions to astronomical supernovae, from the winds on an aircraft carrier's flight deck to the plume of death given off by a chemical weapon in Washington, D.C.'s Lafayette Park.

All these modelers capture the workings of reality in a set of rules—rules that tell the computer how one property of the model changes in response to changes in another property. More of this protein means more growth of that muscle; smoother wind flow means less mixing of gases; greater structural loads than the components' strength can withstand means failure; and so on. The relationship of the properties within these rules is an algorithm (recall that an algorithm is a procedure or formula for solving a given problem).

So, say, if a woman in Hollywood dates a guy for some period until she meets another who's less pompous, more successful, blushes cutely, and has a nice car, a modeler would say she's operating according to this set of rules. And if she meets another guy and another and another, each time comparing and selecting the one who blushes *more* cutely and whose car is *nicer,* then the modeler would say she's an algorithm. The particular situation a model applies to is captured by fixed data, called parameters: Does the new date's car cost $30,000 or $50,000? Does the guy blush dusty rose or scarlet? Change the parameters, and you change the reality being modeled, which is what makes models the tool of choice for almost anyone in the what-if trade.

For a computer, applying sets of rules millions of times over is easy: a computer could calculate our woman's progress through every man in the film industry in seconds. You can even buy inexpensive software that will do that kind of modeling for you—off-the-shelf programs can create a simple world, inhabited by a reasonable number of variables, and make projections about it into the future. Such a program might predict how your garden will look by taking into account where you plant your shrubs and then calculating where the sun and shadows will fall at different times of day and how that changes with the seasons. Because programs that model in this way are frequently used for such planning, they are often referred to as analytic or decision-support tools: they can analyze a situation in a million different ways, since a computer can plug in different parameters and run a model ad infinitum. In fact, sometimes models are used primarily to isolate anomalies, to reveal special sensitivities in a system, sudden departures from expected results. Knowledge of these surprises, of what unlikely parameters could cause the system to malfunction, can go a long way in preventing mishaps.

For the researchers at Sandia, this notion—that a system whose scale is unthinkably vast can be represented inside a machine—is the key to keeping the nation functioning with its infrastructure vulnerable. At Sandia, researchers such as Mark Ehlen and Dean Jones are building models that show multiple infrastructures interacting. "No human," says manager Lillian Snyder, "can solve an eight-dimensional problem on the back of an envelope," by which she means, it takes modeling to bring the management of that mess into the range of human decision making.

Sandia National Laboratories is set inside a dusty fence, itself inside Kirtland Air Force Base in Albuquerque, New Mexico. This is the rocky, dry territory, surrounded by barren mountains, where suspicions of secret military activities and alien visitations run rampant. Security regulations—your escort must be able to see you at all times; no cell phones—make visiting slightly eerie, but it's the history of the place that infuses it with a certain solemnity.

In 1949, Harry Truman wrote to the president of AT&T, offering "an opportunity to render exceptional service" to the nation. The opportunity in question was the chance to manage Sandia, a crucial node in the intellectual and physical infrastructure that was growing up to provide the nation with its nuclear weapons. The bombs were designed in Los Alamos, New Mexico; the uranium they needed came from Oak Ridge, Tennessee, and the plutonium from Hanford, in eastern Washington State. Over time, these sites became the backbone of the Department of Energy's national laboratory system, the citadels of American physics. There are now nine such laboratories, owned by the government but run by outside contractors, and some have no connection at all to nuclear weapons work. When Robert Wilson, the first director of the Fermi National Laboratory outside Chicago, was asked by a congressional committee what the laboratory's vast particle accelerator contributed to the national defense, he replied that it made the nation worth defending.

Sandia, though, contributes a lot more than that. It is one of the three labs at the heart of the process by which nuclear weapons are designed, built, tested, secured, and, when the time comes, decommissioned. The design work goes on at nearby Los Alamos, where Robert Oppenheimer led the team of young scientists (their average age was 25) that fashioned Little Boy and Fat Man, the two bombs dropped on Japan. The Lawrence Livermore Lab, founded in 1952 in Livermore, California, was conceived by Edward Teller as an alternative power base that might supersede Los Alamos in the same way fusion weapons—hydrogen bombs—were superseding the original nuclear weapons. There wasn't a truly pressing need for the second lab, but in a dramatic example of the sort of spare capacity only possible when the government foots the bills, Livermore and Los Alamos have competed with each other, complemented each other, duplicated each other's abilities, sniped at each other's failures, and run through every sort of sibling rivalry imaginable for the last 40 years. A sign on the wall at Livermore used to remind the physicists that "the Russians are the competition. Los Alamos is the enemy."

These two rivals handled the assembly of the bombs' actual explosive materials—referred to euphemistically as "physics packages." The third of the big three, Sandia, focused on the ancillary technology that allowed the bombs both labs built to be transported, maintained, armed, and detonated. By the 1980s, there were 5,000 components in a typical warhead; 200 of them were assembled at Lawrence Livermore and Los Alamos. The rest were up to Sandia. The difference in activities fostered a difference in ethos: Livermore and Los Alamos have been more academic, dominated by physicists and run by the regents of the University of California; Sandia has always been more industrial— run first by AT&T and, since the mid-1990s, by Lockheed Martin— more attuned to engineering and more likely to look for clients outside the world of bomb making.

Throughout the cold war, these three institutions were responsible for amassing the tens of thousands of nuclear warheads stockpiled by the United States, and since then, they've been occupied with stewarding those weapons—which is to say securing, protecting, and maintaining them in what leaders of the programs hope is working condition. But the labs' original purpose has clearly grown less pressing with major shifts in world politics and, especially, with the changing status of weapons testing: In the 1990s, when the United States put a halt to setting off test blasts, one of the main occupations (and entertainments) of the weapons labs—exploding things and studying the results—was taken off the agenda.

Not surprisingly, the labs have sought new pastimes to fill the empty days—and to attract new funding. They study everything from wind energy generation to treating SARS. Engineers at Sandia are working on securing shipping containers, for instance, and it was at Los Alamos that Paul Keim studied various strains of anthrax and Scott Layne hatched his vision of massive automated labs for keeping track of pathogens in the food supply or the hospital or anywhere else.

Computer modeling is an area in which all the labs are continuing to excel. Richly funded for years, they have accumulated more computer horsepower than most anyplace else in the world—employing it while weapons were under development, and today as they are silently tested. What will an explosion look like, how hard will the ground shake, how vast a region will be affected—the idea was to answer such questions, free of actual destruction, inside the machine. So when, in 2003, the federal government conducted its largest reorganization in a half-century to establish the Department of Homeland Security, a handful of projects at the DOE's National Labs were originally selected to join in. One of those was the infrastructure modeling program of Mark Ehlen, Dean Jones, Lillian Snyder, and a few dozen others from Sandia and Los Alamos. The Department of Homeland Security, which Tom Ridge had forged into being and staffed with 170,000 employees and 22 agencies, gathered up this little team and almost immediately put it to work on challenges such as railway damage and chlorine threats.

Until then, for the most part, the team had grown only modestly since 1992, when, after seven years at Bell Labs in New Jersey, commuting from New York, Lillian Snyder was crossing a Manhattan street and found herself slamming her briefcase down on the hood of a taxi that cut her off. It was time to leave the Big Apple. So she returned to her original home, which happened to be New Mexico, where AT&T was still corporate manager of Sandia. Snyder slid easily into the culture of the place. A sucker for systems research, she almost immediately grew interested in the ways infrastructures depend upon one other.

At the start of her tenure, she and three others began tackling the problem of interdependencies, camped out amid the 8,000 workers at the lab. Over the next several years, the pod of researchers grew to eight people, and by 1998, when Bill Clinton signed PDD 63, his proclamation of concern for the nation's physical underpinnings, the group was positioned to help. Prompted by the Oklahoma City bombing and insistent worries about information warfare, PDD 63 established a National Infrastructure Protection Center within the FBI, and the very day that new organization was inaugurated—before it had even furnished its offices—Snyder showed up in Washington, D.C., to pay a call. Her message was simple: We're already out here; use us. After the 9/11 disaster, the team's budget went from $500,000 to $4 million, and then in the text of the USA Patriot Act of 2001, the group was specifically chartered and given its name, the National Infrastructure Simulation Analysis Center (NISAC). Today its budget hovers around $20 million.

NISAC's function is to provide answers—mostly for the Department of Homeland Security—to questions that might run something like this: If we have a budget to fund added security guards at 10 water treatment plants in New England, which sites should we prioritize? If an oil shortage causes slowdowns in 20 different industries among midsized, small, and large companies, and we have a limited budget for rescuing them, who should we bail out first? If two power plants and six bridges are damaged, which facilities should get the speediest repairs? And if a dirty bomb gets smuggled in through a West Coast port, how long can we keep the port closed without causing the region to spiral into recession?

The place that can answer these kinds of questions must have a rare combination of qualities. For one thing, it has to be free of narrow interests. Responding to the terrorist threat has become a concern in many industries, but the work of NISAC may never make it onto these corporate agendas. "There is no business case for any in-

frastructure to look across *other* infrastructures," Snyder points out.
The federally funded lab, on the other hand, can transcend marketplace competition to bring people and data and industries together in
the national interest.

What makes Sandia and Los Alamos even more apt as a home for
NISAC, though, is their resources, and one type of resource in particular. As of November 2003, seven of Sandia's computers were among
the 150 fastest supercomputers in the world, and two machines at
NISAC's partner lab Los Alamos were among the top 10. Sandia's
fastest computer, completed in 1997 and sequestered behind a navy-
gray door in what looks like a spic-and-span locker room for 100 athletes
or so, is now ranked a lowly 27 in the race for supremacy, yet it handles
models involving millions of parameters. In one well-publicized feat, it
predicted the effects of a comet, a half-mile in diameter, barreling into
one of Earth's oceans. (Among the consequences of the great collision:
water vapors splashing up into orbit and tidal waves overrunning
entire states.) With this kind of muscle, Sandia's fastest machine can
take on models of multiple infrastructures with ample power to spare.
Even its slower computers are extraordinarily fast, well out of the
range of most corporate- or university-owned machines. So rather
than spend the four to six weeks on a waiting list for time on the
world's second-fastest supercomputer (which is at Los Alamos),
groups at the labs can turn to dozens of other lesser machines. In fact,
NISAC does most of its modeling on its own less-than-super supercomputer: 24 of Dell's fastest commercial machines lashed together
and working out problems in tandem.

Given that engineers and policy makers have only recently recognized
the need to track infrastructure interdependencies, most researchers
are still just sending out probes. Rather than looking for the best solutions, they are simply trying to understand what's there. Systems
engineers at Argonne National Labs, for instance, have spent a good
deal of time just developing a nomenclature. They've formalized terms
for the types of infrastructure interconnections: physical, logical,
cyber, and geographical. They've categorized types of cross-system
failures: cascading, escalating, and common cause. They've labeled
some characteristics: loose (backup inventory is available) and tight

(one failure follows another immediately); local, regional, national, and international; political, regulatory, budgetary, civic.

It's the various techniques of modeling, though, that will actually expose potential problems. For his work on railroads, Dean Jones used a systems analysis technique invented by a mathematician named George Bernard Dantzig, who was working at the Pentagon in the late 1940s as an adviser to the comptroller of the Air Force. Dantzig's task was to make the logistics of warfare more efficient at a time when the military had no well-grounded means for scheduling tasks and allocating resources. At worst, leaders just took a guess at, say, how many tons of supplies should be shipped to a staging area. At best, they relied on some general's experience or judgment. In the face of a complex system the size of a national economy, Air Force planners were fundamentally trusting their intuition. But Dantzig fixed that, with a modeling technique called optimization.

The best way to understand Dantzig's mathematical invention is to think of a decision-making problem as geometric. Try conjuring up an irregular shape, first in two-dimensional space, as on a piece of paper, then in three dimensions—an object with many smooth facets. (Dantzig's equations can handle hundreds of dimensions, but that's a little tougher to picture.) The surfaces and edges of this shape represent the boundaries of your problem, the constraints you're dealing with. If you're on a diet that tracks vitamin intake, for example, your constraints might be the total number of tomatoes you can eat without getting sick to your stomach, the amount of the vitamin C you can absorb in a day, and so on. The edges represent superlatives: the *largest* number of carrots you can afford, the *least* amount of vitamin A required to keep you healthy, the *maximum* number of green beans at the market. (For Dantzig, these parameters might be such things as numbers of troops or the hauling capacity of transport vehicles.) At any spot outside the edges of this shape lies a combination of factors that breaks at least one rule of your diet: you may be able to afford 10 pounds of carrots, for example, but that amount contains more vitamin C than your innards can use. Inside these edges, on the other hand, lies a combination of elements that breaks none of the rules; here, you're in compliance with every constraint. If you eat three tomatoes, your stomach will be fine, you won't waste vitamin C, and you'll eke out enough vitamin A to stay fit. This regimen sits in what's referred to as the "feasible region" of your problem.

Dantzig's innovation exploits the insight that the best combination of elements in a problem such as this will always reside not just along an edge of this many-sided object, but right at a corner. This fact, that you can ignore the entire thick, dense middle of the many-faceted object and concern yourself only with the corners, is what makes calculating an optimum possible. Such a shortcut is appealing when you consider that, as Dantzig likes to point out, even a simple problem, like matching 70 men to 70 jobs optimally, could take longer than the time span between the Big Bang and the present.

When Dantzig came up with the algorithms that would crack this optimization problem, it would take eight more years for anyone to build a computer powerful enough to actually execute them. Immediately, Dantzig's colleagues saw the potential he'd unleashed, but they couldn't exploit it. At one point during those years, Dantzig tried to determine the least expensive means of deriving the most nutrition from a diet of 77 different variables, such as types of food: using calculators, it took the equivalent of 9 people working 13 days to find the solution.

Probably the most commercially influential use of Dantzig's discovery today is in transportation scheduling. It's his mathematical tool that allows you to find the shortest, cheapest flight from among the entire airline industry's offerings, and that helps trucking companies avoid snowstorms, save gas, eliminate idle time, and get drivers home for the holidays, while still delivering goods on schedule. And it was Dantzig's invention that helped Dean Jones answer all sorts of questions about rail freight. He used basic mapping software (software sold by ESRI, the geographical information systems company in southern California), packed with data about train departures and arrivals, speeds and layovers, tonnage of corn syrup and chlorine. Then he and his colleagues incorporated Dantzig's kernel of wisdom: they built the model to "optimize." As they tweaked the system again and again, the model would calculate the optimum rearrangement of train schedules and paths in response to a change on the map. If a terrorist event caused the Houston railroad switchyards to shut down, say, here's how you would still deliver car parts to Detroit. If Dallas and Beaumont couldn't handle the excess traffic, here's which railcars you should warehouse. Every scenario took seconds, and every scenario provided a clue as to what train hubs mattered most, which ones should be heavily secured, and how to recover if one were attacked.

Protecting the systems that support modern life—the systems that make toilets flush and electricity flow—involves more than running George Dantzig's algorithms. Optimization allows you to isolate a shape, a feasible region, and ignore the world around it. But for the Sandians, that vast region beyond is crucially important. It's out there where the wrong things happen, where processes malfunction, where anything, it seems, can go haywire. And as Lillian Snyder says, "That's a very large problem space." Like the engineers and officials struggling to ensure cargo security, looking for the proverbial needle in the stacks of containers at ports, these infrastructure modelers are trying to find the trouble spots somewhere within a set of possibilities so large the brain flinches—somewhere within the network of networks. "My job," says Jones, "is to find places where you're gonna have problems, and to say, 'Here, pay attention to this, right *here*.'"

Since Isaac Newton, scientists have made models of the natural world—the physics of gravity or, nowadays, the biology of cells. NISAC's models, however, explore a problem space that's a lot less orderly. Its work for the Department of Homeland Security must deal not only with physical phenomena but also with people—consumers and providers, victims and terrorists—and the institutions, policies, and dynamics they generate.

One technique the NISAC engineers use that handles the world of humans as well as the laws of physics is called dynamic simulation. Its invention in the 1960s, by computer engineer and business management expert Jay Forrester, marked a radical breakthrough in the application of models. Indeed, Forrester's contribution to the practice was nothing less than a demolition of its containing walls. Waterfalls, nuclear explosions, even optimum distributions à la George Dantzig—those are based on laws of nature and mathematics, but Forrester asserted that you could also model social groups, government policy, behavior. In an audacious, controversial act of intellectual faith, while teaching at MIT's Sloan School of Management, Forrester quite literally modeled human civilization.

His book on the subject, *World Dynamics,* published in 1971 and later rewritten and published for the general public under the title *Limits to Growth,* modeled the future of the world and its resources. The model itself was a gigantic web of entities whose connections were expressed on paper with arrows but inside the computer as arithmetic relationships. Here's a short list of some of the model's elements: fertility, mortality, pollution, effective health services per capita, food shortage perception delay, land yield and land yield multiplier, fraction of industrial output, fertility control allocation, arable land, jobs per hectare, land fertility, and population under the age of 14. Some of these labels were parameters assigned hard numbers, but many were defined with respect to each other, so that, for instance, "industrial output per capita" was industrial output divided by population. "Need for fertility control" was maximum total fertility divided by desired total fertility multiplied by -1. The model was especially deft at capturing feedback loops in which a change in one element caused a change in another, which would cause a change in the first again, and so on.

So here was the scary part: When you packed Forrester's giant simulator with baseline numbers and ran it over and over, indicators such as child mortality shot up and life expectancy plummeted. By 2100, child mortality quadrupled and average life span plunged below 30. In less than 100 years, natural resources basically ran out. Forrester's highly practical quantification of the human future was utterly dismal. This newfangled model spelled certain doom. *Limits to Growth* became a rallying cry for the burgeoning environmental movement, while for most people in industry it was a piece of unwashed pessimism. The book was glorified and vilified and some say discredited, but the discipline of dynamic simulation had nevertheless been launched.

The engineers at NISAC like to think of dynamic simulation as just another approach to studying the network of networks—another modeling tool in their toolbox. Mark Ehlen's specialty is yet another. If dynamic simulation seems overly deterministic, too top-down to trust, Mark Ehlen's "agent-based" approach is decidedly democratic. Here the rules are not entirely imposed by the modeler, but are produced and adjusted by elements of the model itself; Ehlen, in a way, just sits back and watches the model evolve as elements, or agents, in the system "learn." Each agent—say, an ocean freight carrier or a port employee—will still begin functioning according to rules that Ehlen selects, but those can change in response to other rules that develop dynamically. So, say a model's software code includes a rule that port employees will only work overtime for upwards of $40 per hour. And say that, off in another part of the model, jobs are becoming scarce and the price of consumer goods is increasing. The port employees may cease to follow the original rule and establish another: all the agents representing port employees will now work overtime for $30 per hour. These adaptations, which in turn provoke other adaptations among other agents, give the model, in a sense, a mind of its own. And as millions of separate entities follow their different agendas, the big picture changes slowly. Agents pursue their own interests, jostling their associates along the way, and like a landscape through seasons, the whole model slowly transforms. It sounds pretty much like life. Not surprisingly, agent-based models are often used in studying living organisms and their environments. And not surprisingly, they require huge amounts of computing power. "Agent-based models are software hogs," says NISAC engineer Ralph Keyser, a man with a trim gray goatee and an unflappable demeanor. "To get that level of detail is expensive and complicated."

NISAC's most impressive agent-based modeling project, built by engineers at Los Alamos, began as a tool for analyzing the effects of community transportation habits on air quality. Like other technologies once intended for protecting the environment (or waging the cold war), this one has recently been enlisted to do counterterrorism work through NISAC. It looks at the movements of individuals through cities, forecasting changes in behavior as a result of traffic jams, road closures, and so on. What's extraordinary about this model—or more precisely, this modeling framework, which is called TRANSIMS and can be used to model any geographic area—is its stunning level of detail. Like a fabulous game of dolls for god-like children, it actually tracks and predicts the behavior of every single person living in a given area. So for a model of Portland, where TRANSIMS has been used most elaborately, more than 1 million residents are represented, each as a separate agent (or as the documentation for the project puts it, "One synthetic individual is created for each person in the true population . . . one synthetic household at a time"), along with more than 100,000 transportation links, such as walkways, bridges, roads, and rails. Using a survey of thousands of Portland households, the modelers assign activities to every traveler in the model, activities such as going to work or driving the kids to school, and they instruct each agent to make travel decisions based on numerous criteria. The price of light-rail tickets, the length of a walk, the speed of the bus, or the road repair work that would make driving hell—the model accounts for thousands of decisions. It times the street lights and follows the train schedules. It accounts for shorter travel times for agents using the multi-passenger fast lanes, and it accounts for the changing behaviors of other agents. Set in motion—a typical run of the model might take a day on a good computer—the synthetic city goes about its business, encountering change and adapting in ways that no human could possibly predict.

A favorite axiom among engineers, originally pronounced by a guy named George Box, who was a modeler himself, is that all models are wrong. Most engineers who quote him give just the gist of the rest of his statement: but some models are useful.

The first half of Box's adage is pretty much semantics—a thing that simulates reality can't *be* reality. But it also reflects the profound awareness every modeler has of the acute susceptibility to error of his or her endeavors. (One DOE engineer who monitors the western power grid, when asked what caused the New York City blackout of August 2003, said succinctly, "Our models were wrong.") Just as tiny shifts in a complex system can prove to have potent effects, so can a complex model's seemingly negligible imprecisions. In fact, minor errors can turn results into utter absurdities. George Dantzig tells the story of his attempt to lose weight based on a computer-optimized meal plan: the first dinner menu the model produced included 500 gallons of vinegar. Jay Forrester's critics have joked that if you ran his world model beginning in 1800, by 1900 the projected increase in carriage horses would have buried the streets under 20 feet of manure.

The remainder of Box's adage, though, is where the pith is—that models can be useful.

There are two signs of a good, reliable model: the quality of data it crunches and the degree to which it has been honestly tested. For the folks at Sandia, gathering accurate data is a headache, to put it mildly. Eighty-five percent of the infrastructure in the United States is private property (just consider the holdings of Archer Daniels Midland, Verizon, Citigroup, and Con Edison alone). Not only are the corporate interests who own the data under no obligation to share information, they are in fact loathe to do so. Swapping maps of their networks or data on inventory and pricing would raise both competitive and security issues. Plus, information may want to be free, at least some of the time, but it also evidently wants to be the basis for a lawsuit: infrastructure companies figure they had best keep as much secret as possible.

This is where Snyder and Keyser come in, along with, they hope, the weight of the Department of Homeland Security. It's true that there is clean data available at a price or even free, but much of what NISAC needs is proprietary. It requires prodding and persuading. For one project, the Sandians forged ahead and began constructing their mathematical apparatus even though participants had refused them data. Their output, they demonstrated, would be harmlessly general, and the companies, mollified, handed over the data. Snyder sees the data problem as a national safety issue, which means the higher-ups should impose some authority. "The Department of Homeland Security is looking for ways to get data from private companies," she says, "while simultaneously giving dispensation to protect the sources." In other words, give it to us and we won't let anyone sue you. "Another approach is to disaggregate the information enough so that all players don't see all the data."

More fraught in the struggle to make models useful is the critical need to test them and the scant means to do it. Researchers speak of "tuning" their models, which is an apt metaphor. To tune a guitar, you've got to pluck its strings and listen—the procedure is responsive and repetitive. Similarly, to get a model right you have to make it run, correct it, and make it run again. The most straightforward way to do this is to feed the machine historical data for which you already know the outcome. Tom O'Rourke tests his earthquake models with data from Loma Prieta; if the model fails to mimic what actually happened in San Francisco, he adjusts his algorithms. Then he runs Loma Prieta again. The Sandians use historical data from power outages and transportation disruptions and water shortages, then tighten or loosen the guitar strings many times over.

Modelers also test their creations against their colleagues'. They can model the same system, for instance, using different techniques— agent-based versus dynamic simulation—and run the same data through both of them. At Sandia, they also fall back on outmoded approaches such as talking to people with experience. They show their work to longshoremen or electrical technicians and tune the models with the experts' feedback. For lightweight, desktop versions of their code, they first run the system over and over on supercomputers and then strip out the denser material, leaving the finely tuned basics.

In the end, no model is a crystal ball, as the people who contrive them will tell you many times over. Some bristle at the word "prediction." Still, Ralph Keyser points out that these tools can at least expose assumptions: "Do you *really* need new transmission lines or should you just have more reserve power?" If they can't read palms, models can nevertheless add valuable insight into situations with potentially drastic consequences. As Keyser says, reiterating one of Snyder's remarks and more to the point reprising the mantra of the Sandians as a group: "It's a huge problem space." And one in which human beings need whatever help they can get.

ELEVEN

Cracking Codes

ERIC THOMPSON, a 39-year-old software entrepreneur, had just returned from church one Sunday, settling in for a lazy afternoon with his family, when his cell phone rang. The call might have felt like an annoying interruption to his vacation—he was visiting his parents in Newport Beach, California—if he hadn't immediately recognized the urgent-sounding voice. It was an FBI agent Thompson had worked with before. His company, AccessData, makes the software used by federal law enforcement and intelligence agencies to gather information from confiscated computers. The agent didn't mention what case he was calling about. He didn't have to. "It's not rocket science when you are called a week and a half after September 11," Thompson says, "and they tell you, 'Eric, we've got this really, really important case that you need to come out and take a look at.'" The FBI, it turned out, was analyzing computers that had been used by the 9/11 hijackers, and they needed Thompson's help.

Thompson is one of the world's most proficient code breakers, an expert in retrieving digital secrets through a process known as computer forensics. At a minimum, hundreds, and probably thousands, of criminals—from Bolivian terrorists to child molesters to convicted spy Aldrich Ames—have tried to hide the computer evidence of their crimes, only to have Thompson and his software dig it up. Access-Data's programs can recover e-mails that appear to have been deleted, uncover or crack unknown passwords, and, most significantly, break into files that are protected with a ubiquitous information-hiding process known as encryption.

Encryption is a mathematical procedure used to turn a piece of legible text, referred to in the field as "plaintext," into unreadable random gibberish, called "ciphertext." Ciphertext can be unencrypted only by using the same mathematical algorithm and a specific string of numbers, called a "key," to reverse the process. There are thousands of encryption algorithms in use, ranging from simple (and easily break-able) homemade varieties to the overwhelmingly complex military-grade encryption used by the United States government. Nowadays, widely used software programs like Microsoft's Word and Excel come with the ability to encrypt files. You simply pick a password, click a button, and your file is sealed from prying eyes. That is, unless Thompson gets hold of it.

He recounts the story of the 9/11 computers outside a bagel shop in Tallahassee, Florida, where he lives with his wife and three kids. Thompson runs AccessData from his house, while the rest of the com-pany's 40 employees work in Utah—where Thompson paid his way through college breaking the early encryption on WordPerfect files. Sporting a Hawaiian shirt over khaki shorts and flip-flops, with a trimmed salt-and-pepper beard, he doesn't look the part of a software CEO. He has the casual demeanor of a man who often goes to work in his pajamas. "My management style is very unconventional," he says. "I hire people and try not to get in their way."

The Sunday afternoon call from the FBI, however, was important enough to go straight to him—and it was a sign that something had gone haywire in the AccessData software. "When I don't get calls, I'm doing my job," he says. "When I do get calls, we screwed up." So he flew several of his programmers from Utah out to the Newark area the next morning, and they worked around the clock with the FBI to fix the software bugs. Thompson won't talk about what specific comput-ers were being analyzed—although it is publicly known that the hi-jackers used both Kinko's and public library PCs to make plane reservations, send e-mail, and communicate via instant messenger—or what information the FBI extracted from them using AccessData software. They told him only, he says, that "it was helpful and they got plenty of stuff."

Such is the normal course of business these days for Thompson. When the government needs to excavate information buried somewhere in a computer, anywhere in the world, chances are they are using his software. He often finds out how it's been applied through pat-on-the-back phone calls from government agents, but the details are usually kept mum. "We are breaking a lot of stuff out of Afghanistan," he says, "a lot of stuff out of the Middle East." His software was used to break into files critical to the case against would-be shoe bomber Richard Reid, and to scan the computers of suspects in the slaying of *Wall Street Journal* reporter Daniel Pearl in Pakistan. It was also employed to crack two computers obtained by the *Wall Street Journal* itself in Afghanistan. Bought off of a looter in Kabul by *Journal* reporters Alan Cullison and Andrew Higgins after the fall of the Taliban in November 2001, the computers were loaded with tantalizing documents, potentially the records of Al Qaeda operatives and their past and future plans. But many of the files—which the *Journal* also handed over to the United States government—were in Arabic, and some of them were secured with passwords and encryption.

Thompson's programs—with names like *Password Recovery Toolkit, Forensic Toolkit,* and *Distributed Network Attack*—attack secured files by first determining the length of the encryption key. Short keys are breakable using straightforward encryption-decoding methods, or cryptanalysis. Modern encryption keys are measured in bits, the unit of storage for computer information; in general, the more bits, the harder the encryption is to break. If the key protecting an encrypted file is 40 bits long—like those used in Microsoft Word and Excel—rather than a more secure 56 or 124, Thompson uses raw computer power to try every possible key in what's called a "brute force" or "exhausted key space" attack. To do so, he rounds up thirty or forty computers that can work in parallel, scanning billions of keys over several days. "Then," he says, "ba ba ba boom, we get the file encryption key back, and you are into the documents. Even though the files may have been encrypted with a Chinese or Arabic password."

According to the *Journal*'s accounts, it took a set of computers crunching numbers 24 hours a day five days before the passwords on its Al Qaeda files eventually gave way, revealing the contents of the documents. In essence, the *Journal* was replicating the intelligence-gathering methods of the U.S. government, and the outcome reveals something about what kinds of information agencies gather from digital sources. The revelations that the reporters eventually gleaned from Al Qaeda computers—first reported in two front-page *Journal* articles in December 2001 and January 2002 and later detailed in the *Atlantic Monthly*—were astounding. The computers—in both the encrypted and unencrypted files—held detailed accounts of Al Qaeda operations, including plans to develop nerve gas and conduct biological weapon testing; tracts defending the killing of civilians; memos from Al Qaeda's top lieutenants; and a video file showing the 9/11 attacks, overlaid with sounds of mocking chants and Arabic prayer. They also included, Cullison and Higgins reported, "a primer on the coding and encryption of documents"—further proof of the terrorists' interest in technology-enhanced secrecy.

One of the encrypted files was an itinerary for a man named Abdul Ra'uff, whose travels, the *Journal* showed, matched those of shoe bomber Richard Reid prior to his flight to the United States. According to the documents, he had flown on the Israeli airline El Al from Amsterdam to Israel and surveyed potential terrorist targets in Israel and Egypt before flying on to Pakistan. "A key part of the scouting mission," the *Journal* said, "evidently was to scope out El Al's security procedures." In an acknowledgment of the effectiveness of the behavior detection system Rafi Ron is proposing in the United States, the document told of how Israeli security procedures made the potential bomber's job difficult. But the file also showed how methodically terrorists would study that system in order to try to beat it: included was a transcript of the questions El Al had asked "Ra'uff," and suggested answers for future Al Qaeda operatives traveling on the airline.

Al Qaeda's use of encryption was a clear example of terrorists attempting to turn a dual-use technology to their advantage. In the mainstream media, however, the role of encryption in the *Journal*'s findings was noted only in passing. Less than a decade ago, such a discovery would have fueled calls for the government to pass harsh regulations limiting encryption research and to control the future export of a technology that could hamper the ability of the National Security Agency (NSA) and the CIA to collect valuable intelligence. Indeed, precisely such a debate raged between the government and pro-encryption advocates throughout the 1990s. But by the end of 2001, it was widely accepted that, for good or ill, universal encryption was here to stay. The question today is no longer whether encryption is being used by good guys or bad guys. As Thomspon points out, the answer to that is, both. The real question is whether either side is more secure because of it.

Information is the fulcrum upon which all counterterrorism efforts rest. In the last few years, the U.S. government has redirected entire institutions, thousands of personnel, and billions of dollars in hardware solely to the task of seeking knowledge about terrorist threats. Information is clearly a necessary prerequisite to any effort to break up terrorist cells and stifle their plots. But it is also essential to calculations about our own risks and vulnerabilities, and our decisions about which elements of society are worth the time and money to protect. Intelligence information concerning a threat of attack, in particular, must be not only illuminating but timely. When we are deaf to terrorist communications and plans, we are left with only our last lines of defenses—our airport x-ray machines, subway sensor systems, explosion-resistant architecture, and the like—to protect us. Any such single-point defense can quickly be turned into a fatal vulnerability, a fact that infrastructure expert Tom O'Rourke, airline security guru Rafi Ron, structural engineer Ted Krauthammer, and port security wonk Steve Flynn can all aver.

As Eric Thompson's experience illustrates, terrorists and other criminals will go to considerable technological lengths to try to shield their communications. It's well established that Al Qaeda makes use of digital technology—from the Web to cell phone text messaging—to conduct reconnaissance, transfer funds, recruit followers, and communicate. When Pakistani security officers captured Al Qaeda member and computer engineer Muhammad Naeem Noor Khan in July 2004, for example, they discovered that he was a key node in a global communications network that relied on intricately coded e-mail messages.

But such groups are merely joining a universal trend; secure communication is now a premise underlying countless aspects of modern society. Governments, as they have for millennia, require the ability to exchange information within and among themselves without having enemies intercept it. And preserving the integrity of communications is today equally critical to businesses—whether banks, manufacturers, or retailers—who must secure vast amounts of digital transactions that connect them to other institutions and to their customers. Entire financial markets operate virtually, counting on disparate participants to trust the security of their dealings. Online commerce requires that we rely on digital connections as we hand over our credit cards and personal information in exchange for convenience. Post-9/11 proposals for smart cards and biometric-based identification—if they are to make us more safe instead of less—require that we find a way to secure the information about our fingerprints and irises from identity theft. Around the world, the lives of human rights activists and dissidents depend on shielding their Internet communications from the repressive governments they oppose.

The fate of all of this information rests at least in part on cryptography—the science of securing information. The algorithms used to encrypt messages, also called "ciphers," are a close relative of what we commonly refer to as codes. In a code, as physicist and journalist Simon Singh writes in his 1999 history of cryptography, *The Code Book,* "a word or phrase is replaced with a word, number, or symbol." Some of the Al Qaeda files that the *Wall Street Journal* found, for instance, were "couched in elliptical, coded language," Cullison and Higgins reported. "The Taliban regime, for example, is apparently referred to as Omar & Brothers Company. Mr. bin Laden's Al Qaeda is the Abdullah Contracting Company."

A cipher, in contrast to a code, "acts at a more fundamental level, by replacing letters rather than the whole word," Signh writes. The word "cryptography" derives from the Greek words for "hidden writing," and the beginnings of the field date back to classical times. In the fifth century B.C., Spartan generals wrote their battle plans on a thin ribbon wrapped around a cylinder. When the ribbon was unwrapped, it would read to anyone intercepting it as just a jumble of letters. The intended recipient of the message, however, would have a cylinder exactly the same size. Wrap the ribbon around it, and the message became comprehensible again.

The most well-known and dramatic use of cryptography involved the Enigma, a combination electrical and mechanical machine used by the Nazis to encrypt their messages prior to and during World War II. When coders typed a message into an Enigma, the machine used a complex series of rotors to map each letter to a different one in the resulting ciphertext. A typed letter A might become a Q on one occasion, and then a W on another. Decrypting the message on the other end required another Enigma machine, along with the exact settings for the rotors. Those settings—the Enigma equivalent of a cryptographic key—were changed daily. The Germans believed that the Enigma code was unbreakable, and for much of the 1920s and 1930s their trust in it paid off. In 1931, however, Poland obtained the operating instructions for an Enigma, enabling the Poles to build a replica of the machine. A young Polish mathematician named Marian Rejewsky then developed a brilliant mathematical strategy for determining the Enigma's settings, using intercepted German messages, and the Poles listened to Germany's communications unhindered for several years.

In 1938, when the Nazis added additional scramblers to the machines—the equivalent of adding more bits to a cryptographic key—Rejewsky's method ceased to be effective. By late 1939, however, a collection of mathematicians and cryptographers had taken up the cause for the Allies at the British government's Code and Cypher School, in Bletchley Park. The group, which included computer pioneer Alan Turing, cracked a variety of ciphers from the Enigma, another complex encryption machine, the Lorenz, and those of the Italians and Japanese throughout the 1940s. Their mathematical feats, which helped spawn the birth of computers, are believed to have shortened the war in Europe by years.

Today's cryptography, used to secure everything from e-mail to bank transactions, employs a combination of complex mathematics and computing power. Software-based encryption algorithms perform mathematical transformations, using keys to turn plaintext to ciphertext and back. To explain how it works, cryptographers typically describe a scenario involving three imaginary characters: Alice, Bob, and Eve (person A, person B, and an "eavesdropper"). Alice wants to send a secure message to Bob—an e-mail containing business trade secrets, for example—while Eve is attempting to intercept it. So Alice uses encryption software to generate a long string of random numbers to serve as her key. She then applies the encryption algorithm, which, using the key and a mathematical operation, converts the letters of her message from plaintext into ciphertext (by substituting other letters or numbers for the message's letters, according to the formula). Alice can then send the message to Bob, who uses the same algorithm and key to reverse the process and decrypt the message. If Eve intercepts the message, it will look to her like a jumble of indecipherable letters and numbers. Without the key, she can't unlock its contents.

The beauty of the method, called "symmetric encryption," lies in a principle established in 1883 by the Dutch linguist, Auguste Kerckoffs: The encryption algorithm itself—meaning whatever mathematical formula is used to encrypt the contents of a message—can be made available to anyone without compromising security. The security of the system depends on keeping only one element secret: the key.

The only drawback to symmetric cryptography is that it requires Alice and Bob to each have a copy of the same key. Alice needs the key to encrypt her messages, and Bob needs it to decrypt them. That means that Alice must relay the key to Bob at some point after she generates it. If Eve intercepts the key when Alice and Bob first exchange it, she can use it and the encryption algorithm to decrypt Alice's message herself.

In 1976, three Stanford researchers, Whitfield Diffie, Martin Hellman, and Ralph Merkle came up with an ingenious solution to this problem, an encryption method that didn't require exchanging secret keys—hence it was called "public key cryptography." (The idea was independently discovered a few years earlier by British researchers working at a postwar incarnation of Bletchley Park, but their work was only declassified decades later.) In their landmark paper "New Directions in Cryptography," Diffie and Hellman proposed that instead of using the same key to both encrypt and decrypt a message, an algorithm could be designed to use one key to encrypt the message and a different one to decrypt it. The encryption key could then be made "public," meaning that anyone would have access to it, while the "private" decryption key would only need to be known by the recipient of the message.

Public key cryptography works, essentially, like this: Let's say Bob wants to receive secure messages from Alice. He uses software to produce two keys, and then publishes one of them, the public key, to anyone who wants to send him a message. Alice looks up the public key, and then uses it along with an algorithm to encrypt a message, which she sends to Bob. Bob decrypts the message using the same algorithm and his second, private key, which he has kept secret. If Eve intercepts the message, she can't read it without access to the private key. In the same way, someone who had never met Bob, let's call him Charles, could also send Bob an encrypted message. All Charles would have to do is look up Bob's public key.

Academics and businesses subsequently developed a collection of algorithms around the Stanford group's ideas—the most well-known of which is RSA, invented by three MIT cryptographers and named for their initials. But the situation gets one step more complicated: because public key algorithms are slow, they are unwieldy to use to encrypt entire messages. So instead, cryptographers designed hybrid systems that use traditional symmetric encryption—where the encryption key and the decryption key are the same—to convert the actual message into ciphertext. The public key algorithm is then used to encrypt the symmetric key to send along with the message.

It's an ingenious method of protecting information, but as evidenced by Eric Thompson's flourishing business—which grows at a rate of 100 percent per year—there are ways for the Eves of the world to decrypt messages without knowing the key. Such cryptanalysis techniques—embedded in AccessData's software—are largely of interest to intelligence-gathering agencies such as the NSA, and evidence collection and computer forensics outfits such as the FBI. The brute force attack Thompson uses on short keys is the simplest cryptanalysis method. If a key is n bits long, the number of possible keys would be 2^n. That means that if a key is 40 bits long, "there are about a trillion possible keys," writes cryptographer and security expert Bruce Schneier, in his book *Secrets and Lies*. "This would be impossibly boring for Eve, but computers are indefatigable; they excel at impossibly boring tasks. On the average, a computer would have to try about half the possible keys before finding the correct one, so a computer capable of trying a billion keys per second would average 18 minutes to find the correct 40-bit key." While Thompson recommends setting aside a weekend if you're using 30 to 40 computers working in parallel on a 40-bit, key, the end result is the same. Expanding the key by only 1 bit, though, can double the time it takes to attack the encryption. Cracking a 128-bit key, according to Schneier, would take even the fastest computers millions of years.

When the sheer muscle of a brute force attack is not enough, cryptanalysts turn to more nuanced methods. If they have access to both the ciphertext and corresponding plaintext, along with the encryption algorithm, they may be able to use the matching text to deduce the key. If they know that the word "rain" is represented in ciphertext as "sbjo," to take an extremely simple example, they could deduce that the cipher involved transposing each letter with the one after it in the alphabet. Alan Turing employed a vastly more complex version of this method—called a "known plaintext attack"—in cracking one of the Enigma codes during World War II, relying on commonly used references to the weather in German reports. Sometimes cryptanalysts can even initiate a "chosen-plaintext attack," in which Eve is able to induce Alice to send a particular message to Bob. When Eve intercepts the message, she then has both the plaintext and ciphertext and can work backward to recover the encryption key. The Allies were able to deliberately introduce messages into the German system during World War II—by laying ocean mines and then looking for the coordinates of the mines in a message, for example—and recover matching plaintext and ciphertext. The Bletchley Park cryptographers called it "gardening."

By the early 1990s, the academic science of cryptography was well-established, along with a few commercial products based on the principles of both symmetric and public key cryptography. But it took a man named Phil Zimmermann to bring encryption to the masses. In the process, he helped to fuel a battle over cryptography that would reach to the highest levels of government—and nearly land him in federal prison.

In the decades after World War II, cryptography in the United

States was largely the purview of the government—and in particular one organization, the super-secret National Security Agency. Established by hush-hush presidential order in 1952, the Fort George Meade, Maryland, outfit has long been home to the largest collection of mathematics Ph.D.s in the world. The NSA has been the subject of conspiracy theories and wild speculation for decades, but lately it has allowed itself a slightly higher public profile. Its occupation, however, has remained the same since its inception: to collect the communications, known as signals intelligence or SIGINT, of American friends and foes abroad.

The information is swept up from satellites, radio signals, and the Internet at NSA listening posts around the globe and transmitted back to Fort Meade, where it is processed by the agency's computer systems, code breakers, and translators and then examined by analysts and passed on to other agencies. Just one of the NSA's collection systems—the agency has satellite and ground stations spanning the globe—can intercept more than 2 million telephone calls, e-mails, or radio signals every hour, according to James Bamford, author of the two most authoritative accounts of the agency, *The Puzzle Palace* and *Body of Secrets*. Every three hours the total volume of data pouring in—reportedly filtered using a controversial keyword search system known as Echelon—exceeds the amount of information held by the Library of Congress. Today, in fact, the agency gathers so much information that it struggles to process even a fraction of it.

The NSA is charged with both designing the encryption used to protect U.S. government information—including military communications and nuclear launch codes—and cracking any encryption found in intercepts from abroad. With its concentration of brilliant cryptographers (including Robert Morris, the agency's chief scientist in the 1980s and 1990s and the father of Robert Tappan Morris, author of the infamous Morris worm) and a collection of some of the world's fastest computers, it does both tasks extremely well—according to most, albeit speculative, accounts. But as academic and commercial interest in cryptography heightened in the 1970s and 1980s, the NSA began to worry about the spread of powerful encryption methods—with increasingly complex algorithms and longer keys—that could prove impossible to crack. Academic cryptographers had, for years, voluntarily submitted their papers to the NSA for review, but now the U.S. government began enforcing regulations concerning the export of encryption software. By the early 1990s, both the NSA and the FBI were fretting over the advent of digital telephones, which, unlike their analog counterparts, were difficult to tap. When AT&T announced in 1992 that it would begin commercially marketing a telephone that included an encryption feature to scramble calls, the concerns over digital telephones and cryptography merged, and the agencies pushed the government to place new controls on the development and export of uncrackable encryption products.

The resulting "crypto wars," recounted by *Newsweek* technology reporter Steven Levy in his book *Crypto,* pitted an ad hoc group of academics, libertarians, and computer industry lobbyists, in favor of widespread strong encryption, against the NSA, the FBI, and the Clinton administration. The government maintained that in the hands of criminals, mobsters, and terrorists, strong encryption would destroy the government's ability to collect intelligence and evidence. A 1997 paper by Georgetown computer science professor Dorothy Denning and Science Applications Corporation International vice president William Baugh summed up the argument: "We are at the leading edge of what could become a serious threat to law enforcement and national security: the proliferation and use of robust digital encryption technologies. These technologies will be unbreakable, easy to use, and integrated into desktop applications and network services, including protocols for electronic mail, web transactions, and telephony." They offered examples of how terrorists—from the Japanese Aum Shinrikyo cult, to Bolivian insurgents, to the plotters of the first bombing of the World Trade Center—had already begun to employ the technology. "The impact of encryption on crime and terrorism," they wrote, "is at its early stages." The Clinton administration used similar arguments to justify strict export controls, which allowed businesses to export encryption software only if the cryptographic key was shorter than a certain length. The administration also proposed the development of encryption technology under which the government would keep a copy of all the keys—called a "key escrow" system—which it could use as "back doors" to access encrypted communications for criminal cases and national security.

The pro-crypto side countered, first, that the value of strong encryption—for commerce and privacy, in particular—outweighed the harm of its potential use by nefarious elements. With the increasing influence of networked computers on everyday life, it was becoming possible for the government and corporations to gather more and more information on citizens. In the view of many cryptographers, ordinary people needed a way to keep both criminals and the government from snooping on their private communications without cause or permission. Cryptography would free the citizenry from government intrusion, they said, and provide security for the coming explosion of online commerce. Encryption back doors would be vulnerable to abuse by rogue agencies and undermine confidence in the security of basic communications. Besides, they argued, domestic regulations and export controls would ultimately prove futile. The technology was charging forward, and if the United States didn't export strong encryption, countries such as Japan would. In the end, the NSA would be none the better for letting the advances occur overseas. Such was the position of Eric Thompson, who, even as a code cracker, joined leading cryptographers in calling for loosened restrictions.

It was into this melee that Phil Zimmermann waded, in 1991, with the creation of a piece of encryption software he called Pretty Good Privacy. Zimmermann, a cryptographer and former antinuclear activist, is a bearded man with a soft, friendly face and rectangular glasses. Sitting outside a Borders bookstore in downtown Palo Alto, in jeans and a maroon Hawaiian shirt, he recalls the first indication that his encryption work had run afoul of the law. Like Thompson's post-9/11 run-in with the hijackers' computers, it started with a phone call from the FBI. In early 1993 a U.S. Customs agent called Zimmermann's home in Boulder, Colorado, inquiring about the software that Zimmermann had written two years before. "I thought maybe they encountered PGP in some of their work," he says, "and needed some advice in assessing what they were up against."

PGP, as it came to be universally known, was a relatively easy-to-use application of the concept of public key cryptography, one that could encrypt everything from documents to e-mail. "PGP made it possible for the first time in history to communicate over long distances without significant risk of interception," Zimmermann says. "Even though there was other encryption software before PGP, either that software didn't use public key algorithms—which means you would have to send the keys somewhere, and that could be intercepted—or they did use public key algorithms, but the block cipher"—the symmetrical algorithm used to encrypt the message itself—"wasn't strong enough. It could be easily broken by intelligence agencies, not just the United States but all the major powers." PGP was both comparatively user-friendly and virtually uncrackable (to this day, there is no record of the cipher itself being broken), both of which made it a potential nightmare for the NSA and the FBI. Zimmermann—fearing the passage of stringent export restrictions—had asked a friend to post the software on an online bulletin board back in 1991, and PGP had quickly spread around the world.

Unfortunately for Zimmermann, encryption software was included on the State Department's list of "munitions" requiring a license for export. To U.S. Customs, the release of PGP online was equivalent to Zimmermann having traveled the world handing out surface-to-air missiles.

That first phone call to Zimmermann turned into a visit from two Customs agents, and eventually a three-year investigation in which government prosecutors convened a federal grand jury and subpoenaed his friends and colleagues to testify.

Zimmermann decided, against his lawyer's advice, to take his case straight to the public. "During that three years I talked to the press almost every day," he recalls. His argument—that the government was trying to imprison him for the simple act of providing ordinary people with a way to protect their secrets—resonated, and he quickly grew into a kind of geek icon. One news story after another took a skeptical view of the prosecution and offered up ready-made examples of PGP employed for good, by everyone from Burmese activists to lawyers protecting attorney-client privilege. MIT published the source code for PGP in a 600-page book—further complicating the government's case, since the book itself could be exported. Nonetheless, the investigation, with its corresponding threat of imprisonment, "was all-consuming," Zimmermann says. "I remember talking to him when he was really down and out because Customs was putting the screws on him," recalls Thompson. "I remember him saying, 'If they put me in jail, how am I going to feed my family?'"

After three years, however, Customs finally dropped the case in January 1996. Zimmermann's reprieve would foreshadow the direction of the encryption debate in general. After years of congressional hearings, public forums, frenetic lobbying, and online screeds, the crypto advocates finally won the crypto wars. In 1999, a federal court ruled that government restrictions on cryptography violated the free speech guarantees of the First Amendment.

Meanwhile, the Clinton administration set up a council to determine if changes to encryption export rules were needed, and the group took only nine months to significantly relax the controls on encryption. The move in effect allowed the spread of even the strongest encryption, no matter what the key length. The NSA, meanwhile, had long since foreseen the end and moved on. Encryption technology was out of the bottle, and there was nothing that the agency could do to put it back in. "Number one, the loosening of the availability of encryption has been more good than bad," reflects William Crowell, the former deputy director of the NSA during the 1990s, who chaired the council. "Number two, it hasn't been the end of the world, with regard to intelligence. And number three, it wasn't U.S. technology in the first place. It was very widely spread throughout the world, and was going to continue to spread far and wide." By the beginning of the new millennium it appeared that strong crypto for the masses had finally arrived, along with the security and privacy that it would provide—for ordinary citizens and terrorists alike.

There was only one problem, says Bruce Schneier: "Both sides were wrong." Schneier, the veteran cryptographer and founder of the digital security firm Counterpane Internet Security, had been a general in the crypto wars. He lobbied vociferously on behalf of strong encryption in his *Crypto-Gram* newsletter and in congressional testimony. He had also invented several well-known encryption algorithms, with names such as Blowfish and Twofish, and written a classic text on the topic, *Applied Cryptography*. But the glow of the crypto victory had barely faded when he came to the shocking realization that strong encryption was, in some sense, immaterial to information security. There are so many other weak links in digital security—passwords, lazy humans, and vulnerable networks—that encryption is rarely the part worth attacking. "I and my colleagues, who break security products for a living, we almost always succeed," he says. "No matter how bad the cryptography is, we almost never break the cryptography. There is always something else worse. That is the lesson of the crypto wars: We thought math could save us. It turns out that math is irrelevant, because software security and network security is so bad that math is not the weak link."

The endless effort to strengthen encryption, he concluded, is almost beside the point. "It's like we are trying to keep the enemy out of our town, and we are putting a huge spike in the ground. And we are arguing whether the spike should be a mile tall or a mile and a half tall. Honestly, the enemy is going to go around the spike. It doesn't matter how tall the spike is." A second problem Schneier discovered is that encryption programs typically tend to be less than user-friendly, and both businesses and ordinary people are often too lazy or technically unsavvy to use them. These dual dilemmas formed part of the premise for Schneier's 2000 book, *Secrets and Lies*. Then came 9/11, and suddenly technology was being touted as a necessary means to secure not just information but society itself—through the likes of encrypted biometrics and national IDs. Inspired by a profile in the *Atlantic* magazine, which pointed up the potential importance of Schneier's thinking for a multitude of homeland security problems, Schneier penned another book, *Beyond Fear*. In it, he skewers single-point security approaches, advocating instead for building layered defenses, avoiding brittle technological solutions, and keeping a human-centered approach to security similar to that advocated by Rafi Ron and Paul Ekman.

Beyond Fear doesn't even mention the word "encryption," an absence reflecting Schneier's newfound belief that cryptography, in practical terms, isn't particularly relevant to counterterrorism. But in another sense, both sides of the crypto wars turned out to be right. The math behind strong cryptography might not have saved us, but it has—as its advocates predicted—become an integral part of the economy, enabling electronic commerce of all kinds. A form of public key cryptography called Secure Sockets Layer (SSL), to take one example, is used to secure online transactions such as those at Amazon.com; it's what is happening behind the scenes in your Internet browser when you embark on a purchase and see an "https" materialize in front of Amazon's Web address.

Strong crypto opponents in the government, by turn, were also partly right. Terrorists and other malfeasants, from mobsters to child pornographers, *do* use encryption. "The argument that export controls did matter, because of the impact on law enforcement and intelligence, was true," says Dorothy Denning, now a professor at the Naval Postgraduate School in Monterey, California. During the crypto wars, Denning advocated a key escrow system, but she recognizes why the other side won out. "In the end, that argument wasn't as important as the other arguments about the need to make cryptography widespread for the security benefits."

Fortunately, Schneier's lament—that encryption doesn't provide protection because security is a weakest-link problem—turns out to be just as true for the bad guys as the good guys. "Even if the strongest encryption was deployed widely—which it is not—I do not believe that it would make it impossible for intelligence and law enforcement to do their job," says William Crowell. "Does it complicate it? Yes. But so did the telephone. So did satellites. So did so many other technologies." Many law enforcement and intelligence-gathering encounters with encryption are resolved not through pure technical expertise but with a combination of computer processing power and plain old detective work. The NSA, Cambridge University cryptography expert Ross Anderson told the engineering magazine *IEEE Spectrum* in 2003, has "covered up some quite spectacular successes at breaking into cipher systems . . . by pretending that they were simply better at mathematics and computer science. Whereas what was usually happening was some form of sabotage, blackmail, theft, corruption, or whatever." Even the most famous code-cracking incident in history— the Enigma machine—wouldn't have succeeded without the confiscation of numerous versions of the German machine and its codebooks.

Today, many encryption programs protect their keys using passwords, and to break those passwords Eric Thompson and his software exploit human weaknesses. "As much as technology continues to increase," he says, "a human's ability to remember completely random, complex alphanumeric patterns will not increase. You assign someone a password 17x!42*, they are not going to remember that. They are going to write it down in their Palm Pilot, they are going to put it in a document in case they forget." And when people choose their own passwords, they have to pick something they can remember. A security study Thompson helped conduct of a Dallas company in the late 1990s, for example, found that more than half of the employees used some derivation of the football team the Cowboys, such as "GoCowboys," or "CowboysRule."

So instead of randomly searching possible words—the brute-force version of password breaking, and known as a dictionary attack—Thompson's software instead constructs a biographical profile of the computer user. It scans the hard drive, recovering deleted files, searching through e-mails, and cataloguing names of family members, pets, and hobbies found on the computer. It traces the user's history of Web site visits, finds any passwords that were used on those sites, and also pulls down thousands of key words to permutate and try as passwords—at a rate of millions per second. The software works in all the romantic languages, and versions for Arabic, Chinese, Japanese, and Korean are in development. But in the case of the Arabic world, Thompson has found users using English characters for their passwords to accommodate Western-made software.

Other times, investigators can rely on the ineptness of their targets. Take the oft-cited example of the Aum Shinrikyo cult's use of encryption. It is true that Japanese authorities discovered encrypted files on Aum computers in 1995. But less noted is the fact that the terrorists had left the decryption keys on a floppy disk sitting on the same desk, obviating any need for the Japanese government to crack the cipher. Aum had locked the house, so to speak, but left the key in the front door.

The Al Qaeda computers found by the *Wall Street Journal*, similarly, featured encryption on some files; but the operatives were too lazy or technologically illiterate to use it on others. The computer contained a primer on encryption, but just the fact that using encryption required reading the primer meant that the less disciplined among the terrorists probably wouldn't bother. When Osama bin Laden went into hiding at the beginning of the war in Afghanistan, he reportedly began using only a few trusted couriers to carry handwritten and spoken messages to his network—no e-mail, no satellite phones. When the need arose for completely secret communication, the terrorists went low-tech.

"I'm sure that Al Qaeda uses PGP," Zimmermann says. "There is no doubt about that." Rohan Gunaratna, an expert on the organization, reports as much in his book *Inside Al Qaeda*. So how does Zimmermann feel knowing that terrorists are using the software he designed? "Well, I wish they didn't," he says, with a wry laugh. "Use somebody else's product." It is precisely PGP's popularity—outside of the encryption included in Microsoft programs, it is one of the most-used encryption programs in the world—that makes it as accessible to terrorists as it is to the everyday citizens that Zimmermann first wanted to help. "Damn," he continues. "I mean, it does bother me quite a bit. But I'm not sufficiently clever to come up with a way to not let them have PGP and let everybody else have it. It's easy to say let's just not have crypto." He pauses to correct himself: "Even that's not easy, because someone will make it somewhere else."

Immediately after September 11, the debates of the crypto wars briefly returned. Days after the attacks, Senator Judd Gregg, Republican from New Hampshire, announced that he planned to propose legislation requiring encryption to be built with backdoor keys for the government. His reaction was based in part on speculation that the hijackers had used encryption, and the *Washington Post* even quoted Zimmermann saying that he was "overwhelmed by feelings of guilt" over the possible role of PGP in the 9/11 plot. The rumors, however, were never substantiated, and by October Senator Gregg had received so much pressure from encryption advocates that he reversed his position. Zimmermann issued a statement explaining that he had been misquoted by the *Post*. "Did I re-examine my principles in the wake of this tragedy?" he wrote. "Of course I did. But the outcome of this re-examination was the same as it was during the years of public debate, that strong cryptography does more good for a democratic society than harm, even if it can be used by terrorists. Read my lips: I have no regrets about developing PGP."

These days, Zimmermann sees all the attention paid to the dual-use nature of encryption as somewhat puzzling. He points to the example of GPS devices, which, according to the 9/11 *Commission Report*, the hijackers bought to pinpoint their targets. Yet there was no public call for restrictions on the sale of GPS receivers after 9/11. "I mean, it's a military technology," Zimmermann says. "It was designed to target weapons, and the terrorists may have used it to target a weapon, a big weapon. That makes a good argument for not letting anyone buy GPS receivers. But can you imagine what the effects would be on our economy, if we tomorrow said no more GPS receivers can be bought? There are lots of industries that depend on them, just as today lots of industries depend on encryption, or the Internet."

If the NSA had moved on while cryptographers and the government continued to go at it, so did a young computer scientist enthralled by another kind of information security—one that was quietly advancing beneath the public radar. Neil Johnson had been interested in both computer graphics and cryptography since he was a kid, and he graduated from college with a degree in computer information systems. He went to work for IBM for a few years before returning to school for a master's at George Mason University in Virginia, fully expecting to rejoin the corporate world when he was finished in 1996. In one of his final classes, however, Johnson was searching for a paper topic when he ran across a then-obscure area of information security called steganography. The professor had urged his students to choose little-researched subjects, Johnson says, so "I asked him if he knew anything about steganography. And he said, 'What?' I said, 'It sounds like I have a topic.'" Actually, in steganography Johnson had found not just a topic but a career.

Unlike cryptography, which involves scrambling information to make it unreadable to an interceptor, steganography (or "stego," as it's often called) involves hiding the existence of the message itself. Whereas an encrypted message is designed to remain secure if it falls into the hands of the enemy, a steganographic one is intended to avoid detection by the enemy entirely. The practice, which like cryptography takes its name from Greek, meaning "covered writing," predates the use of ciphers. Among ancient Greeks, it was used to physically conceal secret messages, by carving the message onto writing tablets before covering them with wax or tattooing information onto a messenger's shorn head and then allowing hair to grow back over it.

In its modern, computer-based form, steganography is a less exotic but more intricate process, one that hides information within the makeup of digital media files such as photographs, songs, and videos. Because human senses have a limited auditory and visual range—we have trouble detecting slightly different shades of color, for example, or hearing small amplitude changes in music—we can't perceive minute changes to the makeup of a media file. "Those weaknesses are what are exploited by steganographers," says Chet Hosmer, president of WhetStone Technologies, a New York–based firm that specializes in the investigation of steganography. "They can basically hide information and make very subtle changes to either the visual rendering or the auditory rendering to that piece of digital data, so that—except for Superman or somebody—you can't hear or see the difference."

A digital image file, for example, is made up of strings of ones and zeros (the same bits used to measure cryptographic keys) that store information and combine to form the basic units of color, called pixels. If an image is made of eight-bit color, each pixel consists of eight of these ones and zeros. In the simplest form of steganography, an algorithm hides the information within the least significant of the eight bits—the ones in which our eyes will be unable to detect the subtle changes. In one paper, Johnson offers the example of the letter *A*, concealed within three image pixels. A group of three normal pixels looks like the following:

(00100111 11101001 11001000)
(00100111 11001000 11101001)
(11001000 00100111 11101001)

Hiding the letter *A*—represented in the binary world by the value 10000011—requires changing only the last digit in three of the numbers, and then reading the last number of each set from right to left:

(00100111 11101000 11001000)
(00100110 11001000 11101000)
(11001000 00100111 11101001)

Because a digital photo is made up of thousands of pixels, each with a collection of bits to be manipulated, large amounts of information can be hidden without detection. The equivalent is true for audio and video files. A steganography program does all this hiding behind the scenes, encoding information into the document and then creating a password required to recover the message.

When Neil Johnson began his steganography research in the mid-1990s, the field was a virtual backwater in computer security. "There wasn't a whole lot written about it," he recalls. He started calling up the authors of every steganography paper he could find, and "everyone I contacted wanted to see what I was working on." His master's paper quickly turned into an offer to teach at George Mason and pursue a Ph.D. in the subject. His subsequent thesis, entitled *Information Hiding: Steganography, Attacks, and Countermeasures,* earned him not only a degree but the attention of the NSA, which contacted George Mason and subsequently funded some of his research, as did the Department of Defense's Computer Forensics Laboratory.

Like cryptography, steganography has its counterdiscipline, steganalysis—the science of detecting and uncovering hidden messages in media files. Much of Johnson's research involved finding ways to uncover the existence of steganography, and then crack the messages contained within it. Stego programs leave minute traces of their digital alteration, anomalies that an alert sleuth can uncover. Johnson and others employ a variety of techniques to detect those tiny code tweaks, with software that automatically scans media files for abnormalities. Other tools alter the display of image files, so that abnormalities can be more easily seen by the naked eye. But with the number of media files stored on any given computer—think of all those pictures your digital camera–wielding family and friends have sent you—the process is time-intensive and often fruitless. "If you are looking for stego," Johnson says, "it's like finding the right piece of straw in a haystack. At least a needle is shiny and sharp, so if you accidentally come across it, it's going to poke you. A piece of straw is so much like the hay, you're not exactly sure if you've found the right thing or not." Even if the steganalysis does locate an image that has been altered, the only way to recover the message is through a dictionary attack on the password.

Today, there are hundreds of steganography programs—with names such as Invisible Secrets and MP3Stego—available via the Internet, and the number is growing steadily. Steganography is also employed in digital watermarking, the process of embedding copyright labels within media files. Music and image companies use digital watermarking as a hedge against piracy. But the field largely avoided notice outside of defense circles and hobbyists until 2001, when news reports speculated that Al Qaeda might be using it to send undetectable messages to its members over the Web—by placing them in pictures on eBay and pornographic sites, among other places. (One of the accounts, however, was penned by *USA Today* reporter Jack Kelley, later dismissed from the paper for fabricating stories. The steganography story was one called into question.) Such claims are difficult to prove, but impossible to disprove. Whetstone's Chet Hosmer, who regularly works with law enforcement on potential steganography cases, argues that, since steganography is a tool known to be employed by child pornographers, there is no reason to believe terrorists aren't also hiding communications in plain sight.

On the topic of whether terrorists use steganography, Johnson remains circumspect. "My take is that if an individual or organization has to have secret communication, they will use whatever technology is at hand," he says, "be it crypto, stego, high-tech, or low-tech. Is it being used? Maybe, probably. Is there solid evidence? Maybe, probably." But even prior to 9/11, Johnson became sufficiently concerned that terrorists could be using stego that he stopped publishing his work—much to the dismay of his department at George Mason—for fear that foes, knowing his methods, would engineer programs to avoid them. "Toward the end," he says of his employer's response, "it was, 'Neil, maybe you could put out a paper once a year now?'" So Johnson left to become an associate at the consulting firm Booz Allen Hamilton, where he now plies his expertise developing digital watermarking technology. He still gives training lectures to law enforcement personnel on steganalysis and keeps up his own, unpublished research on the side. When a new program comes out, he can't resist the challenge. "It's kind of like when you look at those stereo grams, the pictures that have a pattern in them," he says. "If you stare at them long enough just right, you see the image. Steganalysis isn't that simple, but it requires that you take a look at things a little differently."

At the height of World War II, the American military became concerned with the inherent vulnerability of traditional communication security methods. There was always a chance that, as with the Enigma, a cipher could be broken by enterprising mathematicians on the other side. So the commanders in the Pacific campaign decided to revert to an even more primitive form of information security: language. They employed Native Americans from the Navajo tribe—whose impenetrable language would be meaningless to the Japanese—to act as radio operators. "The success of the Navajo code," writes Simon Singh in *The Code Book*, "was based largely on the simple fact that the mother tongue of one person is utterly meaningless to anybody unacquainted with it."

In the world of counterterrorism, language provides much the same cover for terrorists. Among intelligence agencies, language translation is a well-known bottleneck to collecting usable information. The head of the FBI, Robert Mueller, testified in the weeks following 9/11 to the "critical need" for Arab, Farsi, and Pashto translators. According to the findings of the Congressional Joint Inquiry on the attacks, "Prior to September 11, the Intelligence Community was not prepared to handle the challenge it faced in translating the volumes of foreign language counterterrorism intelligence it collected. Agencies within the Intelligence Community experienced backlogs in material awaiting translation, a shortage of language specialists and language-qualified field officers, and a readiness level of only 30 percent in the most critical terrorism-related languages." The most dramatic illustration of that backlog came from the NSA, which on September 10, 2001, intercepted two Al Qaeda telephone conversations referring to a significant (but not specific) attack against the United States. One of them, it was subsequently widely reported, said "tomorrow is the zero hour"; the other: "the match begins tomorrow." But the messages were not translated until September 12.

The language problem as it relates to terrorism becomes even more difficult when you examine the range of nations and groups that the United States considers a threat. From militant groups in the Philippines, to Indonesia, to Uzbekistan, to wherever the next threat arises, the range of obscure dialects makes finding human translators a constant challenge. The 9/11 Commission reported that even in Arabic, only 6 university degrees were granted in the U.S. in all of 2002. Meanwhile, a 2004 audit by the Justice Department turned up thousands of hours of untranslated terror-investigation audio recordings.

To confront that dilemma, the government is building the National Virtual Translation Center, designed to handle the translation activities of all federal intelligence agencies. The goal of the center is to centralize and speed up the translation process with a combination of human translation and technological aids. The center plans to staff itself with 300 or more linguists covering languages from Arabic to the Afghan dialect of Dari. For the last few years, meanwhile, DARPA has been funding a variety of translation technologies through its Translingual Information Detection Extraction and Summarization (TIDES) program, a project intended to "make it possible for English speakers to find and interpret needed information quickly and effectively, regardless of language or medium."

Translation and, in a broader sense, the ability of computers to process language, have long been among the most intractable problems in computer science. The birth of computing in the 1950s, coming on the heels of the code-cracking successes of World War II, led some scientists to believe that the new technology would quickly conquer language translation as well. Different languages, they reasoned, were simply "encrypted" versions of the same ideas. Figure out the algorithm in question, and translation would be a cinch. A computer-based translator that could achieve the same results as humans seemed feasible, and perhaps just around the corner. But the initial promise of machine translation, or MT, quickly dissipated as early efforts failed to get even close. Many early MT efforts were unintentionally hilarious—exemplified by the often retold tale of a computer translating "the spirit is willing but the flesh is weak" into Russian, and then back into English as "the vodka is good but the meat is rotten." The story, as it turns out, is apocryphal—but the essential point remained true for decades: Because words can have multiple meanings and be combined into limitless ideas based on context, software was hopelessly lost in figuring out how to translate accurately.

Early MT efforts—and the ones that currently lie behind basic translators such as AltaVista's online Babelfish today—were designed as rule-based systems. The goal was to "parse" languages into their grammatical and syntactical rules and structures and then use those rules to take apart text in one language and reassemble it in another. Linguistic experts in each language developed the rules, which were then programmed into the software. But hard-coding the rules, because of the infinite combinations and exceptions of meaning, proved cumbersome, time-consuming, and limited. Rule-based translation only worked well for small domains—such as a particular company's user manuals—where text was often translated in the same ways over and over again.

In the 1990s, MT scientists began pursuing two new tracks. Both involve using software to scan a set of documents, called a "corpus," that has already been translated between two languages—a novel, for instance, that has been translated from English to French. The first method is known as "example-based" MT, which is also sometimes called "translation by analogy." In simple terms, a software algorithm scans the corpus and extracts both the English phrases and their translations—"the book is on the shelf," say, and "the shelf is next to the table"—and stores them in a database. When it is used to translate a new document, the software scans the text, and when it encounters the phrase "the book is next to the table," it can access the two phrases and create a translation.

The second method is a related technique called "statistical" MT. In it, the software employs complex statistical algorithms to analyze the previously translated text. It develops a model to determine the probability of a particular translation, and then applies that model to new sets of text. The French word "livre," for example, can be translated any number of ways into English: book, document, tome. Statistical MT software assigns a probability—based on how it has been translated before—to "livre" that represents how likely it is to be translated as book or document. It does the same thing at the phrase level—"the book sat on the table"—and then uses complex formulas to work out the translation of new material.

In the last few years, software that is based on statistical MT methods has overtaken rule-based systems as the most accurate form of

MT. The Marina del Rey, California-based company Language Weaver, which began with DARPA-funded research at the University of Southern California's Information Sciences Institute, is developing such software. Language Weaver employs statistical MT methods—originally developed for use in crypography—to analyze corpus text and extract probability-based dictionaries, grammar rules, and language patterns—which it collectively refers to as "translation parameters." It then applies the translation parameters to new texts. The company released its first module, Arabic, in December 2003, followed two months later by Chinese. While the results are not yet close to publication quality, the system is nonetheless able to convey the key idea in each translated sentence—a feature that could be valuable for intelligence analysts. The company is partly funded by In-Q-Tel, the venture capital arm of the CIA. Created in 1999, In-Q-Tel invests in private companies whose technologies have potential applications in intelligence, including everything from GIS software to sensor networks. The CIA connection explains the company's initial focus on Arabic. "Arabic has always been a tough language for translation," says Language Weaver cofounder and chief scientist Kevin Knight. "The output now is certainly much higher quality than was expected."

Eventually, Language Weaver hopes to reach the ultimate achievement in MT: human-quality translations. The long-term goal, says Daniel Marcu, the company's other founder and chief operating officer, "is to eventually produce first-draft human translations—what you would get from a very sophisticated bilingual speaker who just looks at a text and writes the translation." Technology giants such as IBM and Microsoft are also pursuing human-quality MT. But for now, even the best statistical systems are only capable of approximating a translation, called "gisting," to give a human translator a place to start. The National Virtual Translation Center is investing in technologies that will combine gisting with software to digitize and categorize foreign language information for its analysts. But the human brain—which can see meaning where the machine cannot—remains the ultimate arbiter, and the central part of the translation process. Technology is simply an assistant.

•　　•　　•

Eli Abir wants to take the human out of the process entirely. To tackle the problem of multilingual understanding, Abir is building his own software brain. Or, at least, that's what he calls it. "At the end of the day, the way you should think about it is that it really is a brain like the human brain," he says. "So any possible way that you could have told a very smart person to search or to think or to condense information for you, it can do, but much better, in a sense." It's a typically grandiose and perplexing statement from the founder and chief architect of Meaningful Machines, Abir's start-up company, whose offices are on the fortieth floor of a glass-clad midtown Manhattan skyscraper. The first goal Abir plans for his "brain" is the ability to translate with the precision of human translators. After that, he will set it to the task of comprehending and extracting useful ideas from information in a single language, a process known as "text mining." Both tasks rely on what Abir says is a way of teaching computers to comprehend context. He believes he has done nothing less than to unlock the secret of allowing computers to understand ideas. "Everybody else has stones," he says. "We have a sharp knife."

These would be bold claims even if they emanated from the most celebrated computer scientists in the world. Coming from Eli Abir, they at first appear to be the preposterous ravings of a deluded soul. This is because Abir has no formal education in linguistics or computer science, and never graduated from high school. Until 1993, he had never even used a computer.

But the enigmatic Abir has gained some prominent backers, and his company—even with most of its technology still under wraps—is being taken seriously. One of the world's leading machine-translation experts, Jaime Carbonell at Carnegie Mellon University, became so enamored with the possibility of Abir's technology that he joined the company. "Machine translation is my passion," Carbonell says. "I've been working with many techniques, some of which have done very well. Here I saw another one that had a chance to beat all of these—and it looked like it was going to—and that I could help to make it even better. So I joined them."

Meaningful Machines is highly secretive about its methods, but it will divulge that its work relies on a variation of example-based MT. At the most basic level, Abir's technology works by splitting up language into a finite number of phrases and terms, which he calls "language meaning units," or LMUs. In every language, he estimates, there are between 1 and 5 billion LMUs. "These are ideas, they exist in the universe," he says. "Human beings take a language, and they attach it to existing ideas. That's why when I tell you, 'Get out of here,' and the Chinese tell you, 'Get out of here,' they mean the same thing, although we say them in a different way."

"I'll take something more complex," he says, suggesting the sentence "I love you, and I'll see you Monday." He observes that the sentence can be split into a number of LMUs, each of which are separate, but related: I, I love, I love you, I love you and I'll see you, I love you and I'll see you Monday. "Every comprehensive idea," he says, "is made out of either ideas that are smaller, or parts of ideas that are not even coherent when they are apart."

By breaking up language into LMUs, instead of just individual words or short phrases, Abir claims that the system is able to "learn" the translations in much the same way a human does: through context. The software's algorithm takes a statistical measure of the context around the LMU—how, exactly, the company won't say for fear of a competitor swiping its method—and stores it along with the LMU fragment, so that when it appears near similar concepts, the LMU is translated correctly. His method, Abir says, affords the system its most remarkable—and dubious, to those outside the company—feature: it doesn't require corpus text that has already been translated between two languages. Instead, it can translate text using only a separate collection of documents in each language and a basic multilanguage dictionary. Another advantage, he says, is that the system "knows what it doesn't know." When it doesn't have confidence in the translation of a certain LMU, it can highlight that portion of the text. Even if the translation isn't perfect, a human translator can locate the portions that need correcting without analyzing the entire text.

Abir's "brain" has potential applications beyond translation, particularly for text-mining software, which is used to pull meaning out of unstructured data. Structured data is that which lives in a normal database, with rows of records, each record containing a discreet piece of information—name, ID number, phone number, address. Unstructured data is, essentially, everything else, including word-processing documents or Web pages. Unlike search engines, which merely retrieve documents, text-mining software scans through unstructured documents to pull out and categorize data, converting it to a structured format for analysis. One promising area of text mining, for example, involves scanning biomedical literature, looking for information that might represent new diseases. It's a task made difficult for the same reason as language translation: doing it well requires utilizing the context surrounding the terms you are searching.

For Meaningful Machines' text-mining application, Abir has built a kind of thesaurus of ideas, which he calls a "semantic-equivalence generator." For any word or phrase, the software uses a search algorithm to examine a collection of documents—the company currently uses one terabyte of information, the equivalent of one-tenth the amount of information in the Library of Congress print collection—searching for associated LMUs and generating databases of them ranked by how often they are associated. Enter a city name such as New York, for example, and the generator will return with other large cities—London, Tokyo—that it deems are related in the text it examined. Enter the *New York Times,* however, and it returns equivalents such as CNN and Associated Press. "It's as if you are seeing something or hearing something for the first time as a human being," says Meaningful Machines' CEO Steve Klein. "You have the context you find it in when you hear it first, or read it in a book. Context isn't a secret, it's how to measure context that is a secret sauce that we have."

Sitting at a computer in Klein's corner office, Abir types phrases into a rudimentary version of the semantic-equivalence generator, reading off the results with the excitement of a proud parent. When he types in "common knowledge," the screen returns with highly ranked answers such as "obvious," "clear," and "it's fair to say." "Pretty nice, right?" Abir says, grinning. Then he turns back to the keyboard and types in "9.11." "You notice," he says, "it doesn't give me dates. Notice that it puts, 'September 11,' 'tragic,' 'barbaric,' 'terrible.' Why? Because when it reads about it right now, that's the connotation it gets." He types in "Al Qaeda," and the program spits back phrases such as "extremism" and "America is hated."

"Apply this to text mining that the government wants to do for terrorism," Klein says. "An analyst wants to know about Al Qaeda selling anything to any sovereignty, or some sovereignty selling something to Al Qaeda." A Google-like search using standard keywords—called a boolean search—he continues, requires "the keywords to find all the right combinations of how something might be expressed out there, in all the documents. If you can generate all of the alternative ways of saying a concept"—using the semantic-equivalence generator—"you could search every way of saying that concept automatically." In short, the software could help you automate your own thinking.

Formidable computer-science problems such as translation, having foiled computer science and artificial intelligence researchers for decades, tend to attract thinkers from unexpected quarters. Often, the line between an outsider genius and a kook is a thin one—and one that Eli Abir at first glance appears to straddle. People encountering Abir and his ideas for the first time have an almost universal reaction: They have no idea what he's talking about. Abir is small and stout, in his late forties, with spiked gray hair and a goatee. He rides a shiny new Harley-Davidson motorcycle to work, and typically wears jeans and a black T-shirt with, say, a silk-screened Superman cartoon on the front. He has an effusive personality and a playful arrogance about his own intelligence, both of which make him prone to drift from any discussion into his philosophy of the universe. A typical technologist with a Big Idea, he's flanked by Steve Klein, a lawyer by training, who is the founder of Apple Core Holdings, the investment company that has put around $7 million into Abir's ideas. After much cajoling by Abir, Klein says, he agreed to sign on as Meaningful Machine's CEO. Klein, in a button-down shirt and slacks, actually acts as a sort of translator for Abir, attempting to explain his esoteric language in simpler terms—and to offer a tempered view of some of the chief architect's more extravagant claims.

It's when Abir starts talking about his past that the bewilderment really sets in. He was born in Israel, and won't talk about his time there, except to say that he emigrated to the United States in the 1980s, speaking no English, because of political differences. "I've lived a very hard life," he says. "I've lived maybe five lifetimes. But more importantly, more than anything else, I'm a very curious guy. I spend an enormous amount of time reading, from the time I was a kid." One of his previous habits, he says, involved reading up to 10 newspapers a day, clipping articles from each and writing down what he thought would happen in the future on that topic. A year later, he would look back and see if his prediction had come to pass—most of the time, he says, he was either wrong, or right for the wrong reasons—and then tried to figure out why it hadn't. His hobby, essentially, was knowledge.

Abir says he cycled through stints as a restaurateur and a used-car salesman in New York. His introduction to computers came in 1993, when he bought his son a Performa 400 personal computer. It came with a version of the video game SimCity, which allows players to design and build virtual cities. "I locked myself in a room with it and two days later I know that I understand computers better than anybody in the world," he says, with characteristic bluster.

"We almost didn't invest," Klein interjects, with a sheepish laugh, "because of some of these claims."

"What I mean," Abir continues, "is what a computer is and what you put inside it is no different from us sitting in this room now. Think of the whole universe as a computer. I'm just replicating what I see."

Abir had decided by this point to pursue a legal career, but abandoned law school in favor of pouring his savings into buying generic Web addresses such as officesupply.com and DNAanalysis.com, which he later sold off for tens of thousands apiece. The proceeds helped him hire two programmers and build his first software concept, a program called Internet Driver that would translate website addresses from Hebrew to English.

With help from a friend who was well-placed in business, Abir shopped his Internet Driver idea to potential investors, showing up in his usual jeans and T-shirt to announce that he had figured out how to make computers understand language. "Except for Steve, I got kicked out of every place I came to," he says. At first, Klein too was skeptical of Abir's seemingly outlandish assertions. But when the conversation turned to law and Abir began quoting Supreme Court cases, Klein says, "that's when I set out to find out if he was full of shit or not on the technology front." When he determined that there was nothing comparable to Internet Driver on the market, he says, he committed to investing.

In the first week of September 2001, Abir and Klein traveled to

Uruguay for an Internet conference, where they demonstrated the software for the first time. They arrived back in Newark Airport on the morning of September 11, shortly before United Airlines flight 93—which would eventually crash in Pennsylvania—was scheduled to take off. "We were walking out of the airport," Abir reflects, "and the terrorists were walking in."

Abir had been voicing larger plans to build a generalized translation program—"the brain"—that could translate documents between any two languages. September 11 and its aftermath led Abir and Klein to reevaluate the need for translation, and they decided to shelve the Internet Driver in favor of pursuing the translation technology. "We wanted to give it to the government for free," Abir says with a shrug, "but they didn't want to talk to us."

Instead, they took it to Jaime Carbonell at Carnegie Mellon. Carbonell, the founder and director of the university's Language Technologies Institute, is used to hearing from people claiming a breakthrough in MT. "Some people think that they keep discovering the fundamental structure or fundamental meaning or the fundamental translation or the fundamental secret of language. So there are a number of, kindly spoken it would be amateurs, less kindly spoken it would be crackpots, out there. I get a fair trickle of them coming through all the time." But when Carbonell got a call from Klein, he became intrigued with some of Abir's ideas and agreed to meet with them. A simple demonstration and description of the software at Carnegie Mellon, Carbonell says, blew him away. He agreed to conduct a detailed analysis of the system, and after a month of evaluation wrote that "the [Meaningful Machines'] method is clearly the most promising and theoretically important recent MT development in the past several years—and probably since the advent of MT itself."

Carbonell joined the company's board of directors, and eventually came on part-time as Meaningful Machine's chief science officer. Still, even to Carbonell, Abir remains something of a mystery. "I still don't quite understand how Eli's mind works," he says. "He can make connections between things that other people would never even think would be rationally connected." But "the bottom line," he says, "is that it doesn't matter whether language works the way Eli thinks it works. What matters is that this way of thinking about it was leading to a whole class of algorithms for processing language that were radically different from the algorithms that had been investigated before."

It's possible that Abir's very status as an outsider may have contributed to his ability to see language a different way. "I always enjoy working with brilliant people who have a unique perspective on a problem," says Danny Hillis, the pioneer of parallel computing—a method of massively increasing computing power by tying processors together. Hillis, a renowned technologist and inventor, joined Meaningful Machine's advisory board after reviewing the technology. "Often those are people who don't have formal training, because the formal training tends to give you the same perspective as everybody else. Eli fits that pattern, as somebody from the outside who is obviously just very smart and looking at it from a fresh perspective because he is not trained in the field."

Any application of Abir's technology for language translation in intelligence gathering may still be years away—if it ever arrives. The company's first planned module is Japanese, a language that Klein hopes will help them gain a foothold in the multibillion-dollar commercial translation market. But if the company succeeds and Abir's ideas are validated, they could potentially have artificial intelligence applications well beyond translation. If this happens, it may yet remain a mystery as to why Abir saw the puzzle as he did. He recalls the day he first walked into Klein's office and announced that he had solved an age-old problem in computer science. "He said, 'How come you? How come anybody in the world, Microsoft, everybody'" couldn't come up with it? "I said, 'I don't know. I'm interested in things.'"

On October 17, 2002, Lieutenant General Michael Hayden, the director of the NSA, sat before the congressional joint inquiry on the 9/11 intelligence failures. It was only the third time in history that the head of the formerly secret agency had testified in an open session of Congress about the agency's operation, and Hayden had not trudged from Fort Meade up to Capitol Hill to mince words. While acknowledging that in some areas the agency's "performance . . . could have been better," he stridently objected to charges that it had mishandled intelligence information. He also complained about unauthorized leaks of intelligence failures—such as the two untranslated intercepts that warned of an unspecified attack on September 11. What public criticisms of the NSA lacked, he told the committee, was the greater perspective of the process of intelligence gathering. To determine the value of a piece of signals intelligence, he observed, requires answering a host of questions: "Where was the information collected? Were any of the communicants targeted? How many calls a day are there from this location? In what languages? Hazzar? Urdu? Pashto? Uzbek? Dari? Arabic? Is there a machine that can sort these out by language for you, or do you have to use a human? If there is such a machine—does it work in a polyglot place where one conversation often comprises several languages? How long does it take NSA to process this kind of material?—after all, we are not the intended recipients of these communications. Does our current technology allow us to process it in a stream or do we have to do it in batches? When the data is processed, how do we review it— oldest to newest or newest first? And aside from how we normally process it, did the sequence change at 08:46 A.M. on September 11th?"

Hayden's questions—while intended to blunt criticisms against the NSA—also suggested the nature of the true challenge faced by information collectors in an age of ubiquitous communication. The issue is not simply cracking encryption, detecting steganography, or translating obscure dialects—but finding the context in which the resulting information can be understood.

Al Qaeda, Hayden observed, "did not need to *develop* a telecommunications system. All it had to do was harvest the products of a $3 trillion a year telecommunications industry—an industry that had made communications signals varied, global, instantaneous, complex, and encrypted." Hayden's points weren't lost on the 9/11 Commission, which restated his observation in its report, cautioning that "targets of intelligence collection have become more sophisticated." Just as the crypto wars foreshadowed, the very technologies that we design to protect ourselves are the ones which may later surface in a bombed-out building in Kabul.

Eric Thompson, however, knows that the information collection process still comes down to a contest between the hiders and the seekers. From his perspective, the agents and analysts—and their tools— have the advantage: He's seen it in the child molesters put away because of their digital trail, and in Richard Reid—rotting in jail with the help of agents wielding Thompson's software. He knows the hidden flaw that plagues those who are trying to keep secrets. "The fact that there is a human involved in the process is a weakness," he says. "They can't get around that."

TWELVE

The Dangers of Data

"Y OU CANNOT BEAT a roulette table," Albert Einstein is supposed
to have observed, "unless you steal money from it." The same
could be said for any game in a casino (except possibly blackjack),
which is why gambling houses—besides being home to the perpetually hopeful—have always attracted a legion of small- and big-time
frauds and cheats. There are few human institutions more creative
than the criminal mind, and as casino security has grown—with its
eye-in-the-sky surveillance cameras, undercover officers, and automatic card shufflers—so too has the sophistication of those looking to
beat the house by cunning and sleight of hand. Modern gambling
cheaters carry out elaborate schemes, using teams that conduct practice runs, divide responsibility, and defy easy categorization—"the
grandma and the guy with a bunch of piercing working together," as
Las Vegas software entrepreneur Jeff Jonas describes it. "They have
identity packages. They are picking totally different names, social security numbers, and dates of birth, creating fake identities and dis-
guises so that they can come in and not be detected. And they are
walking away with millions."

So the security detail at one riverboat casino was surprised when, a
few years ago, they observed a rather obvious scam by a cheater who

seemed to be working alone. The player was slipping bets onto the roulette table just after the ball had dropped, guaranteeing himself a win. The scam is called "past posting," and without a distraction team the dealer should have easily spotted it. When security did so on the surveillance camera and then swooped in to pick up the cheater, the dealer was rightly aghast. "I'm so embarrassed," she told the security detail. "It'll never happen again."

That would have been that, perhaps, except that a security officer took down the cheater's information—name, address, phone number—and ran it through a piece of software designed by Jonas called Non-Obvious Relationship Awareness, or NORA. NORA organizes and then sorts through information in computer databases, looking for connections between people and linking them through common addresses, phone numbers, and more obscure connections such as geographic location. The principle behind the program, that people can be traced through their connections to others, is commonly referred to as "degrees of separation." Two people who know each other directly are said to be connected by a single degree. If they are connected through a common acquaintance, they are two steps, or degrees, apart. If instead there are two connected people between them, then they are three degrees apart, and so on. John Guare popularized the notion in his play, *Six Degrees of Separation,* in which he played off the premise that all people on Earth are connected by a half-dozen or fewer. The concept has also spawned a number of social networking websites—such as Tribe.net and Friendster—that allow people to connect to others through their friends. The physicist M. E. J. Newman, who wrote "Who Is the Best Connected Scientist? A Study of Scientific Co-authorship Networks," which examined how various researchers are connected, was tracing a scientific version of degrees of separation. The notion even formed the basis of a thought game, called "Six Degrees of Kevin Bacon," in which one tries to connect any given actor, through shared movies, to the industrious movie star.

In the case of the riverboat gambling scam, NORA uncovered a very different game. Taking the cheater's information as a starting point, NORA scanned through databases of fellow cheaters, known gambling felons, and casino employees, looking for matches within a few degrees of separation. In this case, all it took was one: In a matter of seconds, the system came back with a hit. The cheater had shared a home phone number with the dealer; they had been colluding the whole time. Security made another trip back to the table.

Jonas, the gregarious 40-year-old founder and chief scientist of the data analysis company Systems Research and Development (SRD), has a wiry frame honed by triathlons, a goatee, and an unlikely passion for deconstructing databases. "My job is my hobby," he likes to say, when asked about his long hours at the office. The conference room at SRD's Las Vegas headquarters—with its framed magazine and newspaper clippings trumpeting his rise from high school dropout to lauded entrepreneur—is evidence of his sincerity. NORA, Jonas's first big idea, is a variation on a field of computer science called data mining. He built SRD's reputation down the road, on the Las Vegas strip, where NORA uncovers schemes such as the past-posting collusion. Casinos also use it to match names against a list nicknamed "the Black Book," of a few dozen known gambling felons and other unsavory characters, whom they are required by law to keep out of their establishments.

Nowadays, however, Jonas is after a bigger prize than gambling cheats. He thinks NORA's ultimate calling may be to catch terrorists. "You gotta see this!" he says, as he leans over his laptop on the conference room table, clicking through a slide presentation projected on a screen behind him. He stops on a slide that reads, in large text, "September 11, 2001: Connecting the Dots," and looks up with an excited gleam in his eye.

His tale begins with a network. After the attacks in New York and Washington, as facts about the hijackers and their organization seeped into the media, a picture of the 9/11 plot emerged. Valdis Krebs, a Cleveland-based business consultant and expert in the study of the patterns of social interaction, called social network analysis, created a widely reprinted illustration showing the web of links between the hijackers. There were hubs, such as Mohamed Atta and Marwan al-Shehhi, linked to a wide group of nodes: plotters, financiers, and final executors. Much has been made of the terrorists' level of interconnectedness and intelligence agencies' inability to penetrate their plot. Krebs says of his diagram, "It was all built with information afterward, so it was easier to do. But it does show you that there were these connections, and they were fairly obvious."

To Jonas, the important issue was not what the network looked like after the fact, but whether there could have been any way to uncover it beforehand. "The pictures of how people are connected mean nothing unless you know which guy you were looking for to begin with," he says. "The question is, what's the thread?" So Jonas, using the same information that was available before the attacks, retraced the connections, beginning with only two of the hijackers, Nawaf al-Hamzi and Khalid al-Midhar. "What could you find in the network," he asked, "if you start with them?"

The answer, which has since been recounted in the 9/11 Commission Report but which Jonas presents with his particular dramatic flair, is quite a bit. Based on leads from NSA signals intelligence, the CIA tracked al-Hamzi and al-Midhar to a meeting of Al Qaeda plotters in Malaysia in January 2000. Intelligence officials lost track of the two, however, when they entered the United States later that year. (Al-Midhar would leave and re-enter the country again in 2001.) Sometime around August 23, 2001, the CIA finally provided the names to the State Department, to be placed on a joint INS/State Department watch list called TIPOFF, which at the time held around 61,000 names. The FBI office in New York tried and failed to locate the two, and the Los Angeles office didn't receive a request to search for them until September 11. But two days after being placed on TIPOFF, al-Midhar purchased a ticket, in cash, for American Airlines flight 77 on September 11; al-Hamzi did the same on August 27. A simple check of those reservations against the TIPOFF list would have flagged them both, Jonas says, as he clicks the mouse. Two red circles appear around their photos, now projected on the screen.

By pulling the threads of those first two hijackers, authorities could potentially have unraveled much—if not all—of the plot. The address on al-Midhar's reservation matched that of two other hijackers: the ringleader Mohamed Atta and the highly connected Marwan al-Shehhi, both of whom used it as a contact address for their own September 11 reservations. Meanwhile, al-Hamzi's address matched that of a fifth hijacker, his brother. Three more red circles pop up on the screen. A phone number used by Atta on his reservation tied him to five other terrorists, and al-Midhar's frequent flier number revealed a link to another hijacker. Jonas clicks, and now 11 hijackers are circled.

"We've taken one bad-guy list," he says, "and just airline data, and we've found II of them. We don't need banking, medical, what you buy at the store." A search of public records would then have turned up a common address connecting two more hijackers to the network. "Now you are at 13," he says. "They are all connected by four degrees." That left five hijackers unconnected. The next best thread, says Jonas, was Ahmed Al-Ghamdi, who was on a longer Immigration and Naturalization Service watch list for expired visas. Al-Ghamdi was connected by two degrees to the remaining five.

Jonas contends that had NORA been installed to scan the right databases at the proper places—whether at the airlines or in the government—it could have unraveled the plot with nothing more than the TIPOFF list to go on. And Jonas has some credibility in the intelligence community: SRD is partially funded by the CIA's venture arm, In-Q-Tel. Through that connection, SRD works with various other government agencies—Jonas won't say which ones—to apply NORA in efforts to identify terrorists using data. His slide show has been passed around Washington and eventually ended up in an influential report of the Markle Foundation—where a collection of top experts on information technology and homeland security are investigating how the two can be united. The tale has since bounced around among policy makers and become the poster child not just for failed intelligence but also for the prospect of using information technology to uncover terrorists.

Jonas's version of data mining is just one of the many what-could-have-beens to lament. Following September II, 2001, the nation agonized over the question of whether or not the attacks could have been prevented, if only intelligence analysts had been able to synthesize disparate knowledge about the plot. It's a question all the more haunting because of the information we know was available. There was the July 2001 internal FBI memorandum—the famous Phoenix memo—that raised concerns about Middle Eastern men attending flight schools in the United States. There were the two September 10 Al Qaeda telephone intercepts that the NSA translated on September 12. There was the arrest on August 16, 2001, of Zacarias Moussaoui, who had been taking flying lessons in Minnesota. And there were the years of intelligence reports warning of the growing threat of Bin Laden and

Al Qaeda. As the Joint Congressional Inquiry on 9/11 intelligence failures reported, the intelligence community failed to take advantage of
"opportunities to disrupt the September 11th plot by denying entry to
or detaining would-be hijackers; to at least try to unravel the plot
through surveillance and other investigative work within the United
States; and, finally, to generate a heightened state of alert and thus
harden the homeland against attack."

Hindsight, of course, plays subtle tricks on the truth. Prior to the
actual attack, as security expert Bruce Schneier points out, "the crucial bits of data are just random clues among thousands of other
random clues, almost all of which turn out to be false or misleading or
irrelevant." This is important to understanding what it means that the
9/11 Commission eventually concluded the events of September 11
were preventable. Both the post-9/11 congressional inquiry and the
independent 9/11 Commission rightly focused on the institutional reasons for the failure to put all this information together and act on it.
(To take one example, the Federal Aviation Administration's "no fly
list" was completely unconnected to the TIPOFF list.) Much of the
9/11 postmortem has therefore centered on how intelligence agencies
should be organized and how to create greater cooperation among
them—including the sharing of watch lists such as TIPOFF. But no
matter how intelligence bureaucracies are restructured, or how much
power the head of those bureaucracies is given, other questions
remain: how does the overwhelming volume of intelligence information get analyzed in the first place, and what role can technology play
in connecting the dots of terrorist networks? Would software such as
Jonas's help us find the right threads to pull? Could a computer look
at a set of random clues and see the terrorist network that an intelligence analyst might not?

The desire to answer those questions is fueling the government's interest in the mushrooming fields of data analysis and data mining. The
idea that we can search data to catch terrorists is premised on the
recognition—a frightening one to some—that our lives are shadowed
by a growing catalog of digital information about every aspect of ourselves and the world. A myriad of facts, labels, and measures of our
identities are tracked and stored on computers. Every day we leave

our fingerprints in information—when we use an ATM, pay with a credit card, download a document, send an e-mail, or make a cell phone call.

The past decade has seen an explosion in data collection by corporations and the government, through everything from health-care records and supermarket frequent-shopper cards to online book buying and airline reservations. That information is gathered, exchanged, and combined freely in the commercial world, often without our knowledge. Giant data aggregation firms such as ChoicePoint, LexisNexis, and Acxiom make their living obtaining data from both public sources and private companies, such as credit reporters, amassing it, and then reselling it for target marketing and employee background checks.

Terrorists, especially if they are trying to avoid detection by operating within social norms, leave their own digital fingerprints, just as the 9/11 hijackers did on bank accounts, frequent flier cards, and airline reservations. The idea behind using data mining in counterterrorism, from a technical perspective, is to sort through a vast sea of information and find the patterns or clues that point to a terrorist plot—no easy feat. From a political perspective, the problem is even thornier. We're caught today between a widespread fear that our privacy is at stake, and an equally palpable fear that critical information may be lost in the flood of data. The two are not mutually exclusive. Technologies employed in intelligence gathering overseas, for example, may become unpalatable when applied to citizens at home. For many Americans, the prospect of a government scanning through our information looking for clues is itself a terrifying one. But for those designated with the task of thwarting future deadly acts, the greatest fear is missing the essential clues that precede one.

The problem is that we have little or no idea where, when, how, or from exactly whom an attack might come. It is a quandary—as Rao Vemuri, a computer scientist researching data mining for Lawrence Livermore National Laboratory, has observed—analogous to one posed by Plato in his dialogue *Meno*. In Plato's fictional exchange between Socrates and Meno, a Greek aristocrat, the latter poses a dilemma about the nature of inquiry that has come to be known as Meno's paradox. "How will you look for something when you don't at least know what it is?" Meno asks Socrates. "How on earth are you going to set up something you don't know as the object of your search? To put it another way, even if you come right up against it, how will you know that what you have found is the thing you didn't know?"

The question posed to analysts—and to anyone advocating using data mining to detect terrorism—becomes a version of Meno's paradox: How do we find the signs of terrorism that we don't know we are looking for? How can we avoid the problem of "fighting the last war"—looking for truck bombs while terrorists are busy planning to turn undefended airlines into missiles? Instead, we must look for the next plot, the next incarnation. But how will we know when we see it? We may actually have information that would reveal the plot—we may have "come right up against it," in Meno's words—without recognizing what we have.

The potential for data mining to help answer those questions lies at the intersection of the information explosion and the increasing computing power available to process it. Data mining is, at its root, an attempt to take any amount of raw information—from corporate records, to genomic data, to military intelligence—and extract from it meaningful facts. It is a field with vague boundaries and one plagued by labels that don't always fit. The term "data mining" is itself somewhat of a misnomer. "When you are gold mining you are looking for gold," says David Jensen, a computer scientist and data mining specialist at the University of Massachusetts. " 'Data mining' implies that you are looking for data. Well, you're not. You are looking for knowledge, or statistical patterns, or useful predictive knowledge. Not for data. You are looking *through* data." (Many researchers have thus come to favor the term "knowledge discovery in databases," but data mining as a general label has stuck.) Whatever you call it, the goal, Jensen says, is "to identify high-level things—organizations and activities—based on low-level data—people, places, things, and events." In isolation, the low-level data appears void of meaning. But by piecing together disparate pieces of information into a coherent puzzle, meaningful patterns suddenly emerge.

A common dictum among data miners is that "we are drowning in data, but thirsting for knowledge." As our mass of collected information grows—whether about people, scientific research, or business transactions—so too does the difficulty in sifting through it to find anything meaningful. As far back as 1945, the engineer Vannevar Bush wrote the following in a prescient *Atlantic Monthly* article: "There is a growing mountain of research. But there is increased evidence that we are being bogged down today as specialization extends. The investigator is staggered by the findings and conclusions of thousands of other workers—conclusions that he cannot find time to grasp, much less to remember, as they appear. Yet specialization becomes increasingly necessary for progress, and the effort to bridge between disciplines is correspondingly superficial." Such ideas, in fact, were part of the cover for the 1957 founding of ITEK, Richard Leghorn's pioneering spy satellite company.

The problem has grown astronomically since Bush's and Leghorn's time, as can be attested by anyone who has typed a search phrase into Google and turned up millions of documents. Finding precisely what you are looking for—especially if you suffer from Meno's dilemma of not knowing exactly what you are seeking—requires a skillful search and some measure of luck. Often, you are not hunting for a specific website but the answer to some question—what's the closest train station to the Eiffel Tower? Who holds the record for pennies stacked end to end?—and finding that answer, among the vast amounts of irrelevant data that your search inevitably nets, is a challenge.

The government security agencies charged with counterterrorism—the CIA, the Department of Defense, the FBI, the NSA, the National Geo-Spatial Intelligence Agency, and the Department of Homeland Security—suffer from the same type of information overload. Like an ocean drift net that catches 10 fish for every edible one, they gather much more information than they need, and more even than they could possibly analyze. The NSA has its more than 2 million signals intelligence intercepts an hour to contend with, while the CIA and FBI have the masses of information they gather from human sources, confiscated documents, open literature, and investigations. The amount of data multiplied further with the Department of Homeland Security's new multi-billion-dollar US-Visa tracking program, which combines the visa information of U.S. arrivals with digital photographs and fingerprints. A 2002 assessment by the National Research Council summed up these agencies' predicament: "Currently one of the significant problems is managing a flood of data that may be relevant to their efforts to track suspected terrorists and their activities."

Just as in a Web search, however, intelligence analysts are not just looking for particular documents or data. "The more interesting thing that intelligence analysts do, and the more interesting thing that you'd like to do on the Web, is to understand, which is different than to search," says parallel computing godfather and Meaningful Machine's adviser Danny Hillis. Hillis is the cofounder of Applied Minds, a multifaceted research and development company that works on everything from software, to mechanical design, to biotechnology for both government and industry. "If I want to figure out something like what is going to be the next great development in particle physics, it's not like the answer is out on the Web. But on the other hand, there might be a lot of information on the Web that, if I looked at it the right way, would help me understand that thing." Intelligence analysts, similarly, "are often trying to understand rather than to find the document that has the answer." Unfortunately, while humans are excellent at piecing together the right data to gain that understanding, we are slow and limited in the amount of information we can sift through. Computers, in theory, are ideally suited to assist analysts in their search.

There is no shortage of government efforts to develop software that will do so. The Financial Crimes Enforcement Network (FinCEN) of the U.S. Treasury Department has been investigating ways to apply data mining to money laundering since the mid-1990s. The Advanced Research and Development Activity (ARDA), the secretive research and development arm of U.S. intelligence agencies housed in the NSA, has been actively funding data mining projects since at least 2000. In 2002, the National Science Foundation expanded its data-mining research effort with funding from the Intelligence Technology Innovation Center, administered by the CIA. The Northern Command, the homeland security division of the military, has its own data-mining research underway, funded by the Pentagon's Counterintelligence Field Activity (CIFA) division. According to the Los Angeles Times the project is charged with "figuring out a way to process massive sets of public records, intercepted communications, credit card accounts, etc., to find 'actionable intelligence.' " The Multistate Anti-Terrorism Information Exchange (MATRIX), a controversial effort to create searchable law enforcement databases across 16 states, began in 2002 (at least 11 of the states have subsequently withdrawn over cost and privacy issues). The Homeland Security Advanced Projects Research Agency has its own data mining effort underway, and even NASA has gotten in on the game, funding research to scan airline records for potential threats.

Outside the government, the National Association of Securities Dealers uses data mining to detect insider trading and fraud on NASDAQ, and the 2001 USA Patriot Act imposed data-monitoring requirements on banks and securities firms, many of which use data mining to detect money laundering in their accounts. Dozens of companies—with names such as ClearForest, Mantas, and Fair Isaac—identify data mining of various forms as their primary technology, and many of them claim to be supplying unnamed government agencies.

The attempts to apply data mining technology to counterterrorism fall loosely into two categories: subject-based and pattern-based. The former, like Jeff Jonas's NORA software, begins with a subject—a person on a watch list, say—and attempts to trace links from that person to related entities or a hidden network. Pattern-based data mining, on the other hand, takes a model—a set of rules that describes money-laundering behavior, for example—and tries to find data that matches it. Traditionally the model is programmed into the software based on human experience. An anti-money-laundering program, for example, would incorporate rules that federal agents have learned from past experience with the crime: that money launderers tend to use multiple bank accounts and deposit in cash. In the most automated form of pattern-based data mining, the software algorithm discovers rules on its own, by learning from a collection of historical data. In the case of money laundering, agents could feed the software a set of data that includes some bank accounts that have already proven fraudulent, and let the algorithm use that data to determine its own rules.

In everyday terms, the contrast between subject-based and pattern-based applications might be thought of as "pulling the thread" versus "finding the needle." A subject-based analysis begins with a thread, like Jonas's portrayal of the 9/11 plot, and then proceeds to find other subjects connected to it. Pattern-based analysis, on the other hand, is more akin to looking for a needle in a haystack. Or rather, in the words of Ted Senator, a DARPA program manager and data-mining expert, it is like "finding dangerous groups of needles hidden in stacks of needle pieces."

Both types of data mining—as they are envisioned for counterterrorism—take advantage of networks, the critical component of terrorism and the feature that gives it its power. They attempt to help analysts uncloak the terrorist links hidden among vast quantities of information. The larger challenge in counterterrorism, however, is not simply to try and locate suspicious people. It is to do so without generating the twin enemies of any search: false positives and false negatives. Put another way, is it possible to use data mining to connect the dots of a terrorist plot in advance without infringing on the privacy of innocent citizens in the process?

Data mining has its roots in the field of classical statistics, which uses applied mathematics to categorize and describe collections of data. "For 100 years, statisticians have been working on ways of finding patterns in data," says Andrew Moore, a computer science and robotics professor at Carnegie Mellon University. As both the power of computing and the volume of stored digital information grew after World War II, computer scientists began to create software that would seek out those patterns over data collections much larger than humans had previously been able to analyze. The problem in the early days of computing was that those computer scientists had little knowledge of statistics. "Instead of doing the clever things that statisticians do to find *meaningful* patterns in the data," Moore says, "they just found patterns."

In just the last decade, however, the fields of computer science and statistical analysis have converged to create the modern discipline of data mining, in which increasingly powerful computers scan massive amounts of information for useful patterns. "Now," says Moore, "we're in a nice situation where people are doing statistical analysis, finding meaningful patterns in large amounts of data, which previously neither the computer scientists or the statisticians by themselves had been able to do."

As a result, data mining is already employed all around us, both in ways that we rarely object to—such as fraud detection, the study of disease, spam e-mail prevention, and the personal recommendations on Amazon.com—and those, such as the mining of marketing data, that privacy advocates find distasteful. A typical credit card company's fraud detection system, for example, examines each card's activity for possible misuse. The software is built on a set of rules derived from the company's past experience with fraud. One such rule involves flagging a card that is used at a gas pump (which, because it doesn't require a signature, presents an easy place for thieves to check that stolen cards haven't been canceled), and then immediately afterward to buy an expensive item. The software might also scan the data for unusual charges—called outliers—which may signal a change in usage. If purchases suddenly start appearing on your card from Russia, you may get a call from a service representative, the result of the computer flagging your purchase as potential fraud.

These kinds of relatively simple data-mining efforts are widespread, but it wasn't until the first week of November 2002 that the field and its connection to terrorism burst into the American consciousness. That was the week that John Markoff, the veteran technology reporter for the *New York Times,* wrote a story—buried in the paper's Saturday edition—about a new government program called Total Information Awareness. TIA, as it came to be known, was a data analysis research venture funded by DARPA under a new division called the Information Awareness Office, run by former national security adviser John Poindexter.

The Information Awareness Office, established by DARPA in January of 2002, was a collection of research and development projects designed to apply a variety of cutting-edge technologies to counterterrorism. Poindexter oversaw the development of language translation technologies, biometrics, data search and pattern recognition software, and information-sharing technologies. The goal of the Total Information Awareness portion, as defined on the DARPA website, was to combine various parts of the Information Awareness Office into a single prototype that would test whether information technology could be used to uncover the digital footprints of terrorists. As Poindexter explained in an August 2002 speech to a DARPA conference in Anaheim, California, "If terrorist organizations are going to plan and execute attacks against the United States, their people must engage in transactions, and they will leave signatures in this information space. . . . We must be able to pick this signal out of the noise. Certain agencies and apologists talk about connecting the dots, but one of the problems is to know which dots to connect."

Markoff had first broken the Poindexter story nine months before, in a February *Times* article detailing his initial appointment. Public reaction was minimal, however, until Markoff's more detailed November account, which described TIA as "a vast electronic dragnet, searching for personal information as part of the hunt for terrorists around the globe—including the United States." The article was followed by one from Robert O'Harrow at the *Washington Post,* which laid out details of what he called "a global computer-surveillance system." On November 14, *Times* columnist William Safire set fire to the kindling, describing TIA as a Poindexter-led "assault on individual privacy" whose goal was to "create computer dossiers on 300 million Americans." Such dossiers, he wrote, are not "some far-out Orwellian scenario. It is what will happen to your personal freedom in the next few weeks if John Poindexter gets the unprecedented power he seeks." To Safire—and to the legion of editorialists and pundits who subsequently heaped scorn on the program—TIA seemed like Poindexter's master plan for delving into every crevice of American life.

In fact, almost all of the projects funded under TIA predated not only Poindexter, but also 9/11. The core of TIA's data-mining efforts— and the source of much of the angst over the program—was funded through a program called Evidence Extraction and Link Discovery, or EELD. EELD had begun in 1998, when a DARPA program manager named David Gunning organized a meeting at the dedication of the Carnegie Mellon's Center for Automated Learning and Discovery in Pittsburgh. Gunning's idea was to apply research in a new subset of the data-mining field to national security questions. Traditional data-mining methods had always involved looking at each record in a database independently—analyzing, for example, a series of cancer patients and examining the symptoms of each for a pattern that would indicate cancer. In such problems, each patient's status is unrelated to the rest. The EELD program, instead, would examine instances where the entities in the database were interconnected. The process of finding those connections is called "relational data mining." Relational data-mining software might, for example, examine a set of genes to determine which ones produce certain human diseases. To do so, it would analyze not just each individual gene's properties but its interactions with the other genes nearby. The goal of EELD, and of relational data-mining in general, was to investigate complex connections of people, places, and institutions—the exact kind of interactions that DARPA reasoned existed in terrorist networks.

Complex data-mining research has applications in fields well outside of counterterrorism—in everything from scanning scientific literature, to classifying Web pages, to, potentially, finding the patterns in George Poste's Z-chips. And even coming as it did before 9/11, Gunning's meeting drew many of the prominent names in the field. "All of the university researchers were intrigued," says David Jensen of the University of Massachusetts, who gave the opening talk. "Not so much because of the application, but because of the technical challenge, which was an entirely different kind of machine learning than had been done previously."

After preliminary studies of the feasibility of EELD, DARPA funded several academic labs to begin research in mid-2001. Gunning, meanwhile, established another program called Bio-ALIRT, an attempt to build syndrome surveillance software. Then came 9/11. "In the space of three months in late 2001, we have a terrorist attack—where we are told that the U.S. intelligence community didn't connect the dots—and we have an anthrax outbreak," Jensen says. "I wrote [to] Dave at that time and said, 'Congratulations, you did an amazingly good job at predicting the technologies that people would need.'" The September 11 attacks and the subsequent recriminations over intelligence failures created an urgent need to speed up EELD and counterterrorism-related projects. In early 2002, the agency reorganized its data analysis and information-sharing programs under the Information Awareness Office.

From a qualifications standpoint, Poindexter must have seemed a natural choice to run the office. Brilliant and ambitious, he had graduated first in his class from the Naval Academy, eventually rising to the rank of vice admiral. He earned a Ph.D. in nuclear physics at the California Institute of Technology—where his advisor was a Nobel laureate—with a dissertation entitled "Electronic Shielding by Closed Shells in Thulium Compounds." Having served for a year as the national security adviser under Ronald Reagan in the mid-1980s, he also brought extensive policy experience and was well-versed in both technology and DARPA itself. Prior to his reentry into government, he had founded his own software company, and later worked as a vice president at Syntek Technologies, a high-tech company working with DARPA on a project called Genoa. Genoa was a software system designed to improve search capabilities over large amounts of data.

Poindexter, however, was more well known for another part of his résumé: his central role in the 1980s Iran-contra scandal. A federal court convicted him in 1990, and sentenced him to six months in jail for five felony counts of lying to Congress and obstructing the inquiry—a conviction that was later overturned only on the basis that Poindexter had been granted immunity in exchange for his testimony. His history of obfuscation and government secrecy made him a dubious choice for a program that DARPA should have known would ignite controversy over questions of civil liberties and government abuse. He made an obvious villain for privacy advocates: here was an agency largely unknown to the public, seemingly handing the reins of a massive technological snooping engine to a man with a proven history of flouting the law. Groups across the political spectrum quickly mobilized against TIA as Big Brother incarnate. A San Francisco journalist even turned the tables on Poindexter by publishing personal information—including his home phone number and address—which quickly spread across the Internet. Congress, fueled by public outrage, suspended funding for TIA in early 2003.

The public controversy was only deepened by the fact that DARPA and Poindexter seemed given to contradictory claims about the goals of TIA. While in some places the TIA website claimed that the program did not intend to build a centralized system to collect information, other pages described it as an attempt to develop "architectures for a large-scale counterterrorism database." Poindexter himself called for "large scale repositories" of information. In its requests for proposals, the agency referred vaguely to the need to gather a "much broader array of data than we are currently capable of doing." Much was made of the program's ominous logo, which featured an apparently all-seeing eye atop a pyramid (ironically, the same one as appears on the dollar bill), scanning the earth. Below the image read its poorly chosen slogan, the Latin words *scientia est potentia*—"knowledge is power."

More serious critiques came from nonprofits such as the Center for Democracy and Technology and the Electronic Privacy Information Center, which used Freedom of Information Act lawsuits to delve into the complex web of technologies that TIA hoped to bring together. They argued that the tools were intended not just to scan data about overseas terrorists, but to illegally do the same to American citizens at home. To privacy advocates, TIA looked like the most extreme example of a post-9/11 administration spun out of control, ready to cast privacy laws aside to build an all-knowing catalog of the American populace in the name of security.

Public outrage was heightened by oversimplification of the issues. Safire's assessment that Poindexter and TIA would be preparing "computerized dossiers" on American citizens "in a matter of weeks" had been grossly inaccurate, as had much of the ensuing coverage. Few made the distinction that DARPA itself was not an intelligence agency; its role was to foster better technology and hand it over to other agencies. For their part, DARPA officials backtracked in the months following the initial media attention, trying to contain the damage. The agency changed the name of the program to Terrorism Information Awareness, a move that smacked of Orwellian-speak and drew further derision from its critics. In a TIA report requested by Congress, DARPA explained that the program was never designed to build dossiers on Americans, and that the researchers under TIA were going to be working with a set of synthetic data created to simulate realistic databases. "The TIA program," the report stated, "is not attempting to create or access a centralized database that will store information gathered from various publicly or privately held databases."

In July 2003, the media jumped on another Information Awareness Office program, this one designed to enlist policy experts and the general public to wager on the likelihood of major world events, in an online terrorism futures market called FutureMap. The premise was that by allowing people to cash in on their knowledge of potential events, the government could watch the market and gauge the likelihood of an attack. To critics such as Senator Ron Wyden of Oregon— a Democrat and outspoken opponent of TIA who had sponsored the bill suspending its funding—FutureMap was another dubious Poindexter scheme. "A federal betting parlor on atrocities and terrorism," he said, "is ridiculous and grotesque." DARPA canceled the program within days, and less than a month later Poindexter resigned his post and left the government.

Even with its arch-villain banished, however, Total Information Awareness was unable to salvage any public credibility. On October 1, 2003, President Bush signed the Department of Defense appropriations bill, which contained a provision banning the use of funds for TIA and effectively killing all funding under the Information Awareness Office. Privacy advocates celebrated. The legislative counsel for the American Civil Liberties Union declared TIA's shutdown "a resounding victory for individual liberty."

That might have been the end of the Total Information Awareness tale, with the ominous prospect of data mining returned safely to the realm of Orwellian nightmares. But to those looking closely, the defeat of the program was a Pyrrhic victory. First, the bill explicitly allowed four of the more benign components of the Information Awareness Office to continue, including the Bio-ALIRT program and a project to develop software to automate speech recognition. Some of the more controversial elements of TIA, meanwhile, simply shifted quietly to other intelligence agencies—they "went into the black," in defense parlance. In a classified addendum to the appropriations bill, Congress moved the research to unnamed intelligence agencies under the condition that the resulting technology not be used against American citizens. Since DARPA usually used other military divisions such as the Air Force Research Lab to administrate its contracts, the funding could just be supplied from somewhere other than DARPA. From where, nobody seems to know or want to say. "My 'official' funding situation," says Foster Provost, a professor at New York University's Stern Business School who was working under EELD, is that "we will continue to have a contract with the Air Force." The Associated Press reported in February 2004 that some of the research had migrated to ARDA, the intelligence community's equivalent to DARPA. Other reports indicated that some projects may have moved to the Army Intelligence and Security Command, known as INSCOM. "I actually do not know where the funding is coming from," says David Jensen, the computer scientist at the University of Massachusetts. "I'm not entirely comfortable with that. Yes, the DARPA program is done. On the other hand the research, and the terms of the contract—which means I can publish in the literature—have not changed. It's a really strange situation. Unprecedented in my experience."

Its relative merits and eventual fate aside, Total Information Awareness was predicated on a notion that few if any other counterterrorism efforts acknowledged. Namely, that sharing data, translating it, and mining it were all part of the same technological problem. Which is not to apologize for the program's suspect privacy implications or to naysay concerns over its exploitation by whatever agency DARPA may have handed it to. It is merely to note that, in the end, the story of TIA was more about politics than technology. After shutting down TIA, Congress returned quickly to lamenting the pre-9/11 failure to connect the intelligence dots. The pressing questions that had driven the development of TIA—How powerful could data mining be? Could automation help analysts turn raw data into connected dots? Could we use technology to prevent another 9/11 without giving away our privacy in the bargain?—remained unanswered.

The only certainty is that neither the scientific community, nor the commercial sector, nor the government have any intention of stopping their data-mining research. "The notion that TIA was the only data-mining activity ongoing—I don't know if anybody ever believed that, but it is fundamentally wrong," says Jim Dempsey, executive director of the Center for Democracy and Technology. "Data mining is going to be like breathing. To some extent it already is like breathing. It's everywhere."

"Seventy-eight, seventy-nine, eighty . . ." Early on the morning of September 11, 2001, Jeff Jonas pressed his face against the glass of his hotel room at the Marriott Financial Center, peered up at the nearby twin towers, and tried to count the floors. Jonas is what might be called a personal-challenge junkie. He's the kind of guy who, when he's reading a book, stops to look up in the dictionary every word he doesn't recognize; the kind of guy who enters a 100-mile bike race riding a heavy mountain bike with huge knobby tires, just to make it harder on himself. And that morning he was the kind of guy who, having only ever been to New York City a couple of times, tried to see if he could count the floors of the World Trade Center.

After satisfying his curiosity, Jonas headed down to the lobby around 8:30 A.M. and flagged a taxi. His morning meeting, with one of Systems Research and Development's large corporate clients, had been moved the day before from the World Trade Center to midtown Manhattan. But as the cab pulled away, Jonas had a thought: Maybe he could take a quick trip up to the observation deck. He asked the cab driver if he had time to do so. When the driver replied that it would take a full half hour to get to the meeting, Jonas decided to forgo the sightseeing. As the cab headed uptown, he wondered absentmindedly why people in the streets seemed to be staring southward. Only after getting to midtown did he hear the news.

It was a week later, after Jonas had made his way back across the country to Las Vegas by taxi, rental car, and a friend's private jet, that he realized how his invention, NORA, could be used to track terrorism suspects. He saw on CNN that the FBI, in the post-9/11 panic over possible follow-up attacks, was sending out its terrorism watch list to companies, asking them to look for records on potential suspects. "Suddenly I realized that our technology could make a difference," Jonas says. He had founded SRD in 1983 and built a thriving business selling to casinos. But nabbing increasingly sophisticated networks of gambling cheaters, he says, turned out to be a pretty good analog to catching terrorists: Their methods of operation often mimic that of a terrorist network. "Las Vegas has been facing an asymmetric threat for years," he says. "There are these highly organized teams that can completely reorganize themselves every couple of years." And in casino work, he adds, "the tolerance for false positives and false negatives is incredibly low."

Jonas proposed donating NORA to a number of large companies to help them run checks against the watch list, and the FBI accepted his offer. "What we do," says Jonas, "is we take a who—somebody that needs to be found or not let in our country—and then we say, 'Where are they?' And if you find the roommate of Mohamed Atta, he's certainly more interesting than Grandma. That doesn't mean that Mohamed Atta's roommate is bad, by the way. But it's smarter to look at that person, where there is some kind of reasonable cause to be more interested." The watch list was a classic example of the subject-based version of data mining. "It's like it's been trained for that," he says. "So we just plugged it in and turned it on all over the place."

NORA's first premise is that if you want to find out who is connected to whom, first you have to know who *is* who. Even the large-scale private databases of companies such as ChoicePoint and Acxiom are plagued with incomplete records, misspelled names and addresses, and transposed dates of birth. The information collected in intelligence agencies is typically still more problematic. It's a fallacy, says David Jensen at the University of Massachusetts, to "think that if all the databases were made available, suddenly one day, if all legal protection were removed from them, it would be a trivial problem to search all of them. In fact, it is shockingly hard to do this. Even within a single organization where there are no institutional or legal barriers, companies and government agencies have enormous problems doing this well." Just sorting out which records actually point to the same person is a complicated undertaking—recall the 2000 election, when Florida's infamous felons list kept some non-felons with similar names from voting. The problem can be further compounded by the translisteration of Arab names—just look at the list of the 19 hijackers, some of whom were brothers, and see how many names appear similar.

So when NORA is installed, the business or agency first feeds all of its data into the system, and the software attaches an ID number to each record, cleaning up the data as it comes in using "data hygiene" tools that Jonas developed. It links names to their common roots—tying Bill and Billy to William, for example, and matching names against the more than 100 different spellings of Muhammad—and cleans up addresses by checking them against publicly available postal tables for the United States and other countries. It can also add precise latitudes and longitudes to addresses, allowing NORA to search along geographic proximity. Someone who shares a backyard fence with a known gambling felon, in NORA's world, may be a person who invites further speculation.

Then the software undertakes the task of determining which two people or records in the database are the same, a process Jonas calls "entity resolution," and which is known generally in the field as "disambiguation." A Patrick and Patricia at the same address—which the original database may have listed as the same person (Pat) but were actually two—can be split into separate records by examining their social security numbers, dates of birth, or other distinguishing data. Using entity resolution, says Jonas, NORA turns the information overload problem on its head. As the software is fed more information, it can uniquely identify more people. "One of the totally amazing things that we've proven with this—and proven at scale, billions of rows, terabytes of data—is that the more hay you add, the more needles you find," he says. "We have proven that with more hay you can reduce the false positives and false negatives."

That, he says, means that NORA could help eliminate what has come to be known as the David Nelson problem. In 2003, people with the name David Nelson across the country suddenly found themselves stopped for extensive searches, and occasionally turned away at airports. Apparently the name had appeared on a TSA no-fly list, and innocent David Nelsons were caught in the net. It was a classic false-positive problem, but one created by a lack of data, rather than an excess. If the computer system was able to resolve each David Nelson as different from the David Nelson being sought, the false positives would evaporate. Officials at some airports reportedly did just that, using one David Nelson's social security number to distinguish him from the David Nelson authorities were looking for. Post-9/11 watch list mistakes haven't been confined to David Nelsons: Georgia Representative John Lewis and Massachusetts Senator Ted Kennedy were among hundreds of passengers repeatedly stopped at airports because their names appeared on watch lists. According to the *Washington Post,* when Lewis and others added middle initials to their reservations, the scrutiny stopped. NORA, which isn't used by any airlines or the TSA, hasn't had a crack at the problem.

Once NORA has used all the available information to figure out who is who, it can then be given a new subject—a suspect on a watch list, for example—and trace the connections to that subject up to 30 degrees of separation. "In large volumes," says Jonas, "there is little reason to look out beyond a degree or two. It degrades very, very quickly." As a record is added to any of the databases connected to NORA, the software's algorithm churns through the data looking for common fields—names, phone numbers, credit cards—and then drops the record into place. It's the inverse of traditional data mining, Jonas says, where "a data miner stands at the bottom of an ocean of data looking up at it, and asks himself the question: Hmm, I wonder what is in there that I need to know about. I'm going to ask this ocean of data a question. And then he gets an answer back, and he takes the answer and decides if that's a thread he wants to pull." NORA, by contrast, sits on top of the ocean, analyzing each drop of water as it falls in. "Where that drop falls, what it's related to, and what those ripples look like are measured. And if that is connected to anything that is interesting already, it's instantly published to someone who cares."

In working with the largest data sets—one of SRD's clients is ChoicePoint, the giant data aggregation company—NORA can do something even more intriguing: it can tell when people don't exist. Firms such as ChoicePoint have collected so much data on Americans that the fact that someone *doesn't* appear in their database means that they have "never had a driver's license, never been in the phone book, never driven a car, never had insurance, never ever shown up anywhere," says Jonas. In essence, it can measure an identity's "thickness." (A controversial feature of the TSA's canceled CAPPS II program was supposed to perform a related task, using private database information to "authenticate" each name and give it a score based on a "confidence level in that passenger's identity.") NORA also matches data against lists of deceased Americans and their social security numbers, checking whether someone might be reusing the names and social security numbers of dead people. A name that has zero identity thickness has a fair chance of being a fake ID. "It doesn't mean they are bad," Jonas says, "but it does mean they are more interesting."

Back in the SRD conference room, Jonas tells another story, the winding tale of his own past. Two decades ago, he was living in his car in northern California, having run his first computer consulting com-

pany into bankruptcy. He had first become interested in computers as
a kid, when his mother, a lawyer, took him with her to purchase a
computer for her firm. "I remember at 12 going, 'This is what I'm
going to do. This is my thing,'" he says. "I took every class that the
high school had. When they had no more, there was no point in me
going there anymore." After eventually obtaining his GED and
trying a short stint in junior college, he landed a job at a computer
consultancy and by age 18 had started his own custom software business with 21 employees. Unfortunately, he says, "the company was
incapable of really completing anything on time or under budget.
So I ended up at 19 going bankrupt for over $100,000 and sleeping
in my car."

It was from the back seat that Jonas started again a year later with
SRD, writing software for whoever would pay. When the Mirage
casino hired him to help them keep up with the Nevada requirements
for keeping Black Book listees out, he hit on his company's niche. The
first wider notice came in January 2001, when SRD obtained a round of
funding from In-Q-Tel. "It would be hard to be a little company from
Vegas and show up in Washington and say, We can make a difference,"
he says. "It's kind of like, do you want to buy a watch?" he adds, mim-
icking a street vendor opening his coat. But Jonas now finds himself
working with government agencies, think tanks, and the largest corporations to define how data can and will be used by the government
in combating terrorism. He wrote a number of sections in the December 2003 Markle Foundation report, "Creating a Trusted Information
Network for Homeland Security," which describes the potential for
balancing information technology and privacy in counterterrorism.

Later in the day, sitting in the extravagant private club atop the
Mandalay Bay Hotel—where Jonas is a member—looking out over
the lights of Las Vegas, he recalls giving his first presentation in front
of a roomful of government and military experts in Washington, D.C.
He had expected them to respond by telling him that intelligence
agencies already had technology such as NORA, or better. But after
the talk, he says, "I pack my bags and am about to leave, and I get
swamped. The basic message is, this is really advanced, and there's
nothing like this around here. That just blew my mind. I said, you are
kidding me!

"To me it's so blatantly obvious and simple," he says. "To this day I'm suspicious that everybody is already doing this and just not telling me."

Jonas may have mastered the art of finding suspicious people through their connections, but NORA requires a place to start, and knowledge of what kind of links to look for. It doesn't solve the ultimate question, the one posed by Meno's paradox: Can we find someone or something we don't know we're looking for? Is there a way to dig knowledge from data and uncover a network without necessarily having a thread to pull? These are questions that David Jensen is attempting to answer. Jensen was one of the researchers under contract with TIA's Evidence Extraction and Link Discovery program. The goal of EELD research was to find ways to apply pattern-based data-mining software to detect terrorist activity.

Pattern-based data mining consists of two phases: learning and detection. In the former, the software learns a statistical model to represent the patterns it is looking for. The model can be something as simple as "gas station purchase followed by a spike in usage of card," hard-coded into the system based on what human analysts already know. For counterterrorism, the program might be given a model that includes "all airline passengers traveling one-way, buying their tickets in cash." Those rules, in fact, were part of the first version of the CAPPS system employed by the airlines. Such models are essentially behavior profiles, sets of rules that, if met, indicate a suspicious pattern of behavior. As with most fraud and money laundering data-mining systems today, the CAPPS profile was developed by human analysts and programmed into the software.

In data mining, however, the most tantalizing technology is an algorithm that develops its own model, using statistics to analyze a set of "training data" for which the answers are already known. In the case of credit card fraud, for example, a training set would include a database of historical credit card transactions, some of which are already known to be fraudulent. The algorithm would look at those cases, compare them to nonfraudulent charges, and essentially say, "Aha! These are the qualities that indicate a stolen card." The hope with machine learning is that the software would discover not only the gas station rule, but also rules that human analysts had never noticed.

The advantage of pattern learning is that the computer—because it can process large amounts of digital data more quickly and thoroughly than the human brain—can spot patterns in that data that humans would miss. It can also improve its performance every time it looks at a set of data, by learning from its mistakes. Among the EELD projects, says Jensen, "pattern learning was always seen as the most speculative, but also potentially the most revolutionary of the technologies."

Once the software learns the model—whether on its own or coded by human analysts—it can apply it to a new set of data, which is the pattern detection part of the process. Humans do their own pattern detection all the time: we examine a set of facts and, given the model we've developed for how the world works, apply it to any new information we come across. You assume that when the light turns red other cars will stop, because that is the model that you have learned from experience. Rafi Ron's suspicious behavior detection training and Paul Ekman's FACS system are also models, ones which the trainees learn and then apply to detect a set of patterns—in facial expressions or body language—that indicate lying or unease. Ioannis Pavlidis's thermal-imaging technology, George Poste's Z-chip, and Steve Flynn's proposed smart-container network all involve some version of the same concept—creating a model and then employing it to recognize both normal and abnormal patterns.

In data mining, models can be applied to a set of data about a specific person or group of people. In the early 1990s, Jensen constructed a machine learning system to detect patterns that indicated Alzheimer's disease. The data came from a group of 300 patients that had been diagnosed by a panel of five doctors as having mild Alzheimer's, severe Alzheimer's, or none, along with the patients' answers to a series of diagnostic questions. The software scanned the database and created a model that correlated the questionnaire answers to the patient's disease status. Then Jensen applied his model to a new set of patients, this time without knowing their final diagnoses. The system, it turned out, was highly effective in identifying patients that should undergo additional Alzheimer's screening, with one particular exception—it repeatedly labeled one patient as healthy whom doctors diagnosed as having severe Alzheimer's. When Jensen called up the medical school, they told him that the patient had in fact, been misdiagnosed: He was merely an eccentric whose personality had been mistaken for evidence of Alzheimer's. The doctors, not the computer, had been fooled.

But simple data-mining programs such as that one, along with those commonly used for fraud detection or spam prevention, turn out to be poorly suited to counterterrorism questions. The reason is that they don't look for the most important factor in terrorism: connections between people, places, and institutions. Instead, they treat each particular person (a patient, say, or someone depositing money in a bank) or thing (a credit card or a piece of e-mail) independently. The fact that one patient has Alzheimer's does not affect the condition of any other patient, so there is no need to look at connections between them. Similarly, the fact that my credit card is fraudulent tells you nothing about whether someone else's might be.

To take another example, recall the shipping container profiling system designed by the Fogel family at Natural Selection. The goal of the program is to sort the containers according to their potential risk, so that U.S. Customs can decide which ones should be searched more thoroughly. Each container's risk is defined independently, by its own qualities: where it came from, who is sending it, what is supposed to be inside. The fact that a container sent by Honda happens to be placed on a ship next to one with items bound for Wal-Mart doesn't increase or reduce the risk of either one.

In contrast, when you are looking for plots among people, the data is by its nature connected. Searching Mohamed Atta's flight reservations alone for patterns would reveal very little—he flew to Las Vegas, perhaps, and he flew first-class, neither of which alone indicate anything unusual. But mine his connections to other people—namely, Khalid al-Midhar, who was already on a State Department watch list and with whom Atta shared an address—and you have begun to possibly detect a plot. That is where relational data mining—which attempts to discover and learn patterns over this kind of interconnected data—comes in. Whereas Jeff Jonas's NORA would start with one thread and pull it, relational data mining would look at the overall set of data and try to predict potential signals of nefarious networks. There are a variety of ways to do it, and research by Jensen and others is still in the early stages.

When David Jensen started out in computer science, researchers hadn't even begun to look at relational data mining, but an unusual career path positioned him perfectly to be at the forefront. After finishing his Ph.D. in computer science from Washington University in St. Louis, Jensen eschewed academics and went to work for the now-defunct Office of Technology Assessment (OTA), which provided independent scientific advice to Congress. At the OTA, Jensen was enlisted to help study how data-mining technologies could be used by the Treasury Department's Financial Crimes Enforcement Network (FinCEN) to try and uncover evidence of money laundering. FinCEN analysts examine transaction data from the nation's financial institutions, which are required to file "currency transaction reports" for all cash deposits over $10,000. Money laundering often involves networks of criminals operating together, using front businesses and small deposits to elude the reporting requirements. Jensen and fellow researchers determined that using data mining against money laundering was possible, but to do so effectively would require a new type of technology—one that could mine not just data but a network of interconnected data for patterns that reflected suspicious behavior.

Jensen left the OTA to return to academics at the University of Massachusetts in 1995. A few years later, he returned to the problem of relational data mining, "thinking it was probably professional suicide," he says. "Nobody was interested in it. But I thought that it was a fascinating area because of my work on the OTA study." In 1998, a researcher working with DARPA on EELD contacted him about his work. "Someone called me up and said, 'How did you know?' I said, 'Know what?' They said, 'Know that DARPA is starting a program in this?' I said, 'Well, I didn't know, but I know now.' In exchange for helping write the opening presentation for the 1998 Carnegie Mellon meeting, Jensen wrangled an invite to it. Eventually he obtained one of the early grants from DARPA to work on EELD problems.

Jensen's lab has used a variety of data sets to hone algorithms that could make sense of relational data. One was a collection of actors, directors, and other data from the Internet Movie Database (an online compendium of film facts) and the Hollywood Stock Exchange (a virtual exchange on which people trade the "stocks" of actors and films). By examining the conditions that created past blockbusters—the combination of actors, directors, release date, and other factors that made *Titanic* a hit, for example—the system developed a model that was able to predict a new movie's success. His lab has also examined interacting genes, the links between corporate boards of directors, and nodes on the Internet. "Essentially any data that you can represent as a network is something that we can analyze," he says. "We are fascinated at how networks can be analyzed and understood, and how you make predictions about a network's interacting nodes, whether on the Internet; or hosts in a peer-to-peer network; or actors, producers, directors in studios; or, theoretically, terrorist networks that are planning something."

At its height, by one participant's estimate, as many as half of the world's experts in relational data mining were working on similar projects funded under EELD. None of them were using actual citizens' data, and many of the projects had no obvious connection to terrorism at all. Andrew Moore and two colleagues at Carnegie Mellon were using EELD funds to develop a program to discover the hidden networks among a large group of people—suspects under surveillance in the Middle East, for instance. Their strategy is called a "group-detection algorithm." The question, says Moore, is "can you infer from a large set of these observed meetings if there are particular groups of people who are definitively communicating with each other, as opposed to random meetings of subgroups of people." (Moore also has ongoing data mining research for biosurveillance systems.) Daphne Koller, a Stanford University computer scientist, developed a system that could automatically classify Web pages—using their links to other pages—into which ones were created by students or professors. Foster Provost at New York University built a program that determined which character on a list of suspicious people warranted the most investigation, based on their connections. Researchers at the universities of Wisconsin and Texas were using algorithms to analyze publicly available information about nuclear smuggling, hoping to detect patterns that would indicate nuclear materials smuggled out of Russia.

The relational data mining behind all of these projects derives its power, in part, from the fact that entities that are related to one another often have similar characteristics, a property known as "collective inference." Take cellular phone fraud. Another research project by Provost (this one not done under EELD) showed that a simple relational model could help detect fraudulent calls if it took into account which numbers a person dialed. A cell phone that consistently called numbers known to have been previously called by fraudulent accounts, or that suddenly began making calls from an area with a high history of fraudulent calls, was a strong indicator of fraud. The same rule helped exclude numbers from suspicion: if a particular phone never called a number that had received fraudulent calls, it was less likely itself to be fraudulent.

Collective inference is, put another way, a kind of guilt-by-association, or perhaps more accurately guilt-by-connection (these researchers don't use "guilt" in a pejorative sense, as in criminal guilt, but rather in the sense of being more relevant to the search). Terrorists tend to consort with other terrorists—just like movies featuring actors who have previously been in blockbusters are more likely to be blockbusters themselves—and so the best way to find them is to include a guilt-by-association factor in the calculation. "Of course it is not a deterministic rule," says Koller. "If a terrorists goes into a store and buys some bread, that doesn't mean that the store owner is a terrorist. So consorting with another terrorist is only a very weak indicator of being a terrorist." But by examining the interaction between different people in a network of data, you can potentially narrow down the field of interest. "If you use the data relationally," Jensen says, "you can get big increases in accuracy."

When it comes to putting relational data mining for counterterrorism into practice, however, the obstacles quickly multiply. While researchers have achieved progress in small data sets, there is still the complication of scaling the techniques to the massive and disparate amounts of information faced by analysts. One particular difficulty is that the training data for pattern learning is hard to come by, especially in unclassified work. Moreover, says former Stanford computer science professor Jeffrey Ullman, a database expert, in many cases "we are looking for events that we haven't actually seen before. Artificial intelligence people will always tell you that it is much easier to train a system where you have positive and negative examples. When you have no positive examples, it is almost impossible to train a system."

Even if the data were available, however, a system just trying to predict patterns of attack would have to confront the endless variability in the activities it was seeking to detect. Meno's paradox arises again. "My first question to you is going to be, What does a money launderer look like?" says Chris Westphal, CEO and founder of the data-mining company Visual Analytics and an expert on applying software to intelligence questions. "What does a terrorist look like? What does an insider trader look like? If you can tell me that, I can give you some pretty good answers, in terms of encoding it. Unfortunately, nobody can answer those questions."

That's why the most promising relational data-mining applications wouldn't be fully automated systems that conduct random scans of massive databases looking for "terrorists." Rather, they would be those that incorporate human intelligence and use collective inference—which itself involves already knowing some suspicious people and then tying them to a pattern. The goal, harkening back to J. C. R. Licklider's idea of human-machine symbiosis, is to use data mining as an appendage to human analysis. "A common myth is that the models produced by data-mining algorithms will replace human analysts and decision makers," David Jensen said in a 2002 presentation on data mining and counterterrorism. "However, the last two decades of work with artificial intelligence systems—including data-mining systems—have shown that these systems are usually best deployed to handle mundane tasks, thus freeing the analyst to focus on more difficult tasks that actually require his or her expertise."

The idea of creating an automated system that constantly churns through our data, looking for "abnormal" occurrences—such as people who have traveled to Afghanistan who are studying at flight schools—is a distant dream. Such a system would collapse under the weight of millions of false positives. "If you tried to generally search for abnormal behavior, then it's completely hopeless," says Andrew Moore. "If someone was going to invent a system to do a really 1984 type scan of all of the activities going on in the United States, looking

for people that seem to be doing patterns of activity that are different
from normal, I'm sure you will match 20 percent of the population.
Most people will have something about them which is really atypical."

It was privacy concerns that eventually scuttled the Total Information
Awareness program, and any government data-mining program—
assuming the public is aware of it—will have to address the question
of who is going to get to see our data and what they can potentially use
it for. The public fear that attached itself to TIA, of a single, vast database that would contain all our personal information, was not a realistic one; data mining doesn't require such a centralized system. (It's
also a bad idea, since centralizing information makes it more vulnerable to security breaches.) But the privacy concerns remain as realistic
as ever. With either subject- or pattern-based analysis, algorithms can
search across multiple public and private repositories. Law enforcement agencies already regularly access commercial databases of companies such as ChoicePoint and Acxiom looking for evidence. The
CAPPS II airline screening system would have mined private sector
data extensively. The question of how that information gets shared—
and who can access it—lies at the crux of the privacy dilemma.

Even within the government, different agencies are unwilling or
unable to share information for fear that the data will fall into the
wrong hands. The result is what's known as "stovepiping," the inability of connected parts of an organization to communicate with each
other. One of the most commonly voiced criticisms of the intelligence
community has been the failure to share information within and
among agencies. The FBI, the *9/11 Commission Report* says, "lacked
the ability to know what it knew." The agency is hoping to solve that
problem with a multi-million-dollar advanced search and retrieval
program from a Virginia-based company called Convera. Convera's
software lets agents conduct automated searches across databases,
documents, and audio and video files in 45 languages.

Then there is the problem of sharing data across agencies. Before
9/11, the government maintained 12 separate watch lists among nine
agencies. But despite the relatively simple technical solution, the bureaucratic process proved difficult to overcome. "It's not a lot of data,"
says Dan Prieto, a fellow at Harvard University's Kennedy School of

Government, a member of the Markle Task Force and research director of the Homeland Security Partnership Initiative at Harvard's Belfer Center for Science and International Affairs. "The commercial sector deals with stuff that big every day of the week.". Prieto points out that a whole new system isn't needed; the technical challenge is relatively simple—database connections can be built on top of existing infrastructure. What holds back the integration of information is a resistance on the part of each agency based on its concern over who will see the data. It was not until late 2003 that the administration finally created the Terrorist Screening Center (TSC) to begin integrating the lists. The goal of the TSC is to determine the criteria of who gets on and off watch lists (itself a remarkably complex problem, as Senator Kennedy's appearance on a "no-fly" list illustrated); merge the disparate lists into a central location; and distribute it to other law enforcement agencies that might need to see it. The trouble remains how to get sensitive information to people who need it—without revealing classified secrets. "When it comes to applying new technologies to homeland security," says former NSA deputy director William Crowell, "the most important ones are those that allow seamless sharing of information across many jurisdictions and silos—everything from water and disease control to nuclear and weapons of mass destruction."

The concerns over privacy and those over information sharing are, in some sense, parallel: both involve how either human analysts or data-mining systems access information. The problem is not just one of policy but of technology. Is there a way to mine networked data without allowing it to fall into the wrong hands? Whether confronting privacy questions or for interagency turf wars, finding a way to exchange data while keeping some of it secret is a critical technical issue.

Teresa Lunt, a data privacy expert at the Palo Alto, California-based Xerox PARC laboratory, began examining that particular problem 15 years ago. As an expert in databases and computer network intrusions at a nonprofit research center called SRI in the early 1990s, she built a database system for the government that allowed access to different database records and fields based on the security clearance of the user. A person with a "top secret" clearance might be allowed to see certain fields in a database—say, about classified weapons projects—

that those with only a "secret" clearance would not. But Lunt soon found that sometimes the lower clearance users would be able to deduce the data they weren't supposed to see, something called an "inference problem." So Lunt developed a system that examined the relationships between data to try and predict those "inference channels." If the analyst was supposed to see data field A but not data field B, the software looked at links to determine whether they might be able to use a combination of A and another field, C, to discern B.

After a stint as a program manager at DARPA, Lunt landed at PARC with little intention of doing further government research. But when she saw an Information Awareness Office announcement offering funding for data privacy research, she says, "it occurred to me that you could apply this whole inference-control idea to the privacy problem. If you are trying to protect the identity of people, it's not just enough just to withhold identifying information such as their name, their address, their phone number, their social security number, their credit card number. Because there are other things that you can use to infer something." The problem is even more thorny given how difficult it is to anonymize personal data. Research by Latanya Sweeney, a computer scientist at Carnegie Mellon, has shown that 85 percent of the population can be identified using only their zip code, gender, and date of birth.

Lunt obtained a TIA grant to build a privacy appliance that could be placed in front of databases, to anonymize data against the inference problem. She also began working on a system that would record all the searches done by analysts, to protect against abuse of the system. It's called an "immutable audit trail." "What you don't want is some analyst who is trying to game the system in some way," she says. "If they know that their queries are going to be blocked if they ask direct questions, they may try to ask some very roundabout questions. You should be able to see patterns of activity like that. So you could do real-time analysis of the queries that are being asked. You could also, if some citizen feels that harm had come to them, ask for the audit trail to be opened and look at the records for a particular analyst."

"Immutable audit is very useful," agrees Jeff Jonas, "I like that a lot. I want to know when someone has been searching for something." But Jonas, characteristically, thinks he has an even better answer to the data-sharing problem. He thought of it, he says, while in the shower one morning. He knew that the government was concerned about the safety of its watch list data. (When the FBI sent out the list after 9/11, a source told him, a copy of it somehow made its way onto the website of a water treatment plant.) He also knew—from the TIA controversy and the firestorm of criticism over airlines such as JetBlue giving passenger data to the government—that Americans are becoming increasingly skeptical of corporations handing over their personal data to the government in the name of security.

The answer Jonas came up with is a means to anonymize information but still allow it to be searched for links. "I had this vision," he says. "What if you could take data that was too sensitive to give to anybody and change it into anonymized values that were irreversible. And then you could take somebody else's data and run it through the same algorithms that make it anonymous, and then you compare that." It's what he calls ANNA ("NORA's sister," he likes to say). ANNA, says Jonas, is the answer to "how to know everything about everyone without knowing anything about anyone."

ANNA works like this: The software takes a set of data and applies a mathematical encryption formula that converts each piece of data— a name, an address, a phone number—into an indecipherable string of characters. The name al-Midhar, for example, could be transformed into cbd034409c22929518fa494f99dc9964. Called a one-way hash, it's the same type of function being used by Internet virus traceback researchers to identify the routers as a data packet passes through, and a close relative of the encryption algorithms used to secure information. In the case of ANNA, the hash function serves to create an anonymous version of the information stored in the database. Each string of numbers is unique, so if two pieces of data differ by even a letter or a comma, the resulting hash will be completely different. ANNA also takes the common data errors found by NORA—misspellings of names, transposed birth dates—and hashes them as well. Then it does the same for the names and other information on the watch list (which might include birth dates, addresses, or social security numbers). Once all the data is hashed, NORA or another system could search for matches between the unique numbers without ever revealing the underlying data.

Let's say the government is looking for a particular suspect, John Doe, and wants to find out if certain companies have any data about him. It runs a hash on "John Doe," his birth date, social security number, and any other information it has on him. The result is a string of letters and numbers. It then hands that string over to the companies, which have run the same hash function on all of their data. Then the company simply looks for matching strings in its database. If it finds one, it alerts the government, which then could obtain a court order to un-anonymize the data.

Because a hash cannot be unencrypted, there is no way to use the number to discover the data it stands for—you have to have the original. That means that the government could in theory hand its whole encrypted watch list over to companies without having to reveal it, and companies could potentially send large amounts of data to the government without violating the privacy of the subjects involved. The data would simply be a database full of strings of numbers. Software such as ANNA, Jonas argues, provides a technical fix for similar data-sharing problems in health care, in private sector to government data sharing, and in communication between government agencies. "ANNA," says Jonas with a smile, "may be the most important thing that has ever come out of my own head."

While Lunt's and Jonas's solutions don't solve all of the privacy and information-sharing problems, they would help provide some basic protections. When Congress killed the TIA program, however, it also killed funding for Lunt's initiative, forcing her to abandon the project after building an initial prototype. Her new research involves methods of preserving anonymity in wireless traffic data. Jonas, too, had been in talks with the TIA brass about the concepts behind ANNA. Lunt, for one, doesn't see data privacy work burgeoning without a push from the government. "I would like to see DARPA fund more of this stuff," she says. "I just don't see it in the cards, because they don't want to get burned again."

While the privacy work evaporates, however, the data-mining research is proceeding apace, and increasingly out of the public eye. "Now that the Information Awareness Office programs have gone away," says David Jensen, "we are much more likely to have those algorithms developed either as proprietary research by companies or in classified research programs. Technically I think that's a bad idea. I think that's a bad idea in terms of the confidence that we can have in the technology and the public knowledge about how well those algorithms work."

Researchers in the data mining community have their own concerns about how the technology gets used, but they are less inclined to frame them in terms of Big Brother scenarios. "The ones that actually do frighten me are the commercial things like data mining to decide who gets a loan or who is a good insurance risk," says Andrew Moore at Carnegie Mellon. "I believe that in a sensible world, that would be the kind of thing that people like William Safire would really be complaining about. Those are the things that could really have an effect on people right now. As opposed to a theoretical threat of a bad government sometime in the future." These concerns of Moore and others might at first seem odd, given the post-TIA stigma that data-mining research has taken on in the media. But data mining, after all, is only a method for drawing meaning out of information—any information. Like genetic engineering, encryption, and a host of other technologies, data-mining research has multiple uses. For every Big Brother scenario you can concoct, there is its inverse—a biomedical application, say, with the possibility for improving diagnosis and saving lives.

In the end, only a wide-ranging public debate can resolve whether the potential for abuse is outweighed by the value for security. Data miners themselves like to point out that the technology will march forward regardless, and holding back research into what is possible could eventually result in less protection, not more. "If anything I think that the public should *want* more sophisticated data analysis methods developed," says Stanford's Daphne Koller. "Because the fact is that people are going to be looking at this data anyway, and if you give them better tools then they are going to have a lot fewer false positives. And that's exactly what you want."

If we do decide that data mining should be part of our national security strategy—something that is more than likely already happening—we shouldn't expect that a computer will somehow uncover the next 9/11 on its own, though it might increase our analysts' chances of doing so. How successful their effort will be depends, in part, on the other half of the data-mining debate—how much information we want the government to access. But any utopian hope for data mining—that it alone can solve Meno's paradox and find for us what we don't know we are looking for—is a mirage.

In Plato's *Dialogue,* Meno asks Socrates: "How will you look for something when you don't at least know what it is?" Either we know the thing already, he conjectures, or we won't know it when we find it. But Plato's fictional Socrates—after proclaiming the question "a trick argument"—has an answer. Because the soul is immortal, he says, it has already learned all the knowledge we need to know. That knowledge is stored up within us, all we have to do is unlock it. To demonstrate his idea, which is called the "theory of recollection," Socrates conducts an experiment with Meno's slave boy. Using a series of questions—the famous Socratic method—he coaxes the boy into solving a mathematical formula, even though the boy has no previous teaching in mathematics. Socrates' lesson is that the boy already knew the answer. He just didn't know that he knew it. "All nature is akin," Socrates tells Meno, "and the soul has learned everything, so that when a man has recalled a single piece of knowledge—learned it, in ordinary language—there is no reason why he should not find out all the rest, if he keeps a stout heart and does not grow weary of the search; for seeking and learning are in fact nothing but recollection."

The same could be said of the knowledge we seek from our vast databases and networks, whether about terrorists, spam e-mail, blockbuster films, or genes. The answers, we hope, are already contained somewhere within the data, waiting to be unlocked. The question is whether data mining can unlock them, and at what price. "This research is going to continue on because it's not just about counterterrorism," says Jensen. "It never has been and certainly isn't now. It's about a world full of these networks, these interconnected things. Previously our techniques couldn't really address them, and now it can. That's exciting, and it opens up a huge range of possibilities for us to understand the world better."

THIRTEEN

The Power of the People

IN JANUARY 1995, a magnitude 7.2 earthquake hit the city of Kobe, Japan, setting off scores of fires, destroying 136,000 houses, and killing more than 6,000 people. Some 20,000 residents were crushed or trapped by falling debris, and millions were left without water and electricity. Officials in the region were poorly prepared for such a disaster. The nation's fears of seismic catastrophe had long been focused on the Tokyo area, while the Kobe region's emergency plan anticipated only much smaller events. Fire departments were overwhelmed, and the national government was slow in committing resources. As officials struggled to get the rescue and recovery on track, however, throngs of ordinary Japanese citizens dove in to help.

Japan is not a nation known for its volunteerism. According to Kathleen Tierney, one of the world's foremost experts on the sociology of disaster, after World War II the Japanese people relied heavily on a growing bureaucracy responsible for citizens' welfare. So the out-pouring of citizen aid in the Kobe catastrophe was unprecedented, and in ensuing years 1995 became known throughout Japan as "the first year of the volunteer." In the months following the quake, some-where between 600,000 and 1.3 million people offered aid sponta-neously to those in need—even though, in Japanese, no word for "volunteer" existed. To describe the sorting and delivering of goods, the search-and-rescue assistance, the child care and soup cooking and counseling provided by citizens with no expectation of reward, the Japanese were forced to use a transliteration of the English word.

It's a story to make you marvel at the power of humanity. A million individuals proffering their strengths to be used as needed: what other show of peaceful force could be so great? Indeed, the potential of the public to protect or restore itself in times of calamity is astounding. Nearly without exception, when confronted with man-made or natural disasters, communities have exhibited what sociologists such as Tierney call "pro-social" behavior—altruism and morale on a breathtaking scale. Tierney and some of her colleagues have gone as far as to appropriate the term normally applied to firefighters, police, and emergency medical technicians (EMTs) and give it to the unpaid populace: they are the real "first responders."

The renaming is accurate: Consistently, victims trapped in collapsed buildings are more likely to be rescued by their family, friends, and neighbors than by official emergency workers. Following the 1980 earthquake in southern Italy, 90 percent of those saved were extricated by untrained, uninjured survivors, using axes, shovels, and their bare hands. A study of a 1992 gas explosion in Guadalajara, Mexico, that killed roughly 300 people found that the majority of the 43 victims liberated from the debris in the first two hours were rescued by ordinary citizens. Similarly, disaster victims taken to hospitals are more likely to be transported by family, friends, and neighbors than by ambulances—as was the case after a well-documented tornado in Alberta, Canada, in 1987: out of more than 300 injured, 30 percent were taken to the hospital by a relative, 20 percent by a stranger, 18 percent by public bus, and only 16 percent by ambulance. Some research even suggests that full-on reversals take place after such calamities: *anti*-social behavior, such as crime or membership in exclusionary groups, actually declines.

Yet throughout emergency officialdom, from every branch of the military to the upper echelons of the Federal Emergency Management Agency (FEMA) to thousands of municipal police chiefs, in times of great danger the public is viewed largely as a nuisance. And with reason. The masses are an unruly bunch, and when lives are threatened in a just-bombed building or by a release of noxious gas, safety professionals strive for a baseline of sanity from which to make difficult choices. The flood of citizens' goodwill is often undirected and uninformed. Donations, for instance, can come in by the trainload or shipload and prove to be a huge burden. "Radio stations," says FEMA Voluntary Agency Coordinator Ben Curran, "are notorious for generating scores of 18-wheelers gathering plastic bags and toys." On the receiving end, he explains, there may be no warehouse space available, and to be of any use, goods have to be sorted and labeled and stacked on pallets. The sheer volume of all this impulsive generosity can create a disruptive ripple effect of its own. After 9/11, Curran notes, "we had to keep thousands of tons of donations *out* of New York City."

Curran is careful not to offend, but his position is plain. "We can't just allow volunteers to flow into a site willy-nilly and interfere with the first responders."

This paradox—that the public is at once a profoundly effective element in an emergency and a harmful distraction—is all the more knotty in the context of terror. By nature changeable and diverse, terrorist behavior requires particularly rigorous study, and only experts and professionals have time for that. The effects of nerve gas, the trajectories of explosives, the symptoms of a virus, or the secret plans of a well-surveilled terrorist cell—ordinary citizens are unlikely to possess the kind of knowledge about terrorist acts that would seem to be a prerequisite to being useful. Average Joes are far more ignorant about suicide attacks or synthesized pathogens than they are, say, about wild fires they've witnessed repeatedly in the western United States.

Yet the anytime-anywhere logic of terrorism also undercuts the value of specialized knowledge. Experts are concentrated in a few places: universities, diplomatic missions, Washington, D.C. But like potential terror acts themselves, ordinary people are everywhere. No night watch is as vigilant as the light sleep of the citizenry, and no rescue team, no cop, no molecular biologist, is ever going to be as nearby as one's family and neighbors. Terrorism targets civilians, so why should civilians not be the first line of defense and the advance guard of recovery? At this moment in the evolution of science and public life, civilians may be more appropriate for those roles than ever before.

Among the trends influencing the ways we respond to national and global events today, four form a particularly interesting cluster: the growing attention paid to threats of terrorism, an emphasis on networks in science and technology, a rising level of education among rank-and-file professional first responders, and greater connectivity within the public at large. These trends—new worries and continuously evolving human and technological resources for addressing them—are coalescing, perhaps nearing a "tipping point" that will change our collective experience of major calamities as dramatically as the Kobe quake changed Japanese attitudes, from vocabulary on up.

First, the anytime-anywhere properties of terrorism today mirror key properties of our most transformative technologies. Terrorism shatters the structures of traditional warfare; it is cheap, dispersed, and coordinated through multiple decentralized nodes. Now think of cell phones: cheap, dispersed, and coordinated through multiple decentralized nodes. It's a novel illustration of the expression "fighting fire with fire" that when the passengers on Flight 93 prevented the aircraft from destroying a fourth building, they were armed with mobile phones.

The illustration is novel and tragic, but the phenomenon is neither. Over and over, in surveying antiterror technologies, you encounter the notion that cheap tools and information distributed widely, may be the smartest defense—that for all sorts of purposes, abundance and pervasiveness matter as much as, if not more than, concentrated strength. In the area of sensor technologies, the smart-dust gurus advocate convincingly that creating many tiny detection devices versus a few sophisticated ones is the most promising solution to monitoring large areas for invisible weapons. In airport security, training lots of people to look for a few clues beats complicated lie-detector tests. In microbial forensics, the scientific push to sequence the genomes of lots of organisms, from the tiniest microbes in ordinary soils to every patient who visits an emergency room, is the best way to understand the biowarfare threat. In each case, the idea is to draw information and participation from the greatest possible number rather than from what's conventionally perceived as the significant few.

Meanwhile, an educated community of professional responders and a more networked public are already participating in unpredictable incidents in increasingly active ways. In 1986, Steve Flynn enlisted the participation of the day's leisure boaters when the missile cruiser USS *Yorktown* steamed into the Chesapeake Bay. Since 1991, when the police beating of Rodney King in Los Angeles was captured by an onlooker's personal video camera, the world has been watched and guarded by a random amalgam of witnesses. Today, in Boston, clam diggers are given free cell phones and enlisted to help keep watch on Logan Airport's perimeter. And it was a message posted in a chat room by a worried teacher in Guangdong that alerted the volunteer members of the ProMED network to the onset of SARS.

You might see the open-source biology advocated by Roger Brent as the ultimate expression of such ideas. Making the developments in advanced biology available to anyone who is interested, he argues, will encourage robust, reliable science. This view might seem obviously silly—after all, terrorists, too, fall into the category of "anyone interested" (and it may be that not all clam diggers are nice people either). But Brent believes that open systems are far better understood in the end than secret and proprietary ones, and that deeper understanding is ultimately a force for good. For Brent, if all is equal—if terrorists and innocents have the same access to the same advantages—the innocents, far greater in number, will surely win.

Of course, this vision of technological populism can't be a codified, across-the-board objective. Ben Curran at FEMA has *real* problems on his hands when the 18-wheelers roll in with hastily gathered contributions, and there are no data-mining techniques to handle the immense telephone surges—the Washington, D.C., sniper rampage alone produced 70,000 calls—when the public is solicited for tips. (Isn't that lack itself rather telling, though? Why is more effort put into scanning data people haven't seen fit to offer up than data they have?) There's also the problem that not all "help" that might be offered would necessarily be "pro-social"—or acceptable to the community at large. When the Department of Justice announced its Operation TIPS in July 2002, which would provide letter carriers, utility workers and millions of others the opportunity to report "suspicious" activities, civil liberties groups were horrified. The prospect of a citizen army making judgments about other citizens' behavior was unacceptable (Salon.com's story on the subject carried the tongue-in-cheek headline, "When Neighbors Attack!"), and the program was quickly canceled under pressure from many quarters.

But in other areas of life and science, complexity can be a boon as well as a scourge; why should broadening the participation of volunteers in the challenge of safety be any different? Instinct may tell us to contain, conserve, limit, and simplify systems, but through the work of network scientists such as Albert-László Barabási, we're finding that these impulses are not the only source of strength. "It seems," writes Barabási in his book *Linked*, "that nature strives to achieve robustness through *interconnectivity*." What we've learned from the Internet, which is so terribly complex and yet so peculiarly reliable, and what we are learning from cell biology and systems engineering and numerous other fields, is that redundancy and interweaving—complexity—can be the source of power and resilience.

This is not a romantic vision of anarchy; it's a call to understand what Barabási calls "the architecture of complexity," and to use that understanding to build a large, inclusive structure that supports us, rather than collapsing, when a link is cut. Such re-engineering is not something to rush into. Increasing the role of an empowered populace in combating disaster is something that will happen in slow increments. In the case of emergency response, it will be firefighters, cops, and EMTs—today more informed and educated than the generations of safety officers that came before them—who will first seize upon and circulate information previously accessible only to mayors and chiefs. Later, however, some of this same information and some of science's new skills and practices will reach into the realm of the public—not just as a matter of policy, or a matter of public demand, but as one of the unavoidable consequences of dual use. In an age where we all value the technologies of communication, innovation will inevitably offer new ways to tie us together in strange circumstances.

Secretary of Defense Donald Rumsfeld was unreasonably derided for his use of the terms "known unknowns" and "unknown unknowns" in discussing terrorist threats. The point was a worthy one: it's important to recognize the extent of your own ignorance. If Rumsfeld erred, it was in not going far enough. David Guston, a professor of public policy at Rutgers University, likes to point out that Rumsfeld's taxonomy left out a whole quadrant of the world: the "unknown *knowns.*" High-ranking defense and safety personnel know a lot, but they are often oblivious to what *others* know, including ordinary people. The 1994 earthquake in Northridge, California, offered a graphic illustration of what can happen when the knowledge of ordinary citizens is far more important than the expertise of officials. Sixteen of the 58 deaths caused by the event occurred in a small, three-story apartment building. While neighbors were heroically pulling those who were still alive from the rubble, the Los Angeles Fire Department drove right past the complex. Because the entire first floor had collapsed, the official responders saw an intact two-story building. The neighbors, of course, knew otherwise.

When it comes to disasters, which are by nature unusual if not unprecedented, experts who've learned their fields through careful study of the past will often be thwarted by a version of Meno's paradox: How do you learn to cope with problems when you don't know what they'll be? When this happens, some hard decisions and responsibilities could be up for grabs—handled no better by FEMA directors than by EMTs, who may, in fact, perform no better than rank amateurs.

The "architecture of complexity" that is most often applied in cases of emergency is the organizational structure referred to by the military as "command and control." Although varying definitions of the phrase litter the footnotes of military scholarship, the general notion is this: "command," the orders given by a leader, is informed and executed by "control," the mechanisms for taking action and for reporting the effects of that action, and the context for future commands. Implicit in the term is also an organizational hierarchy, in which each player has a single leader to whom he or she reports, and each "parent" has command over the control system below. By the time of the cold war, a third C had been alliteratively added: command, control, and communications, or simply C3. The value given to that last C reflects the growing realization, early on in the age of electronics, that the quality of communications mattered a lot. Nazi tank divisions excelled at blitzkrieg not because of bigger guns or more powerful engines, but because they had two-way radios that allowed far better situational awareness. British and French tanks, on the other hand, started World War II able only to listen, not to talk.

The military, the United States' largest employer, spends billions of dollars on the maintenance and fine-tuning of its command-and-control topology. But public safety groups such as fire and police departments are also highly consistent practitioners. Throughout the late nineteenth and twentieth centuries, as municipal safety services became institutionalized—boasting good salaries, standardized training, and government appointments—they drew from the model of wartime defense. As far back as the 1870s, large cities such as Chicago began organizing fire departments in terms of battalions and companies. Top-level decision makers, who earned their roles as fire captains, lieutenants, and chiefs through years of training and experience, sent filtered information down the ranks, and subordinates responded. During the post–World War II decades, new layers of emergency services bureaucracy, usually referred to as civil defense, peopled their ranks with experienced, frequently retired, members of the Armed Forces. This disciplined top-down constitution of professional bodies continues to reproduce itself.

Such architectures are what scientists like Barabási call "tree structures," and throughout history, warriors, bureaucrats, bishops, and everyone else desiring clear lines of authority have attempted to get better and better results by these means. One way has been to perfect the connections between orders and actions—in a sense, to tighten the structure's joints. Rather than toppling the chief and introducing radical socialism within the fire department, you improve methods of communication between trunk and branches or heighten discipline throughout.

In some cases, perfect tightening of command-and-control interactions can be achieved by taking people and their decision-making abilities out of the situation altogether and hardwiring the controlled elements in such a way that they have no choice but to follow this invariant command. This was what Ed Lee was about to consider on September 12, 2001, as he made his way to the University of California's Berkeley campus to teach his class on systems engineering. A computer science professor with a solemn manner and a dark, close beard, Lee had spent the previous day, like so many others on the planet, absorbing the horrible reality that the unimaginable can happen: that even when you think the tree of command is intact and someone is safely in charge—airlines, say, pilots, traffic controllers— your security can suddenly be demolished.

That Wednesday morning, Lee felt he couldn't teach the lesson he'd prepared. It now seemed impossibly irrelevant. Instead, he presented his class of some 200 undergraduates with a challenge that was disturbingly germane: create an aviation technology that would prohibit an airplane, even under the control of a hijacker, from crashing into a skyscraper. What the class came up with, and what Lee has been refining ever since, was a means of irreversibly imposing no-fly zones on the operations of an aircraft. He calls the concept Soft Walls.

Many control capabilities on modern planes are achieved through electronics rather than traditional mechanics or hydraulics. These systems are easier for pilots to manipulate, and the miniaturized electronic components reduce the overall weight of the aircraft. They can also be programmed to recognize and react to certain triggers, just like ordinary desktop computers. For instance, on some planes, if data suggest there's a danger of colliding with another aircraft, a cockpit alert sounds. An audio function will say, literally, "Traffic, traffic, traffic!" followed by instructions on some planes, such as, "Climb, climb, climb!"

Soft Walls would build on this capability. Like collision-avoidance systems, it would collect information about the plane's status, compare that with other information, and draw a conclusion about what should happen next—all without the pilot's input. But rather than compare the plane's location with that of another aircraft, Soft Walls would compare it with fixed points on a geographical information system. Loaded into the memory of the airplane's computer, this 3-D map of the ground, hills, valleys, and city skylines would demarcate forbidden areas. The airliner's electronics would determine if the craft was getting too close to, say, Wrigley Field in Chicago or Disneyland in Anaheim or encroaching on any premandated perimeters.

And here's where the soft walls come in: as an airplane approached a no-fly zone, the electronic controls that steer the craft would gradually—softly—resist the pilot's attempt to steer. At first the pilot would experience a light resistance in the steer mechanism, maintaining a portion of control, but if he or she continued to head into a prohibited area, the resistance would get stronger. Ultimately, the power of the computer would, if necessary, supercede the human directive.

Absolutely. Which is what makes Ed Lee's proposal controversial: as he envisions it, the system would not permit overriding. After all, if the pilot were allowed to disable the system, so could a suicide hijacker. No, Lee's system would be mandatory, imposing absolute control from above; it would tighten the civilian chain of command into an unyielding rod.

Not surprisingly, for pilots among others, this vision is dangerously simplistic. Ingrained in the culture of aviation is a belief that the human element is primary and indispensable; when unpredicted weather looms or an unthinkable predicament arises, a human must be in the loop, if not in charge. Most of the navigation on airplanes today is accomplished with *recommended* actions. Instructions from a person in the control tower, electronic guidance such as collision avoidance—these are suggestions that the captain can respond to at his or her discretion. "Pilots have always said they *must* have on-off switches for everything," says Lee. "The classic view is that if these systems were mandatory versus advisory, they'd undermine the pilot's control and take over." There are broader reasons for this insistence on discretion, which in a sense is a form of "slack." Whereas in tightly coupled systems a change in one component has an immediate and singular effect on other components, loosely coupled systems, such as those with redundancy built in, allow for a variety of reactions. Like the ductility of explosion-resistant buildings, which helps disperse excessive loads that might otherwise topple a structure, a system that assigns some discretion to pilots (or traffic controllers or airport security guards or passengers) can diffuse damaging effects.

The question of the pilot's role has long engaged the flight community and forged a sort of philosophical split between the two megacorporations in the industry, Boeing and Airbus. Airbus has implemented some nondiscretionary controls, such as those in what it proudly calls its "flight-protection envelope." Another variation on collision avoidance, this system prevents pilots from making maneuvers that violate certain safety specifications. Boeing, on the other hand, leaves even seemingly risky moves up to the captain, the belief being that a thinking human is better than any machine at weighing unanticipated options. As it turns out, this distinction between the two manufacturers' equipment was not lost on the planners of the September 11 attacks. According to the *9/11 Commission Report*, Mohamed Atta was explicit in his preference for using Boeing over Airbus aircraft, "which he understood had an autopilot feature that did not allow them to be crashed into the ground." All four of the planes hijacked on September 11 were made by Boeing.

So, it's not difficult to understand why Lee falls firmly on the Airbus side of the debate. For instance, he says, "I see absolutely no reason for a pilot to turn off the transponders," referring to the devices that allow traffic control to locate planes from the ground. On 9/11, the terrorists did disable the transponders, and as a result, controllers lost track of the four overtaken craft, making interception impossible. And Lee feels just as strongly about no-fly zones: there is absolutely no reason that a commercial pilot would need to be within 100 yards of the Sears Building or, for that matter, a mountainside. In a frequently-asked-questions document that Lee has posted on the Soft Walls website, he goes as far as to suggest firing pilots who insist on having override capability: "A phase-in plan," he writes, "will have to take into account the burden on the airlines to replace the pilots that refuse to accept the change."

This idea for tightening command-and-control through technology is not the only one of its kind. Every year, on U.S. roads, about 2,500 accidents occur involving hazardous materials shipments, close to 800 of those causing chemical spills. Corrosive, explosive, flammable, and toxic materials, all hauled across the country every day in huge tanks, can spill or ignite even under carefully monitored conditions. With a suicide hijacker at the wheel, the potential for destruction is obviously enormous. The U.S. Department of Transportation advises hazardous-materials ("hazmat") drivers not to leave their rigs while the engine is running and to be wary of strangers at intersections, but since September 11, the department is also considering a more definitive approach—much like Soft Walls.

With the help of the same GPS satellites used by aircraft, hikers, soldiers, and engineers experimenting with securing shipping containers, trucking companies today can track the movement of their fleets. With new features, they could also remotely shut down a vehicle if necessary. The idea that someone at headquarters could disable a rogue vehicle, much as if it had run out of gas, is appealing to shippers who move volatile goods. It's likely to be less attractive, though, to truck drivers, who value their discretion as pilots do and have resisted other curtailments to their autonomy that new technologies have made possible. An operating restriction that would serve to stop a terrorist would also have to take drivers out of the loop—that is, there'd have to be no way to override the device. (Similar techniques are under discussion for combating straightforward car theft by either immobilizing cars reported stolen or limiting their speed to allow police to overtake them.)

Using technology to take the slack out of command and control doesn't necessarily disempower people; it could save them. Many firefighters, for instance, wear a safety alert device that contains a motion sensor and produces a loud alarm when it detects that the wearer hasn't moved for a certain interval and is therefore possibly trapped or unconscious. After firefighters in San Francisco were found disabling the alarms because they produced irritating false positives whenever a group stood around resting or conferring, they were issued new devices that had no on-off switches; there was no longer a way to shut down the devices for convenience.

All of these solutions spring from the same desire to strengthen command and control. The trouble with locking in fixes, however, and using automatic controls is that they diminish the strength of the network's individual nodes. The unique aerial perspective and expertise of the pilot—one such node—is impeded in favor of the will of the system designer. The ability to adapt—the true strength of complex systems—is edited out. Again, in some situations that may make sense; one really can't imagine certain things as desirable outcomes under any circumstances, and the ability to fly into Wrigley Field or dump a hazmat cargo where it shouldn't be may fit into that category. In general, though, hardwiring the limits of your imagination into a system may not be the wisest way of facing up to the unexpected.

•   •   •

Another way to improve the performance of a traditional hierarchical network, if you choose to maintain the basic tree topology, is to add links. You can at least link some of the leaves of the tree structure to each other, and more links mean more interconnectivity.

Seth Rubenstein wouldn't dream of tossing out the tree. Like most firefighters, he has a passionate respect for the chain of command. With ten years in the San Francisco Fire Department, Rubenstein is a team player, a solid, deeply committed member of the rank and file. He has sandy hair and a slight overbite and wears his creased office uniform (those white shirts—a surprising hint that old-fashioned starch might still be on the market) with the pride of a sportsman in the team cap. Surrounded by other crisp uniforms in his newly renovated office space at headquarters, he works on special projects, with an emphasis on information systems.

Yet Rubenstein epitomizes two trends that appear to be fostering a revolution in our response to emergencies: one, that technologies are enabling all kinds of networking and, two, that first responders are ever more informed. A computer engineer in small high-tech companies before joining the department, Rubenstein has found himself floating among the ranks to implement new networking technologies and improve systems that the organization has let linger on for years. Some days, he longs to go back to the station, slide the pole (yes, they still have them—what better technology for getting downstairs fast?), and race into burning buildings. "I joined the fire department trying to get away from the screen," he laments, "and here I am back in front of it." Mostly, though, he's happy to be fixing kinks that are hindering operations unnecessarily. Rubenstein is a do-it-yourself emergency responder, a regular guy with smarts, who's changing the nature of public response simply by doing what makes sense to him.

What makes sense to him? For one thing, creating a link between fire inspections, which are done regularly to ensure that buildings are up to code, and fire suppression, which is what the heroes with axes do daily. San Francisco has an entire infrastructure—budget, staff, permits, regulations—for inspecting the city's 150,000 buildings. To Rubenstein, this ought to make firefighters better informed, but it doesn't. Inspectors search residences and businesses for toxic substances, blocked egresses, flammable materials, but they collect all that data on paper. "Then it gets put into an old Wang computer's database, where firefighters never see it," says Rubenstein. These communication cul-de-sacs are partly the result of outdated technologies and partly of cultural differences. It turns out that 24-hour cultures, such as fire departments and hospitals, suffer a sociopsychic divide from the 9-to-5 world of inspectors. Rubenstein plans to bridge the chasm. "When a suppression truck is dispatched," he asserts, "I want it to get clear, concise information while en route. I should know if there's a chemical lab in a nearby building."

In most large departments, fire vehicles already have computers on board, bolted down in the cab. They're typically used, however, for a very few basic tasks. "Right now, we really only use four buttons," says Rubenstein's colleague, Lieutenant Michael Thompson. "One to say you've heard from dispatch, one to say you're on the scene, another to say you're leaving the scene, and a fourth to say you're back at the station." This isn't meaningless information—it helps dispatch keep tabs on units, and it also stores records of response times and movements for future analysis. But it's mostly information a GPS-enabled truck should be sending out without having to be told—and it's a far cry from what the technology is capable of doing. "If a fire inspector has written a citation for possession of a dangerous chemical," says Thompson, "and that night there's a fire in the same building, my crew wouldn't know it."

There's more that the firefighters might wish for, too. Perhaps the most fruitful, in terms of using the networks already out there, would be applications of mobile-phone systems that know where their users are. How about allowing fire-crew dispatchers to locate passersby with mobile phones near the fire and ask them to send in pictures that allow the extent of the blaze to be assessed? Or, for that matter, to phone people known to work in the building and find out if they got out? The systems to do this—to turn onlookers and potential victims into a sensor network on the fly—don't exist yet. But the technologies and the attitudes that would make them possible are advancing.

In Rubenstein's vision, as an engine peels out and its sirens wail, an officer in the passenger seat will see the computer screen fill with critical facts about the team's destination. Exactly what sort of facts will matter: in redesigning the inspectors' database, Rubenstein is inventing a scheme for labeling the fields so that either a human dispatcher or the computer itself can select certain categories of facts for sending.

This function, the ability to pick and choose what information gets sent, is far more important than it might seem. Information overload might be a minor hassle when Googling your blind date or shopping for CDs online, but for emergency responders speeding to a job, poor presentation or excessive data may affect lives. "Lots of things in the database are irrelevant," Rubenstein says. "You don't need to know if the exposed building failed its fire extinguisher test on one floor." Maybe even more important, if firefighters think something's too fancy or superfluous, they won't use it at all, and Rubenstein knows that from experience.

It's this appreciation for optimal simplicity and functionality that makes do-it-yourself responders such as Rubenstein a true asset in a world where technology developers from the outside, enamored with bells and whistles, are constantly pressing their wares. "One of the coolest things about the fire department," Rubenstein says, "is that if a computer doesn't work, it gets thrown out the window." He loves the tough demands of fire department R&D and contrasts these qualities with his previous experience programming games: "Here, we can't afford fluff at all. This work is so function-specific and time-dependent, the stuff really works. If you're in the private sector and want to see how functional something really is, ask the fire department." And by "functional," he doesn't just mean that the tool does what it's supposed to when its buttons get pushed. For technology used by emergency workers, everything matters, from ruggedness to power requirements, from glare on the screen to the size of the knobs. Lieutenant Thompson likes to tell the story about the first time his computer screen on the fire engine went blank: "One of the firefighters told me to hit it, and it would come back," he remembers. "One punch shattered the screen." And try using a cell phone while wearing those gloves! "These are giant people," Thompson adds, "like 6-foot-3 and 260 pounds." Big, busy, and brutally goal-driven. Rubenstein puts it like this: "If you try to replace firefighters' gear with some new gadget that doesn't help, they'll react by cascading back through the history of technology till they're using cans on a string or yelling out the third-floor window to the fifth."

•   •   •

When it comes right down to it, Rubenstein's inspection-suppression project is all about the hyphen: it's about connection. His goal is to link the guy climbing around construction sites holding a clipboard with the gal pulling a hose off a truck. So whether or not orders come down from above, advice and information can shuttle horizontally, adding slack, options, and useful complexity.

In many areas of on-the-ground terrorism protection, forging new pathways in which "knowns" can flow freely is key. In a Department of Justice survey of law enforcement officials from a range of agencies after 9/11, for example, access to intelligence information across geographies and levels of government was first on their list of needs. The up-down exchange of information intrinsic to any command-and-control structure is not always an impediment, but it is certainly only one way to communicate.

Nowhere is the pressure to bust up this highly constrained circuitry as urgent as in the real-time, on-site drama of response. Today, it's well understood that 9/11 responders suffered crippling lapses in their ability stay in contact, and the widely read McKinsey report that analyzed the performance of the fire department after the disaster is filled with observations about communication failures. Hard as it is to believe, in an age when a Texan can call Lagos and feel as if he's having a conversation with someone down the street, in virtually every major city in the United States, police and fire departments, EMTs, and county, state, and federal response teams cannot talk to each other over their radios. A dozen different police departments may arrive at the site of a plane crash and be unable to coordinate with each other via radio. Now consider that easily 25 fire departments are involved at an earthquake, and at a terror event, sheriffs' departments, highway patrols, the Forest Service, and National Guard will likely leap in, along with ambulances, bulldozers, helicopters, fire engines, command vehicles, pump trucks, and hazmat trailers. How will they pool their strengths? They'll use runners—as in, "Run down there and tell the chief to get the dozer here." In the minutes after the Oklahoma City bombing, fire commanders had reason to believe another undetonated bomb was on the site (much like the feared third plane at the World Trade Center) and ordered all personnel to leave the building. But only firefighters got the warning. All the other emergency responders were on different radio channels.

Getting devices to talk to each other is called "interoperability"— a term used about as frequently these days as "love and peace" used to be. The challenges of interoperability stem from the strange character of that fifth dimension: radio airwaves or electromagnetic spectrum. Though the metaphor eventually breaks down, the easiest way to think about the spectrum is in terms of real estate, with different frequencies parceled, leased, and auctioned off. Acres of land lie side by side; developers get dibs on various lots from the FCC; some parts are set aside as parkland. The Donald Trumps of this Spectrum Town are cell phone companies; the mom-and-pop homeowners are local TV stations, taxi dispatchers, utility companies such as Con Edison that communicate with their people in the field, and even fast-food companies that use radios to relay orders at the drive-through window. Where you live or work in Spectrum Town—what frequencies you are allocated—determines what sort of radio receivers you can use. This is where the trouble comes in for firefighters and cops: different frequencies, different receivers. When the police arrive at a scene, they come with their trusted walkie-talkies. (Real users call them "Handie-Talkies.") These thick bricks of electronics, which cost about $4,000 a pop, are designed and manufactured to handle a given range of signals—say, the 33 MHz range. Now, what about the firefighters? Well, their heavy handhelds may be designed for transmitting and receiving not at 33 MHz, but up, let's say, around 460 MHz. Why? The answer takes us back to Spectrum Town: One of the maddening drawbacks of living there is that public safety bands—just like national parkland—are set aside by the government for free, but they're doled out only as small plots spread all over the spectrum.

So why not make a radio that can jump around? Why not just extend the range and make the dial turn further? Well, in certain cases you just aren't allowed to; where the military talks is off-limits, for instance. It can also be very difficult to do; inside those Handie-Talkies are bits of electronics and wires that shuffle around the electromagnetic waves, ushering them in to be processed and routed and transformed into the fire chief's orders. The components responsible for guiding those oscillations, typically made of metals or ceramics, are chosen according to their properties—their tendency to conduct or resist charges, for instance, or their deftness at moving little waves versus big long ones. These distinctions turn out to matter a lot, a fact that was expressed beautifully by the famed early computer scientist and Navy Admiral Grace Hopper, who, it's said, used to carry around a piece of wire cut to just over 11 inches. She called this teaching prop her "nanosecond," because that's the amount of time—one billionth of a second—it would take for an electron to travel from one end of the wire to the other. For radio waves, measured in time intervals a lot like Hopper's, how you wire a device clearly has an impact. To create a radio that can manage signals in different frequency ranges requires separate sets of components, which are bulky and costly. Such radios do exist, but so far, virtually no safety agencies, burdened as they are by their legacy antennas (which also handle only specific frequencies) and extremely pricey Handie-Talkies, have made the investment.

It may be just as well. The problem of communication among emergency responders is multifold—interoperability is just one facet—and it may be that technologies will soon evolve that can address the need to tune into many channels much more easily. The coming idea here is to develop radios that suck up signals at a wide range of frequencies and use ever more capable software rather than hardwired components to sort them out. For now, disparate agencies are applying quick fixes to the technical rift between frequencies and the resulting logistical rift between firemen and cops. They swap a bunch of radios, for example, and their chiefs carry several at once—one to talk to the Forest Service, one to talk to the state cops, one to talk to the commissioner. A jurisdiction in Pennsylvania reports having mounted seven different radios in an ambulance—the cost of which approached that of the whole vehicle. Many regions are investing in converters, which involve about 4 feet of gear that they haul out to an incident. These can receive audio at several frequencies and send it back out at others, a trick not unlike aiming a tape recorder microphone at the CD player's speakers to transfer an album onto cassette. Using drag-and-drop icons on a graphic display, an operator at the scene can mix and match airwaves. In his book *After: How America Confronted the September 12 Era*, Steven Brill tells the story of Raytheon's attempts to capitalize on the new security fervor, one of which was to whip up an "interoperability vehicle" called the First Responder. The revamped Chevy Suburban would contain not only the radio-frequency converter but also satellite connections, video feeds, and other high-tech gear—all for under a quarter of a million dollars. Brill deftly recounts the elated reactions of homeland security officials, including Tom Ridge, when the Raytheon team showed off the tricked-out truck. Subsequently, Raytheon took the thing off the market for lack of sales: what departments want is the converter, minus the truck and embellishments.

And it may be that these basic, somewhat clumsy workarounds are really the best bet. Heavy spending on real multiband radios could further trap agencies in a communication platform that will never significantly advance. After all, there's got to be something radically better coming down the pike.

Consider the status quo: Not only are emergency responders trapped in their limited radio frequencies, they're also trapped in the dark world of sound. They work blindly, translating where they are, what they're doing, and whatever they see, by voice. Anyone who's ever gotten computer tech support over the phone knows just how inefficient that is.

Go to the list with a file called Bumble-Bee.
I don't see it.
It should be about halfway down.
Oh, I see it.
Double-click on it.
OK.
Now what do you see?
It says "Error."
What else does it say?
It says, "You'll never fix this."
OK, close that window.

David Kehrlein, from the Governor's Office of Emergency Services in California, rails against safety workers' reliance on voice: "It's heresy!" he says. "You can only listen to one person at a time! Plus, it's not persistent," by which he means that once a transmission's been sent, it can be so easily forgotten. There's no means for posting those voice messages somewhere to refer back to them or pass them along to the next shift.

Meanwhile, for a twentieth of the price of a Handie-Talkie, today's cell phones already accommodate two other media: text and imagery. If you've ever wondered what to do with the photos on your cell phone, imagine the benefits of capturing the underside of collapsed construction and sending it to a commander formulating a rescue strategy. "Emergency workers can't drive a fire truck faster," says Kehrlein, "they can't run up a hill faster. The only area for improvement is in coordination!"

But never mind the pictures; there's an even more basic function rescue workers will want: the ability to track their whereabouts seamlessly. Today, 60 percent of radio traffic at an emergency is composed of officers reporting where they are.

D2, give me your location.
At the rear of the first floor.
Is that the street-level floor or the floor of the main entrance?
The floor of the main entrance.
Copy.
Now I'm heading to the roof.
Copy. D4, give me your location.

Meanwhile, cheap cell phones are being equipped with electronics that can calculate geographic coordinates automatically. On the subject of location devices, Kehrlein says: "Even Burger King's 99-cent hamburger deserves to be tracked!"

Among all the unbearable tales of 9/11, a particularly painful one turned up in the accounts of the Port Authority of New York and New Jersey, released in the late summer of 2003. It concerned the fate of more than a dozen Port Authority employees on the sixty-fourth floor of the north tower who never made it out, while hundreds of victims on higher floors survived. Finally, with the release of the reports, the mystery surrounding their predicament was explained. The engineers had stayed put inside the building for nearly an hour and a half—too long to make it down the stairs when they finally tried—because they had been told to. "Stand tight," they were instructed by a sergeant at the Port Authority's police dispatch, early on in the mayhem. "Stay near the stairwells and wait for the police to come up." When one of the men pressed—"They will check every floor?"—the dispatcher reassured him. When the same man called again, shortly before the tower collapsed, the reply he got was different: "Try to get out."

The heartbreaking anecdote resonates in several ways. First, it points to the isolation imposed upon so many people that day because elevators, land lines, and radios didn't work.

Radio contact was disrupted not least because it was coping with 110 floors of two colossal towers. As if the problem of communicating across frequencies isn't enough, there are also difficulties with sending communications through space and matter. Like the interoperability bind, this is a physics problem: most radio waves, though they can dance around the planet in rather surprising ways, can't quite make it through steel. Concrete isn't easy either—the density of the material works much as paint on glass blocking light from passing. So, every day, firefighters venture into chambers of smoke and fire and disappear. Usually they reappear, but the striking reality of firefighting technology is that when they are inside blazing buildings of more than a few floors, there is no reliable means for communicating with them. They often can't see what's ahead of them and often have no way to report back. On their stomachs on the floor, where the air may be somewhat clear, they feel their way along walls with the backs of their hands (burning the front would make holding or gripping things later impossible) and use their ears (heat-sensitive and among the few spots on their bodies not covered) as sensors. Even if geographical tracking information for every vehicle and firefighter in the city were blinking brightly on the fire chief's map, it would be of no help to the people indoors, where satellite signals don't penetrate.

Like Seth Rubenstein, Fire Captain Mike Stein is a guy from the trenches who has taken on a citywide problem. An ox of a man, built like a pro quarterback with the bangs of a Little League catcher, Stein is another do-it-yourself emergency responder. He has spent 25 years in the New York City Fire Department, but hasn't forgotten his skills from "outside": he's a licensed electrician and plumber. In the words of his boss, "He's got a lot of experience as a captain, but he likes to tinker." So along with about a half-dozen others in the department, he handles product evaluations. "We test the harnesses and belts," he says politely, one lazy eye permanently lost in thought while his other looks as alert as a young scientist's. "We test all the ropes. Anything new in the field, we look to determine if it's safe."

And he invents things, all sorts of things. It was Stein who figured out that if you're on a water rescue, you can attach a fire hose to your oxygen tank (all firefighters carry breathing gear that delivers pressurized oxygen to their face masks), inflating the hose; then you can drop the long buoy to float on the water so victims have something to grab hold of. It was also Stein who worked out how to turn a rope into a blinking cable stored on a reel: teams will sometimes unfurl a rope on their way into a burning structure so that they have a guide in the darkness on the way out; making that rope light up using battery power may seem trivial until you consider that the rudimentary "glow strips" placed in the World Trade Center stairwells prior to 9/11 were credited by some with saving their lives. It was Stein who helped initiate the Guardian Angel Project, an enhancement to the protective motion sensors that set off alarms if their wearers stop moving; this version would attach a small voice chip to all firefighters' masks. And it was Stein who got to work on the problem of inside-the-building communications. "Our Handie-Talkie was having trouble getting through walls," he explains, underplaying the point as if, because he solved it, the problem must have been minor. "We were using relays: guys stationed every 10 floors, passing messages along. But it was like a game of telephone." The method was inefficient and error-prone. "So we built this command-post radio."

Stein's insight was simple—so simple it's difficult to imagine what took mankind so long to have it—and his execution has gotten approval from the world's toughest focus group: all battalions in New York now have the devices; Boston's fire department has bought several; and Jersey City has established a requirement that all high-rise building owners purchase them and have them on-site. While similar technology has been used for years in rural settings, where distance rather than steel is an impediment, Stein had the brainstorm to bring the technique to the city—to take the ability to send signals horizontally across large expanses and turn it upright. The idea is to use brute-force power to drive voices right down through the floors of any one of the 2,000 high-rises in New York City. While cell phone signals, which also suffer from poor propagation through large buildings, can head out to the sky in search of a friendly antenna, emergency radio messages aiming to connect officers on distant floors often have no options but to bore through concrete and die trying. One solution is to plant devices called "repeaters" throughout buildings so that signals have something to bounce off to get where they're going. And in fact repeaters, which had been installed following the 1993 World Trade Center bombing, proved useful to commanders on 9/11. But this kind of infrastructure is vulnerable to whatever perils are already assaulting a site. It also costs money. Some day soon building owners may be required (or lured by tax breaks) to build plenty of stepping stones around their property for emergency radios to use, but Stein wasn't planning to wait for that.

Instead, he upped the wattage of a typical handheld 10 times. This would give the waves the heft to travel up and down through dozens of floors of steel and concrete or 5 miles out across the city. But there were two details he needed to work out. First, batteries are big and heavy, and firefighters already carry about 100 pounds of covering and breathing apparatus. It took choosing a lunch-box-sized battery intended for powering jet skis to make his radio portable. Second, the signals that could now penetrate steel and concrete would also be blasting out into the open air—where they would increase the noise congestion already plaguing the city's safety channels. So Stein built the radio to use a different set of frequencies than the ordinary Handie-Talkies. When the commander in the lobby spoke into his 20-pound command radio, he could reach the chief at the incident command station specifically, 80 floors up in the building. With $1,000 worth of parts, Stein had found a fix.

The innovations that Stein and others are pursuing might have made an enormous difference to those Port Authority workers on the sixty-fourth floor of the north tower, who, on the basis of their sole communication link with the world, made what turned out to be the wrong decision. Yet their fate resonates beyond the awful isolation they experienced in the last hour of their lives. Equally moving is the fact that they were loyally following directions. These dozen people went to the authorities for instructions, and got them. It's an inherent weakness of the command-and-control model that when those up the pyramid are uninformed, the ghost of the structure still stands; information still flows along the intended axis, except that the information is wrong. Anyone in charge can give bad orders. What that agonizing outcome suggests is not that because the brass can make mistakes we shouldn't pay attention to them, but that a single channel of communication significantly raises the stakes of getting the message wrong.

Art Botterell tells the story of an important meeting held at the Mitre Corporation, home of 4,700 scientists and engineers who do research for the Department of Defense and the FAA, among others. Botterell is an expressive guy, with animated features set inside a round pillow of a face, who is notably pugnacious and extremely smart about the technology and the politics of emergency response. Mitre, he recounts, was hosting a board meeting for the Partnership for Public Warning, a group that was formed in 2001 to support the development of better public-warning systems throughout the United States. Its membership includes representatives of what you might call the heavyweights of warning: alarm companies such as Federal Signal, several state offices of emergency services, Departments of Commerce and of Homeland Security, the FCC, and a few technology companies such as RealNetworks and Intel. At some point during the meeting, a fire alarm sounded. Here, midstory, Botterell does an imitation of the 25 or so men and women at the meeting: his head dips between his shoulders, his eyes widen with surprise, his jaw goes slack, he sits motionless. "Everyone stared at each other!" he remembers. No one seemed to know what to do. And then the punchline: "We didn't evacuate until someone came to the conference room door to say we really had to leave."

Along with the pleasures of irony it affords, Botterell likes how well this story illustrates a crucial point: "People," he says, emphatically, "want corroboration before acting." Social scientists who have studied disaster response have documented this behavior numerous times: even when faced with potentially grave danger, people check information against more information before doing anything. Which is quite logical, after all; on serious matters, it makes sense that you'd get a second opinion, two sources at least or even two mediums of communication. Human beings want some redundancy in their networks.

To achieve a weave of robust and useful signals across a nation in crisis—especially when rescue workers on the same scene can't yet communicate with each other—is a big task. But Botterell has a fairly optimistic attitude. He thinks a smart, doable system is out there. "It's like before we broke the sound barrier," he says. "We knew a bullet could do it, but we didn't know how." In the meantime, he thinks there's some groundwork to do that we know will be worthwhile regardless.

Like Mike Stein and Seth Rubenstein, Botterell is a do-it-yourselfer. He's a guy who's never scaled professional heights or managed large numbers of people, but he's accomplished an extraordinary amount—without worrying too much about command structures. In fact, Botterell began on this track as the quintessential, self-motivated amateur—"a ham radio kid," he says. "I'd already played around with a shortwave radio that, best I can tell, my father looted from the Air Force after World War II." A classic nerd, he liked to tune into illicit, distant broadcasts, such as Radio Moscow and Radio Cubana, and felt most comfortable being sociable when the interface was a machine. "I had a tendency to deal with people through a keyhole," is how he puts it today. "I had a thin skin and a long antenna." By age 11 he had his FCC operator's license, which required that he take a course in the rules of spectrum usage and the mechanics of transmitters.

What he got from amateur radio, though, was more than just a cursory education in radio science. "It brought me into contact with two worlds," he explains. "One was radio broadcasting and one was emergency services." And these passions kept cropping up. Now nearing 50, he has worked as a radio newscaster ("I covered the first Larry Flynt obscenity trial"), a machinist, a hot line operator ("First it was a lot of bad acid trips; then it was suicides"), and a software programmer. But from the time he was 12 years old and worked as a ham radio volunteer on his first disaster, he was most drawn to the heat of emergencies. "I saw people being just like they ought to be," he says, "genuine, constructive, honest—and I fell in love with it." The pro-social behavior that sociologist Kathleen Tierney has identified so consistently in her research was evidently enough to drag Botterell out from behind his keyhole.

Today, he is drawing from this array of experiences to help address national-level problems in disaster communications. For Botterell and many of his colleagues, Washington, D.C.'s color-coded "threat advisory system," which some prefer to call the "national mood ring" or the "rainbow of doom," is woefully inadequate, and they've begun taking serious steps to improve matters. The idea is to develop a truly robust network for engaging with the public and among agencies in a crisis, and Botterell's thinking is based squarely in these same notions of interconnectivity, redundancy, and the related perception that complexity can be a virtue.

A neat example: Botterell envisions a world in which your fire alarm doesn't only alert the fire department, but also sets off the alarm of your neighbor, who is more likely to save you than anyone else. More channels, more nodes, more connections.

In his emergency response work, Botterell is motivated by the national emergency system that wasn't. When San Francisco suffered its 1989 earthquake, he was shocked to discover that the Emergency Broadcast System (EBS), which, along with its predecessors, had been interrupting our TV shows since the 1950s, wasn't any help at all. It was activated at some point after the quake, but after the few tones that signal emergency, there was nothing else—no explanations, no instructions, nothing. It was actually less efficient than a nineteenth-century town crier. The EBS, from its origins, was intended to facilitate a single authority with a limited amount of information and a single message applicable to everyone regardless of location or circumstances. It would fit into a one-minute audio message after those tones. In short, says Botterell, "it was designed for the president to get on and say good-bye to us." Post-quake, when rumors spread that the water in a small Bay Area jurisdiction had been contaminated, the town's officials searched for a way to announce that in fact the water was fine. EBS didn't allow for announcements from local agencies. When emergency organizations needed to broadcast fairly complex information about food and shelter, the EBS format was much too short. "The earthquake itself was self-notifying," says Botterell wryly. What was needed was some airtime for the plethora of necessary follow-up alerts.

So Botterell took Mike Stein's approach and began gathering spare parts. What was supposed to be merely a prototype for a new statewide emergency alert system turned into the real thing, based on ordinary radios he scrounged up and a primitive modem used by ham radio enthusiasts. Drawing on his background in newsgathering, Botterell and his colleagues created a text message format similar to the one followed by the Associated Press wires, transformed their formatted messages into digital packets like the Internet's, and sent them over the public safety airwaves. Then he persuaded TV and radio newsrooms around the state to purchase the $150 modem and set up a radio next to their wire feeds. The state office of transportation agreed to send out his messages over its traffic network so that they reached rural towns everywhere, and soon after, California agreed to let Botterell's network piggyback on the state's new satellite service, a backup telephone network for its 59 emergency managers. Botterell got a little slice of the satellite for his data messages.

Today the Emergency Digital Information Service (EDIS) serves 100 agencies authorized by the governor's office to promulgate warnings, from law enforcement organizations to county governments and public health officials. During the Northridge quake, 2,000 alerts were sent out, most of them from FEMA, with information about the recovery of services. In less dramatic times, the National Weather Service is the most voluble user. During the 2001 California energy crisis, the state required utility groups to broadcast warnings of outages on EDIS, and in the course of one week in 2004, the system delivered a warning of a salmonellosis outbreak traced to a Los Angeles–area restaurant and an assurance that the smoke wafting from the L.A. airport that morning would be part of a fire department training exercise. On the receiving end, newsrooms, government officials, as well as the law enforcers and health groups who are senders themselves, pick up messages they want, filtering out the ones of lesser priority or distant geography. Ten percent of the 15,000 EDIS subscribers receive the commonly formatted messages on their pagers.

Thirteen years later, Botterell continues to host EDIS without pay. "There never was and still isn't a budget line for EDIS," he says, without any visible pique; for a belligerent guy, he can be remarkably generous. Then he concedes, "Before Y2K, they bought me a new computer to run it on." No longer a state employee—he's now one of the "genteel unemployed," he says, meaning a consultant—Botterell continues to maintain the system. And he is working on realizing EDIS on a much bigger scale: a standard format for use nationwide that allows messages to be fed directly to sirens, cell phones, satellites, television news crawls—anywhere digital information can go, these alerts can travel.

Who will be authorized to send these alerts is yet to be determined, but Botterell, not surprisingly, favors an open approach—many sources. As his fellow meeting attendees at Mitre can attest, people will need corroboration.

In *Are You Ready?: A Guide to Citizen Preparedness,* FEMA spells out a series of sensible steps for anticipating all sorts of events—tsunamis, landslides, accidents at nuclear plants. It also covers national security emergencies. In that section—12 pages of warnings about bombings, suspicious parcels, and so on—the 2002 version of the manual includes advice to "Listen to your radio for instructions" or "Follow the instructions given" or "Pay attention to official instructions"—*11* different times.

This is not an indication of ignorance on the part of national leaders or a sign of failed imagination. It is an assertion that terrorist events demand an ability to improvise, but that only those at the top can improvise successfully. It cleaves to the model of command-and-control even in the absence of strategy, plans, or procedures. It steels itself against the onslaught of people undirected and unassigned, against the panic that, according to decades of research, won't actually arrive.

As the multiple trends that mark our historical moment continue to coalesce, however, even the highest officials, whom the old tree structure has benefited most, are shifting their outlook. In its materials released in early 2004, the Department of Homeland Security has taken a markedly different tack from FEMA's earlier directive. In *Preparing Makes Sense: Get Ready Now,* the department uses the phrase "common sense" a half-dozen times, a phrase that appears nowhere in the full 102 pages of *Are You Ready?.* This change is striking in that it ascribes new value to the "common" individual and assumes that citizens will possess their own "knowns" even when the authorities are uncertain. Along with its injunctions to listen to the radio and "seek medical advice," the newer document now includes recommendations such as "assess the situation" and "be prepared to improvise."

This devolution of responsibility in an emergency is not without its risks and ethical shortcomings—the most powerful being that some people have naturally crappy judgment, and shouldn't suffer for it. Public officials have as much of an obligation to blunderers as to the happily astute. But if current trends are driving us toward a completely new approach to disaster, pretty soon the general notion that each of us is an active node in the network of response won't seem farfetched. Kathleen Tierney and coauthors of a book called *Facing the Unexpected* have little doubt that we're headed there and little praise for where we've been: "The 'command and control' approach," they write, "which never was appropriate for managing disasters, represents a thoroughly outdated way of thinking about crisis response. Instead, policies and plans should conceptualize disaster response as a loosely coupled set of activities carried out by a highly diverse set of entities."

So, it's not surprising that along the way to reaching every leaf of the command-and-control tree structure, these ideas have already begun influencing the intermediate branches: not the public but those who are close to the public—the firefighters and EMTs and other emergency personnel who spend their lives on the scene, on the ground, among the victims. These do-it-yourself responders are already fixated on adding links to the networks that can inform their judgment. In their own deeply practical way, they are searching for a new architecture of complexity that will better support them and protect us. And they are hardly alone. Engineers who model infrastructures, data miners who troll for connections among facts, detectives who study the genetics of seeds to solve crimes—all of these experts share a conviction that within complexity there exists the potential for failure but also for success. There are dangers hidden in our chaotic world, but there is also the possibility of remarkable resilience.

And if current trends continue, the robustness of increasing connection will not end by benefiting merely first responders. It will eventually mean greater determination for those who offer aid after an earthquake; better judgment for those who fill truck convoys full of donations; more confidence for those who visit chat rooms in Guangdong, and more courage for those who study *Preparing Makes Sense,* getting ready to improvise in a crisis. It will mean more resilience, more hope, more safety, for everyone.

INDEX